Foundations of Space-Time Theories

Foundations of
Space-Time Theories

Relativistic Physics
and Philosophy of Science

Michael Friedman

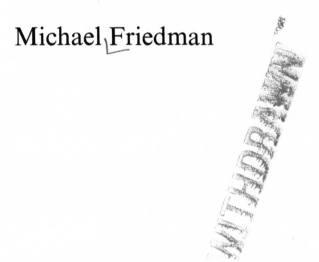

Princeton University Press
Princeton, New Jersey

Publication of this book has been aided by the Whitney Darrow Publication
Reserve Fund of Princeton University Press
This book has been composed in Times Roman

Clothbound editions of Princeton University Press books are printed on
acid-free paper, and binding materials are chosen for strength and
durability. Paperbacks, while satisfactory for personal collections, are not
usually suitable for library rebinding.

Printed in the United States of America by
Princeton University Press, Princeton, New Jersey

In memory of my grandfather,
Joseph Nathanson (1889–1981),
teacher of physics

Contents

The use, which we can make in philosophy, of mathematics, consists either in the imitation of its methods or in the real application of its propositions to the objects of philosophy. It is not evident that the first has to date been of much use, however much advantage was originally promised from it . . . . The second use, on the contrary, has been so much the more advantageous for the parts of philosophy concerned, which, by the fact that they applied the doctrines of mathematics for their purposes, have raised themselves to a height to which otherwise they could make no claim.

Kant, "Attempt to Introduce the
Concept of Negative Magnitude into
Philosophy" (1763)

Preface

To whom is this book addressed? Ideally, of course, to readers very like the author: to scientifically inclined philosophers and philosophically inclined scientists who have read and pondered the literature on Einstein's theory of relativity with a mixture of awe and puzzlement. This theory, it is clear, effects a profound transformation in both our understanding of nature and our conception of that understanding itself, that is, our conception of what a scientific understanding of nature is. In particular, the theory casts an entirely new light on the role of geometry and geometrical structure in our physical world picture. The problem is that very different, if not incompatible conceptions of geometrical structure emerge. On the one hand, the development of relativity appears to lead to what one might call a physicalization of geometrical structure: to the view that the geometrical structure of space or space-time is an object of empirical investigation in just the same way as are atoms, molecules, and the electromagnetic field. As such, geometry describes the furniture of the universe. Indeed, on some extremely ambitious developments of Einstein's ideas it describes the entire furniture of the universe, for everything else is to be constructed from geometry. Yet, on the other hand, relativity is often taken to embody what one might call an idealization of geometrical structure: a view on which geometry is conventionally imposed on the physical universe as a device for conveniently describing the concrete distribution of matter, which alone really exists.

The tension between these two conceptions gives the development of relativity theory its peculiar philosophical fascination. For the conflict does not surface only in critical reflection about the theory; it pervades the theory itself, from basic physical ideas such as the principle of equivalence and the various relativity principles to central mathematical ideas such as the notions of covariance and invariance. This book is the outcome of a long attempt to clarify the conflicting interpretations put on these ideas in the scientific and philosophical literature. I have learned much by writing it, and if it helps even one reader to do the same, I shall deem it a success. I should say at the outset, however, that I have not remained aloof from the central conflict: I come out squarely in favor of the tendency toward the physicalization of geometry. Nevertheless, I think that the contrary tradition has much of value to teach us.

The reader will perhaps obtain a better sense of the content of the book from a brief description of how it came to be written. I first started studying relativity theory seriously as a graduate student at Princeton, where I was fortunate enough to attend philosophy of physics seminars taught by Clark Glymour. I read and reread the classical papers on the theory, the standard textbooks, and the influential philosophical commentaries by Hans Reichenbach and Adolf Grünbaum. This experience left me both deeply intrigued and deeply confused. I was confused by the general conflict mentioned above, of course, and also by a specific mathematical puzzle about how the traditional relativity principles are to be formulated. These principles are standardly presented in terms of our freedom to choose any member of a certain class of coordinate systems to represent the laws of a given theory. The Newtonian principle of relativity is expressed in our freedom to transform coordinate systems by a Galilean transformation, the special principle of relativity in our freedom to transform coordinate systems by a Lorentz transformation, and so on. But this way of putting the matter makes the traditional relativity principles appear trivial. Are we not always free to transform coordinate systems by any transformation whatsoever? Will not any theory, regardless of its physical content, be *generally* covariant? If so, what physical content do the traditional relativity principles express?

While puzzling over these questions I read *Principles of Relativity Physics* by J. L. Anderson, which was published in 1967. It was clear to me immediately that this book marked a major advance in our understanding of relativity principles and, by implication, of the fundamental interpretive conflict noted above. Anderson draws a sharp distinction between covariance on the one hand and invariance or symmetry on the other. He shows in detail how all the traditional space-time theories, Newtonian mechanics, special relativity, electrodynamics, and general relativity, can be formulated in a generally covariant way; and, nevertheless, he shows how they differ in their invariance or symmetry properties. On Anderson's account, these theories have the same covariance group, the group of all admissible transformations, but different invariance or symmetry groups. The symmetry group of Newtonian mechanics is indeed the Galilean group, the symmetry group of special relativity is the Lorentz group, the symmetry group of general relativity is again the group of all admissible transformations. Moreover, Anderson views all the theories four-dimensionally: as theories about a four-dimensional manifold with various geometrical structures, vector fields, tensor fields, and so on defined on it. Their differing content results from different choices of the basic geometrical structures.

Thus, Anderson's book clarified the mathematical aspect of the development of relativity considerably. Similar clarification was contributed by A. Trautman and P. Havas; indeed, the basic approach can be traced back to the work of Kretschmann, Weyl, and Cartan in the early years of the century. At the same time, this four-dimensional approach to the mathematics appeared to fit in very well with a physicalizing or "realist" attitude toward geometry; and, in fact, the philosophers John Earman and Howard Stein took up the defense of this attitude, explicitly opposing the more standard "conventionalism" of Reichenbach and Grünbaum, in articles published around 1970. I was greatly impressed by the mathematical clarity of the four-dimensional approach; it harmonized nicely with my "realist" prejudices; and so I undertook to write a dissertation that cleaned up some of the mathematical machinery introduced by Anderson and attempted to show how it contributed to a "realist" interpretation of the various space-time theories. This dissertation, completed in 1972 under the direction of Glymour, is a remote ancestor of Chapters II–V of the present book.

Unfortunately, the defense of "realism" found in my dissertation was extremely dogmatic and inadequate from a philosophical point of view. To be sure, an alternative to "conventionalism" was clearly articulated, but there was no real attempt to respond to the arguments and motivations of the opposing tradition. I was helped to see this by reading Lawrence Sklar's *Space, Time, and Spacetime*, published in 1974. Sklar emphasized the simple point that, although the four-dimensional approach contributes much mathematical clarification, it does not automatically resolve the fundamental conflicts in interpretation at one stroke. In particular, it does not automatically do justice to the methodological moves embodied in the development of relativity theory itself. If good scientific methodology leads us to reject absolute space, absolute simultaneity, gravitational acceleration, and so on, why are we not moved to reject other pieces of "unobservable" theoretical structure as well? Is not a "conventionalist" interpretation of geometrical structure the right one after all?

In the years 1975–1977 I set out to revise and expand my dissertation so as to develop a genuine response to the "conventionalist" tradition in general and Sklar's challenge in particular. I looked to the development of relativity to find a methodological property sufficient to distinguish "good" pieces of theoretical structure from "bad." Above all, I wanted to understand the difference between those aspects of geometrical structure retained by general relativity, such as the space-time manifold itself and the metrical tensor field defined thereon, and those aspects of geometrical structure abandoned by general relativity, such as absolute space, absolute time, gravitational acceleration, and so on. I thought I found an answer to these questions in the notion of "unifying power," which, as I later discovered, is a throwback to the notion of "consilience" introduced by William Whewell in 1840. "Good" pieces of theoretical structure contribute to "theoretical unification," or what Whewell calls "the consilience of inductions." "Bad" pieces of theoretical structure do not. Moreover, the traditional relativity principles are themselves best understood as methodological mechanisms for eliminating theoretical structure deficient in "unifying power."

I developed these ideas in extensive revisions of Chapters II–V and in ancestors of Chapters VI and VII, which I then circulated to friends and colleagues. I received many useful comments and criticisms on this manuscript, especially from Hartry Field, Roger Jones, David Malament, and Sklar himself. The present version, written in the years 1978–1981, is a revision and expansion of the 1975–1977 manuscript. I tried to tighten up and simplify the argument, greatly assisted, of course, by the critical reaction I had received; I also tried to add a bit of historical perspective to the discussion in Chapters I and VII. In writing this version I had no direct stimulus comparable to the books by Anderson and Sklar, but Hartry Field's *Science Without Numbers*, published in 1980, provided helpful tools for articulating the final mathematical framework.

The present book makes heavier mathematical demands than the reader may be used to in the philosophy of science. Unfortunately, given the way in which physical and philosophical problems are intertwined with mathematical notions here, especially with the notion of general covariance, there is no practicable alternative. In particular, one has to see in detail how one goes from generally covariant formulations to the more usual formulations and back again if one is to attain any real clarity about the various relativity principles. Of course, these formulations and methods are in no way original with me: I have taken them, with minor alterations, from the works of Anderson, Trautman, and Havas mentioned above. In any case, I have striven to make the treatment as intuitive as possible via numerous pictures and diagrams, and I have included an appendix outlining the geometrical techniques employed. All that is absolutely presupposed is an acquaintance with the basic ideas of analysis and linear algebra on the level of an undergraduate course in advanced calculus. (Explanatory references to the mathematical appendix are scattered throughout the text: thus, (A.10) refers to equation (10) of the appendix.)

In writing this book I have naturally acquired many intellectual debts, some of which I have already indicated above.

My first and greatest debt is to Clark Glymour. He directly stimulated my original interest in the subject and, as the director of my dissertation, helped me immensely in formulating the seminal ideas of the book. I inherited from him both a general sense of what it is to do philosophy of physics and a particular view of what a "realist" interpretation of physical theory requires: namely, an epistemological or methodological account of how one chooses between "empirically equivalent" theories. In pursuing this kind of account we have since taken rather different paths, but the similarity of underlying motivation in our work is both obvious and deliberate.

John Earman was the second reader of my dissertation and provided a source of stimulation second only to Glymour's. Indeed, it is not too much to say that Earman's original work on relativity theory in the early 1970s set a standard of rigor, insightfulness, and philosophical boldness that has served as an inspiration to a whole generation of philosophers of science.

David Malament provided constant encouragement and constructive criticism, both early and late. He read and commented on every draft, patiently explaining numerous mathematical, physical, and philosophical complexities I had overlooked. I can truly say that without his encouragement and advice the book would not have been completed.

Roger Jones and Lawrence Sklar supplied the most extensive and helpful comments on the penultimate draft, comments that provided the most important stimulation for the writing of the final version. Jones, in particular, subjected my version of Anderson's interpretation of relativity principles to searching criticisms that I have still not answered fully.

A debt of a special kind is owed to Burton Dreben, who for many years has urged that I should not only concentrate on finding "clean" versions of the traditional relativity principles but also attend to the physical, mathematical, and philosophical history within which the more usual "messy" versions are embedded. It has taken me a long time to see the full force of his complaint, and I have endeavored to be (slightly) less ahistorical in this final version of the book. I hope to delve more deeply into the history in future work. (I am also indebted to Dreben for the technical terminology on p. 289.)

I was fortunate enough to be able to try out ideas in conversations with Scott Weinstein, who is extraordinarily perceptive and helpful with even the most inchoate beginnings of a thought. Warren Goldfarb read several sections of the final manuscript and provided very useful advice on mathematical expression and notation, especially in Chapter II. I am indebted to Zoltan Domotor and Robert Wachbroit for suggestions of a similar nature.

I am grateful to the American Council of Learned Societies for a fellowship in the academic year 1976–1977 and to the University of Pennsylvania for a semester's leave with pay in the fall of 1978. Without their generous financial assistance I would not have had the time serious writing requires.

Foundations of Space-Time Theories

I

Introduction: Relativity Theory and Logical Positivism

The relationship between the development of relativity theory and twentieth-century philosophy of science is both fascinating and complex. On the one hand, relativity theory, perhaps more so than any other scientific theory, developed against a background of explicitly philosophical motivations. As is well known, both Leibnizean relationalism and Machian empiricism figured prominently in Einstein's thought. On the other hand, twentieth-century philosophy of science, and logical positivism in particular, is almost inconceivable without relativity, for relativity theory was second only to *Principia Mathematica* as an intellectual model for the positivists.[1] It appeared to realize all their most characteristic ideals, from a general distrust of non-observational, "metaphysical" notions (such as absolute space and absolute simultaneity) to a specific program for dividing science into "factual" statements on the one side and "definitions" or "conventions" on the other. In short, Einstein himself was influenced by positivist and empiricist philosophy, and the logical positivists used both this fact *and* the theories that resulted as central sources of support for their philosophical views.

This book is an attempt to sort out some of the many different elements in this complicated relationship of mutual influence. To what extent does relativity theory actually realize Einstein's philosophical motivations? And, consequently, to what extent does relativity support the philosophical ideals of logical positivism? My answer is that both Einstein and the positivists were wrong. The

---

[1] Of course, quantum theory was also grist for the positivists' mill; see, e.g., Reichenbach [88]. Nevertheless, it does not rank with relativity as an *inspiration* for positivist philosophy, largely because quantum theory did not take on a characteristically positivist flavor until Heisenberg's work in the middle and late twenties.

theories Einstein created do not vindicate his Leibnizean-Machian motivations. Relativity theory neither supports nor embodies a general positivistic point of view. But this conclusion is of only secondary interest to me. The moral I would most like to convey is this: one cannot hope to understand adequately either the development of relativity theory or twentieth-century positivism without a detailed account of their reciprocal interaction. In particular, we cannot hope to advance beyond the positivists if we ignore the roots of their doctrines in the scientific practice of their time. Logical positivism's intimate involvement with twentieth-century physics is one of the main sources of its considerable force and vitality. Any alternative philosophy of science must come to terms with this physics in an equally vital way.

It seems to me that contemporary criticisms of logical positivism suffer because they ignore this moral, and this is especially true of popular attacks on the observational/theoretical distinction. That is not to say that these attacks are unwarranted. The positivists' use of the observational/theoretical distinction *was* often naive: they assumed to be clear and unproblematic a distinction that is at best both vague and highly complicated; what is worse, they appeared to be unaware of the many ways in which changes in theory can alter what counts as observational and non-observational (in popular parlance, they ignored the "theory-ladenness" of observation).[2] The trouble is that all this is much too easy to say. We know that the observational/theoretical distinction is shaky and problematic, but we want to know more about its role in the positivists' general epistemological program and in the revolutionary physical theories that inspired that program. Criticisms of positivism that avoid these matters are liable to appear superficial and unsatisfying. We can throw away the observational/theoretical distinction if we like, but we will have no idea how to construct a better epistemology or a better account of twentieth-century physics.[3]

---

[2] As I shall indicate below, even these criticisms do not apply to the *early* (pre-1930) positivists. See, for example, the introduction to Reichenbach [85], which contains a quite sophisticated discussion of the distinction, including a clear acknowledgment of "theory-ladenness."

[3] This is one reason Quine's attack on logical positivism is so much more satisfying. Quine concentrates on the philosophical program in which the observational/theoretical distinction is embedded. In fact, Quine thinks that this distinction is one of the few elements of the program that can be salvaged; cf. [82], 84–89.

4

If we look at the actual development of relativity theory, the observational/theoretical distinction appears to play a central role. It is a key element, in fact, in Einstein's arguments for both the special principle of relativity and the general principle of relativity. In outline, these arguments go as follows. Classical physics makes use of *absolute motion*, that is, motion of physical bodies with respect to absolute space. But only *relative motion*—motion of physical bodies with respect to other physical bodies—is observable. Therefore, appeals to absolute motion should be eliminated from physical theory. Thus, the special principle of relativity allows us to dispense with the notions of absolute velocity and absolute rest. All systems moving with constant velocity are "equivalent," and velocity makes sense only relative to one or another such system. Accordingly, Einstein criticized classical electrodynamics for invoking something more than relative velocity:

> The observable phenomenon here depends only on the relative motion of the conductor and the magnet, whereas the customary view draws a sharp distinction between the two cases in which either the one or the other of the two bodies is in motion. ([26], 37)

On the other hand, the special principle of relativity does not go far enough, because special relativity still makes use of an absolute distinction between *inertial* systems (systems moving with constant velocity) and *accelerated* systems. Absolute velocity is indeed eliminated, but absolute acceleration is retained. Einstein again criticized this distinction by appealing to what is observable:

> But the privileged space R_1 of Galileo [= an inertial frame], thus introduced, is a merely *factitious* cause, and not a thing that can be observed. ([28], 113)

The general theory of relativity is supposed to overcome this defect in the special principle by dispensing with the distinction between inertial systems and accelerated systems, just as special relativity dispenses with the distinction between systems at rest and systems moving with (nonzero) constant velocity. According to the general principle of relativity, all states of motion are "equivalent."

In short, the development of relativity theory appears to have the following structure:

5

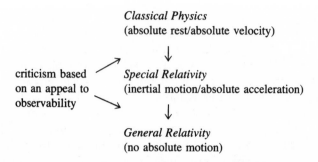

This development appears to illustrate the fruitfulness of both the observational/theoretical distinction and the positivists' general distrust of nonobservational entities and properties. It is no wonder, then, that the positivists were so attracted to the picture expressed in the following passage from Reichenbach:

> The physicist who wanted to understand the Michelson experiment had to commit himself to a philosophy for which the meaning of a statement is reducible to its verifiability, that is, he had to adopt the verifiability theory of meaning if he wanted to escape a maze of ambiguous questions and gratuitous complications. It is this positivist, or let me rather say, empiricist commitment which determines the philosophical position of Einstein. It was not necessary for him to elaborate on it to any great extent; he merely had to join a trend of development characterized, within the generation of physicists before him, by such names as Kirchhoff, Hertz, Mach, and to carry through to its ultimate consequences a philosophical evolution documented at earlier stages in such principles as Occam's razor and Leibniz's identity of indiscernibles. ([89], 290–291)

Similarly, we can also understand why the positivists were attracted to conventionalism. The essence of conventionalism is a doctrine of "equivalent descriptions": alternative, seemingly incompatible theoretical descriptions are declared to be equivalent when they agree on all observations. Consequently, theoretical assertions are not "objectively" true or false: they have truth-value only relative to one or another arbitrarily chosen "equivalent description." This doctrine also appears to be borne out by relativity theory, where alternative ascriptions of motion are not "absolutely" true or false but are true or false only relative to one or another arbitrarily chosen reference system. Moreover, all reference systems are "equivalent"; they provide equally good descriptions of the same observable facts.

This way of looking at the story is still relatively superficial, however. It is not just that relativity theory provides a model or exemplar for the empiricist and verificationist prejudices of twentieth-century positivism; rather, the historical development of the former plays an integral role in the historical development of the latter. Logical positivism did not start out as a version of empiricism or verificationism *à la* Hume or Mach. Contrary to popular wisdom, the influence of Kant was much more important. Twentieth-century positivism began as a neo-Kantian movement whose central preoccupation was not the observational/theoretical distinction but the form/content distinction. Without arguing the point here, I believe that this theme comes through clearly in Carnap's *Aufbau* (1928), in Schlick's *General Theory of Knowledge* (1918) and *Space and Time in Contemporary Physics* (1917), and, most explicitly, in Reichenbach's *The Theory of Relativity and A Priori Knowledge* (1920). Central to all these works is the idea that natural knowledge has two different elements: a *form*, which is in some sense conceptual or mind-dependent, and a *content*, which is contributed by the world or experience. Moreover, all these works side with Kant against empiricism in emphasizing the necessity and importance of such formal elements in our knowledge of nature. Where Kant went wrong was in characterizing these formal elements as unrevisable, synthetic *a priori* truths; they are better described as "conventions" or "arbitrary definitions."

Kant's version of the form/content distinction drew much of its strength from eighteenth-century mathematical physics and in particular from the Euclidean-Newtonian picture of a space-time framework or "container" filled with matter (content) obeying deterministic laws of motion. Here, according to the early positivists, the Kantian view is limited by its too intimate connection with outmoded mathematics and physics. The task of twentieth-century philosophy is to develop a version of the Kantian form/content distinction that is properly sensitive to developments in modern mathematics and modern physics.[4] A brief review of these develop-

---

[4] Carnap, unlike Reichenbach or Schlick, bases his reinterpretation of the form/content distinction almost wholly on modern *logic*. For Carnap, the formal is understood in terms of logical form, and notions from mathematical physics actually play very little role. An important unresolved question (at least in my mind) is how to reconcile the linguistic or logical interpretations of "formal" or "conventional" typical of Carnap's work with the interpretations drawn from mathematical physics typical of Reichenbach's work.

ments will give us a preliminary sense of the issues and complexities involved in the relationship between positivism and relativity theory. We shall see also how the form/content distinction actually becomes transformed into the theoretical/observational distinction.

The most important mathematical development from this point of view is an extreme generalization of the concepts of space and geometry. This process begins with Gauss's theory of surfaces of 1827. But here already there is a twofold generalization that is the source of much later confusion: a generalization in the use of *coordinates*, and a generalization to non-Euclidean *geometries*. First, Gauss considers the use of general (Gaussian) coordinates x_1, x_2 in place of the familiar rectangular or Cartesian coordinates. In Cartesian coordinates the "line element" takes the familiar Pythagorean form

$$ds^2 = dx_1^2 + dx_2^2. \tag{1}$$

In arbitrary coordinates it takes the more complicated form

$$ds^2 = g_{11}\, dx_1^2 + g_{12}\, dx_1\, dx_2 + g_{21}\, dx_2\, dx_1 + g_{22}\, dx_2^2 \tag{2}$$

where the g_{ij}'s are real-valued functions of the coordinates. (For example, in polar coordinates $ds^2 = dr^2 + r^2\, d\theta^2$; so $g_{11} = 1, g_{12} = g_{21} = 0$, and $g_{22} = r^2$.) Secondly, however, Gauss also considers the geometry of arbitrary curved surfaces (for example, the surface of a sphere or of a hyperboloid of revolution). He shows that the geometry on such a surface is in general non-Euclidean. For example, the geometry on the surface of a sphere is a geometry of positive curvature in which there are no parallels; the geometry on a hyperboloid of revolution is a geometry of negative curvature in which there are many parallels.

These non-Euclidean surfaces can also be covered by Gaussian coordinates x_1, x_2, and they also have a line element given by (2). Moreover, as Gauss shows, the coefficients of the line element in (2)—the g_{ij}'s—give us complete information about the curvature, and hence the geometry, of our surface. *It is extremely important, however, to distinguish a mere change in coordinates from an actual change in geometry.* In particular, the more complicated form (2) of the line element does not necessarily signal a non-Euclidean geometry: we can use non-Cartesian coordinates on a flat, Euclidean plane as well (as our example of polar coordinates clearly shows). What distinguishes a flat, Euclidean geometry is the *existence* of

8

coordinates satisfying (1): on a Euclidean plane we can always transform the more complicated form (2) into the Pythagorean form (1) (for example, by transforming polar coordinates into Cartesian coordinates). By contrast, on a curved, non-Euclidean surface it is impossible to perform such a transformation: Cartesian coordinates simply do not exist.

We should distinguish, then, between *intrinsic* and *extrinsic* features of a surface. Intrinsic features characterize the geometrical structure of a surface—its curvature, Euclidean or non-Euclidean character, and so on—and are completely independent of any particular coordinatization of the surface. Extrinsic features, on the other hand, correspond to particular coordinatizations of the surface; accordingly, they vary as we change from one coordinate system to another. Thus, the actual form of the line element (2)— obtained by substituting particular functions for the g_{ij}'s—is an extrinsic feature of the surface, for these functions change when we transform our coordinate system. The connection between intrinsic features and extrinsic features is this: a given intrinsic feature of a surface corresponds to the *existence* of coordinate systems with certain extrinsic features. For example, a flat, Euclidean structure corresponds to the existence of Cartesian coordinates in which $g_{11} = g_{22} = 1, g_{12} = g_{21} = 0$; it corresponds to the simple form (1) for the line element.

In the remainder of the nineteenth century these ideas underwent further generalization. The high point of this process was Riemann's Inaugural Address of 1854. Riemann considers arbitrary spaces or *manifolds* of any number of dimensions: points in such manifolds can be uniquely represented by *n-tuples* of real numbers. Given a particular coordinatization $x_1, x_2, \ldots, x_n$ of an arbitrary n-dimensional manifold, we can define an n-dimensional line element or *metric tensor* by

$$ds^2 = \sum_{i,j=1}^{n} g_{ij}\, dx_i\, dx_j \qquad (3)$$

subject to the conditions of symmetry ($g_{ij} = g_{ji}$) and positive-definiteness ($ds^2 > 0$). (If we drop the positive-definiteness requirement, we obtain a *semi-Riemannian* metric, which is important in relativity.) Riemann shows how to define the notion of *curvature* for such a manifold and shows that the special case of a flat or Euclidean manifold is characterized by the existence of coordinates in which the matrix of coefficients (g_{ij}) in (3) takes the form

9

diag(1,1, . . . ,1): that is, $g_{ij} = 1$ for $i = j$, $g_{ij} = 0$ for $i \neq j$. (If our metric is semi-Riemannian and flat, this matrix of coefficients can be put in the form $(g_{ij}) = \text{diag}(\pm 1, \pm 1, \ldots, \pm 1)$.)

Once again, however, we must distinguish changes in coordinates from changes in geometrical structure, extrinsic from intrinsic. A mere change in coordinates from $x_1, x_2, \ldots, x_n$ to $\bar{x}_1, \bar{x}_2, \ldots, \bar{x}_n$ results in a change in the coefficients in (3) from g_{ij} to $\bar{g}_{ij}$, but, of course, the line element itself is preserved:

$$\sum_{i,j=1}^{n} \bar{g}_{ij} \, d\bar{x}_i \, d\bar{x}_j = \sum_{i,j=1}^{n} g_{ij} \, dx_i \, dx_j.$$

If, on the other hand, we change our geometry—from a flat, Euclidean structure to a curved, non-Euclidean structure, for example—we change to a new metric tensor $d\bar{s}^2$ given by

$$d\bar{s}^2 = \sum_{i,j=1}^{n} g_{ij}^* \, dx_i \, dx_j.$$

If our new metric $d\bar{s}^2$ is indeed non-Euclidean, then there will exist no coordinates in which $(g_{ij}^*) = \text{diag}(1,1, \ldots ,1)$.

Riemann's work allows us to see all the different kinds of geometrical structures—Euclidean and non-Euclidean; constant curvature and variable curvature; two-, three-, and higher dimensional spaces—as particular instances of the very general idea of an n-dimensional manifold. Each particular type of space results from a given choice of n and a given choice of line element (3). So we can view an n-dimensional Riemannian manifold as an abstract form or schema that can be "filled in" by various metric tensors to yield various concrete geometrical spaces. Antecedent to a particular choice of metric tensor, an abstract n-dimensional manifold has no geometrical structure: it is "metrically amorphous." The only structure possessed by such an abstract manifold is the locally Euclidean *topological* structure given by the use of n-tuples of real numbers as coordinates. This locally Euclidean topological structure is common to all particular Riemannian manifolds, regardless of their different geometries.

Further, given any particular geometrical space or manifold, we can distinguish various levels of structure. Consider an n-dimensional Euclidean space, for example. The lowest level of structure is the *metrical* structure given by a Pythagorean line element:

$$ds^2 = dx_1^2 + dx_2^2 + \cdots + dx_n^2. \tag{4}$$

This structure gives us a *length* for every curve, a *distance* between any two points, a notion of *straight line*, and an *angle* between any two intersecting straight lines. But some of these notions are more general than the metrical structure (4), so they really belong on "higher" levels. Thus, for example, the *affine structure*, or class of straight lines, is given by the condition

$$x_i = a_i u + b_i \tag{5}$$

where the a_i and b_i are constants and u ranges through the real numbers. The class of coordinate systems satisfying (5) is wider than the class of (Cartesian) coordinate systems satisfying (4), so different metrics can give rise to the same affine structure. Similarly, the *conformal structure*, or notion of angle, is also more general than the metrical structure (4): if Ω is any real-valued, positive function on our space, then $d\bar{s}^2 = \Omega\, ds^2$ yields the same angles as ds^2. Finally, the highest level of structure is just the manifold structure itself: the locally Euclidean *n*-dimensional *topology*.

An extremely elegant perspective on these different levels of structure was provided by Klein's Erlanger Program of 1872. Klein considers the groups of transformations, or *automorphisms*, that preserve the different levels of structure, where an automorphism is a one-one mapping of an *n*-dimensional manifold onto itself taking each point $\langle x_1, x_2, \ldots, x_n \rangle$ onto a second point $\langle x_1^*, x_2^*, \ldots, x_n^* \rangle$. Thus, the metrical structure is preserved by the *rigid motions*, or *isometries*, which in a Euclidean space take the form

$$x_i^* = \sum_{j=1}^{n} \alpha_{ij} x_j + \beta_i \tag{6}$$

where the α_{ij} and β_i are constants and the matrix (α_{ij}) is *orthogonal*: that is, $\sum_j \alpha_{ij}\alpha_{kj} = \delta_{ik} = 1$ if $i = k$ and 0 if $i \neq k$ (so the Euclidean group (6) is just the group of rotations and translations). The class of straight lines, on the other hand, is preserved by the wider group of *affine transformations*, which in a Euclidean space take the form (6) with arbitrary constants (so the affine group is just the full group of linear transformations). The notion of angle is preserved by the *conformal transformations*, and the topological structure is preserved by *arbitrary transformations*:

$$x_i^* = f_i(x_1, x_2, \ldots, x_n) \tag{7}$$

where the f_i are sufficiently continuous.

11

In short, as we move to more and more general levels of geometrical structure, the associated groups of transformations become wider and wider.

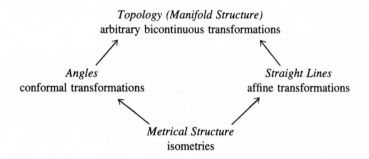

Topology (Manifold Structure)
arbitrary bicontinuous transformations

Angles
conformal transformations

Straight Lines
affine transformations

Metrical Structure
isometries

Note also that there is a close connection between Klein's groups of transformations and the existence of classes of privileged coordinate systems. If we interpret (7) as a *coordinate transformation*, then, for example, a Euclidean transformation (6) maps one Cartesian system satisfying (4) onto another, a linear transformation maps one affine system satisfying (5) onto another, and an arbitrary transformation (7) maps one arbitrary coordinate system onto another. Under this interpretation (7) is the group of all admissible (one-one and sufficiently continuous) coordinate transformations.

The most important development in physics has been an evolving relativization of the concept of motion and a corresponding relativization of the concepts of length and simultaneity. Newtonian physics pictures material objects or bodies as embedded or contained in an infinite, three-dimensional Euclidean space. The primary concept of motion is *absolute motion*: change of position with respect to this Euclidean embedding space. *Relative motion*—change of position of one material body with respect to another—is conceived of as induced by, or composed of, absolute motions. Thus, for example, if we cover Newton's three-dimensional Euclidean space by a Cartesian coordinate system x_1, x_2, x_3 and define the *absolute acceleration* of a body by

$$a^i = \frac{d^2 x_i}{dt^2}$$

we have Newton's Second Law

$$F^i = ma^i \qquad (8)$$

where F^i is the "impressed force" acting on our body and m is its mass. If, on the other hand, we interpret a^i as the *relative acceleration* of one body with respect to another, then (8) will not in general be true: additional accelerations (for example, centrifugal accelerations) will have to be included.

More generally, all Newton's laws of motion are true when the motions considered are interpreted as absolute motions but not when the motions considered are interpreted as relative motions. Nevertheless, one class of relative motions does satisfy Newton's laws. Suppose we take our original Cartesian coordinate system x_1, x_2, x_3—which is of course at rest with respect to Newton's Euclidean embedding space—and impart to it a constant absolute velocity v. That is, we change to a new coordinate system by the transformation:

$$\bar{x}_1 = x_1 - vt$$
$$\bar{x}_2 = x_2 \qquad (9)$$
$$\bar{x}_3 = x_3.$$

Our new system $\bar{x}_1, \bar{x}_2, \bar{x}_3$ is called an *inertial frame*; (9) is called a *Galilean* transformation. Now let $\bar{a}^i$ be the acceleration of a body with respect to such an inertial frame: $\bar{a}^i = d^2\bar{x}_i/dt^2$. It is easy to show that $\bar{a}^i = a^i =$ the absolute acceleration and, therefore, that Newton's Second Law (8) holds also for accelerations with respect to an arbitrary inertial frame. More generally, all Newton's laws of motion are true when the motions considered are interpreted as motions with respect to an arbitrary inertial frame. This statement is the Newtonian, or classical, *principle of relativity*.

That Newtonian physics satisfies a relativity principle is the source of familiar epistemological problems, problems that were first articulated by Leibniz. Leibniz objected that Newton's three-dimensional Euclidean embedding space gives rise to distinct but *indistinguishable* states of affairs if one simply changes the absolute positions of all material bodies while preserving their relative positions. The classical principle of relativity gives precise expression to this kind of indistinguishability. All inertial frames, regardless of their position, orientation, or velocity in Newton's three-dimensional embedding space, yield the same laws of motion. So it is impossible, according to Newton's laws themselves, to determine which inertial

13

frame one is in. Absolute position and absolute velocity appear to have no physical significance. Why, therefore, should we believe in a three-dimensional absolute space? Should we not use the principle of parsimony to reject such "metaphysical" entities?

On the other hand, although Newtonian physics satisfies a relativity principle for absolute position and absolute velocity, it does not satisfy a relativity principle for absolute acceleration and absolute rotation. Newton's laws of motion hold only in *inertial* frames, all of which have constant velocity in our three-dimensional embedding space. The laws of motion are not true in absolutely accelerating or rotating frames. Thus, for example, an absolutely rotating object will experience distorting effects—due to centrifugal forces—that will not be experienced by a merely relatively rotating object (compare Newton's "rotating bucket"). Absolute acceleration and rotation, unlike absolute position and velocity, do appear to have physical significance. Therefore, we are faced with a dilemma. Some types of absolute motion—acceleration and rotation—have observable significance and are distinguishable by the laws of our theory. This speaks for the reality of absolute space. Other types of absolute motion—rest and velocity—have no observable significance and are not distinguishable by the laws of our theory. This speaks against the reality of absolute space.

Maxwell's electrodynamics of 1864 appeared to resolve this dilemma in favor of absolute space. Maxwell's theory implies that electromagnetic disturbances are propagated in the form of waves and that

> electromagnetic waves propagate through empty space
> with a definite constant velocity c (10)

Yet (10), unlike (8), cannot hold in all inertial frames: if we subject the velocity c to a Galilean transformation (9), we of course obtain the velocity $c - v$, not the velocity c. Unlike (8), (10) allows us to distinguish a proper subclass of inertial frames, all of which are at rest relative to one another. It is natural, then, to identify the reference frames in which (10) holds with the Newtonian rest frames, that is, the inertial frames at rest in our three-dimensional embedding space (so these frames are related by the Euclidean rigid motions (6)). Hence, Maxwell's electrodynamics appears to give observable significance to absolute rest and absolute velocity, and Newton's absolute embedding space appears to be fully vindicated.

14

The problem here, of course, is that the combined theory of Newtonian space and time plus Maxwell's electrodynamics turns out to be false. A series of experiments conducted at the end of the nineteenth century—the most famous being the Michelson-Morley experiment of 1887—resulted in the conclusion that we cannot distinguish one inertial frame from another by electromagnetic processes after all. The classical principle of relativity, originally formulated for Newton's mechanical laws, appeared to hold equally for Maxwell's electrodynamical laws. Once again, absolute motion, and therefore absolute space, appeared to have no observable significance. But the situation is even worse, for the fact that the law (10) holds in all inertial frames in highly counter-intuitive, even paradoxical. If something has velocity c with respect to one inertial frame, how can it possibly have velocity c with respect to a second inertial frame in motion relative to the first? Imagine, for example, that the second frame has velocity $.999c$ relative to the first!

Einstein pinpointed the source of this aura of paradox in 1905. It stems from our tacit assumption that the Galilean transformations (9) do not affect *time*, that the time coordinates $\bar{t}$ and t of our two frames are necessarily identical. Einstein saw that the simplest way to implement a relativity principle for electrodynamics is to retain Maxwell's laws intact while changing the transformations connecting the inertial frames. These new transformations, the *Lorentz* transformations, are given by

$$\bar{t} = \frac{t - \frac{v}{c^2}x_1}{\sqrt{1 - \frac{v^2}{c^2}}}$$

$$\bar{x}_1 = \frac{x_1 - vt}{\sqrt{1 - \frac{v^2}{c^2}}} \tag{11}$$

$$\bar{x}_2 = x_2$$
$$\bar{x}_3 = x_3.$$

The transformations (11), unlike the transformations (9), preserve the laws of Maxwell's electrodynamics. (Thus, for example, a simple calculation shows that the velocity c in (10) is preserved.)

15

Einstein had one further problem in 1905. Experience teaches that our relativity principle holds equally for electromagnetic and mechanical phenomena, but the Lorentz transformations (11) do not preserve Newton's laws of motion, which are instead preserved by the Galilean transformations (9). So Einstein showed how to construct new laws of motion that are preserved by (11). The key move here is a modification of the concept of *momentum*. Einstein's finished theory, the special theory of relativity, satisfies a relativity principle for both mechanics and electrodynamics. In this theory the concepts of rest and velocity have been thoroughly relativized. Furthermore, and this is immediate from (11), the concepts of *simultaneity*, *duration*, and *length* are relativized as well: they have significance only relative to one or another inertial frame but no absolute significance.

Perhaps the clearest perspective on Einstein's new theory was provided by Minkowski's geometrical interpretation of 1908. On this interpretation, the special theory of relativity describes a four-dimensional, semi-Euclidean manifold, with line element given by

$$ds^2 = dx_0^2 - dx_1^2 - dx_2^2 - dx_3^2 \qquad (12)$$

where $x_0 = ct$. The inertial frames are just the Cartesian coordinate systems for this line element, and the Lorentz transformations take Cartesian coordinate systems onto Cartesian coordinate systems; that is, they preserve the form (12) for ds^2 (so the Lorentz transformations are analogous to the Euclidean orthogonal transformations (6)). Furthermore, the trajectories, or *world-lines*, of free particles are straight lines or *geodesics* of the metric (12): they are four-dimensional curves of extreme—in fact, maximal—"length" according to ds, and they satisfy the linear equation (5) in all inertial frames.

Einstein's answer to the dilemma of absolute motion, then, is this. There is indeed no three-dimensional, Euclidean embedding space, but there is a four-dimensional, semi-Euclidean *space-time* in which all physical events are embedded. Within this space-time manifold, a privileged class of straight lines, or *inertial* trajectories, represents the world-lines, or histories, of (possible) free particles. A body is moving inertially (neither accelerating nor rotating) just in case its world-lines are all four-dimensional straight lines; a body is moving noninertially (either accelerating or rotating) just in case its world-lines deviate from four-dimensional straightness. Thus, acceleration

16

and rotation are essentially four-dimensional notions, whereas rest and velocity are essentially three-dimensional. In moving from a three- to a four-dimensional "container" or embedding space, we eliminate the latter notions while retaining the former.

This way of putting things shows quite clearly how far the special theory falls short of the relationalist ambitions of Leibniz and Mach. We still have absolute motion, and we explain absolute motion in terms of a class of highly abstract and nonobservational entities: our four-dimensional inertial frames. Moreover, these four-dimensional inertial frames are no more likely to be "occupied" or "anchored" by concrete physical bodies than is Newton's original three-dimensional Euclidean space. From the point of view of a thoroughgoing relationalism, therefore, they are just as objectionable and "metaphysical." This was Einstein's point of view in 1916 as well. As we indicated above, one of Einstein's principal motivations in developing the general theory of relativity was to eliminate *all* absolute motion, to implement fully a relativistic conception of motion.

How was general relativity to accomplish this task? Einstein had two central ideas in 1916. The first was to eliminate the privileged class of inertial reference frames and to formulate laws of motion valid in arbitrary reference frames or coordinate systems. Such laws of motion would be *generally covariant*: they would be preserved by all admissible transformations (7), and not merely by the Lorentz transformations (11). The point, of course, is that such generally covariant laws appear to implement a thoroughgoing "equivalence" or "indistinguishability" of all states of motion and to extend the classical and special principles of relativity to a truly *general* principle of relativity. Second, Einstein was attracted by Mach's reply to Newton's "rotating bucket." The idea here was to account for the distorting centrifugal effects experienced by "absolutely" rotating objects in terms of their relative rotation with respect to the total distribution of matter in the universe. We no longer appeal to such objects' absolute rotation with respect to some unobservable inertial frame—inertial frames no longer exist.

Here is how Einstein's 1916 theory actually works. We start with a flat, Minkowskian manifold with line element given by (12). We then impart a *variable curvature* to this manifold, where the degree of curvature in a given region depends on the distribution of mass and energy. Since our new manifold is not flat or semi-Euclidean, we can no longer use the simple form (12) for the line element. Instead,

17

we need the more general form

$$ds^2 = \sum_{i,j=0}^{3} g_{ij}\, dx_i\, dx_j \tag{13}$$

where the g_{ij}'s are not constant. Thus, inertial or Cartesian coordinate systems no longer exist. Now consider the geodesics—curves of maximal "length"—of this new metric. We interpret these as the world-lines, or histories, of possible *freely falling* (gravitationally affected) particles (so the g_{ij}'s in (13) are in a sense the "potentials" of the gravitational field). Finally, we observe that our new, non-Euclidean straight lines provide us with four-dimensional notions of acceleration and rotation, just as their semi-Euclidean counterparts did in special relativity. The difference is that these lines, unlike their Minkowskian counterparts, are affected and "shaped" by the total distribution of matter—hence the connection with Mach.

As we noted above, the early positivists—Reichenbach, Schlick, and, to a lesser extent, Carnap[5]—wanted to redraw the Kantian form/content distinction in light of post-Kantian developments in mathematics and physics. Like Kant, these writers thought it essential to distinguish between those aspects of our knowledge of nature contributed by the mind or reason and those aspects contributed by the world or experience. Unlike Kant, however, they did not think that the formal or rational components of knowledge could be characterized in terms of unrevisability, necessity, or *a prioricity*. After all, if the developments in post-Kantian mathematics and physics show anything, they show that one central Kantian formal component—the Euclidean-Newtonian picture of space and time— is clearly not *a priori* or unrevisable. The positivists were quick to draw the conclusion that *nothing* in our knowledge of nature could be *a priori* and unrevisable in Kant's sense.[6]

If the apparent certainty and necessity of Euclidean geometry does not provide us with a model or paradigm for the formal or

---

[5] See note 4 above. One way in which Carnap's interpretation of "formal" (in the *Aufbau*) differs significantly from Reichenbach's is that for Carnap, like Kant, the formal = the objective. For Reichenbach, the formal = the subjective. One place where Carnap's ideas touch on a Reichenbachian account is §§146–148 of the *Aufbau* [8], where Carnap discusses intersubjectivity. In effect, Carnap's "intersubjectivizing" function is just a coordinate transformation from the reference frame of one observer to the reference frame of another. (Burton Dreben pointed this out to me.)

[6] See, e.g., Reichenbach [84], 77–87.

mind-dependent, what does? The new paradigm suggested by the positivists is our use of spatial and spatio-temporal coordinates. An arbitrary choice of one admissible coordinate system represents the subjective contribution of reason; what is invariant in a class of admissible coordinate systems represents the objective contribution of reality. According to Reichenbach,

> Just as the invariance with respect to the transformations characterizes the objective nature of reality, the structure of reason expresses itself in the arbitrariness of admissible systems. Thus it is obviously not inherent in the nature of reality that we describe it by means of coordinates; this is the subjective form that enables our reason to carry through the description . . . . Kant's assertion of the ideality of space and time has been precisely formulated only in terms of the relativity of the coordinates. ([84], 90)

Thus, an arbitrary choice of admissible coordinate system provides the positivists with their first (and best) example of a convention or arbitrary definition. This is also the genesis of Reichenbach's notion of "*coordinative* definition."

The idea, then, is that the Gauss-Riemann theory of manifolds and coordinates, especially as embodied in the theory of relativity, provides us with a new method for "teasing out" the formal or subjective elements of knowledge. Again Reichenbach explains:

> The method of distinguishing the objective significance of a physical statement from the subjective form of the description through transformation formulas, by indirectly characterizing this subjective form, has replaced Kant's analysis of reason. It is a much more complicated procedure than Kant's attempt at a direct formulation, and Kant's table of categories appears primitive in comparison with the modern method of the theory of invariance. But in freeing knowledge from the structure of reason, the method enables us to describe this structure; this is the only way that affords us an understanding of the contribution of our reason to knowledge. ([84], 91–92)

Note, however, that there is one important respect in which this new method for drawing the form/content distinction reverses the Kantian conception: for Kant, the formal components of knowledge are the most stable and objective elements; now they are the most arbitrary and subjective elements.

19

How does this new method for isolating the contribution of reason apply to the development of physics? Consider first the Newtonian-Maxwellian version of mechanics plus electrodynamics. The laws of this theory hold only in coordinate systems at rest relative to absolute space: the conjunction of (8) and (10) is preserved only by spatial translations, spatial rotations, and temporal translations. So the choice of admissible coordinate system here is arbitrary only up to an element of $O^3 \times T$, where O^3 is the group of Euclidean "rigid motions" (6) and T is the group of time-translations. Hence, the notions of absolute rest and absolute velocity, the notion of absolute simultaneity, and the three-dimensional Euclidean spatial geometry are all invariant or objective; they are factual not conventional. The only conventional elements in our description of nature are the choices of spatial origin, spatial orientation, and temporal origin.

As physics evolves, the form/content distinction—now the conventional/factual distinction—evolves as well. Thus, when we move from classical electrodynamics to special relativity, we widen the class of admissible coordinate systems to include all inertial systems: the laws of Einstein's 1905 version of mechanics plus electrodynamics are preserved by spatial translations, spatial rotations, temporal translations, and the Lorentz transformations (11). So the choice of admissible coordinate system here is arbitrary up to an element of $O^3 \times T \times$ the Lorentz group. Hence, the notions of absolute rest, absolute velocity, and absolute simultaneity are no longer invariant or objective; they are now conventional not factual. In general, then, as we widen the class of admissible coordinate systems, we narrow the class of objective or factual elements in our description of nature and widen the class of subjective or conventional elements.

Although the special theory of relativity has fewer factual or objective elements than classical physics, it still draws the conventional/factual distinction within the theoretical realm. We still have objective notions of absolute acceleration and absolute rotation, and, as we saw above, these notions are generated by a class of unobservable semi-Euclidean geodesics. Further, in any single inertial frame we still have a perfectly objective and determinate Euclidean spatial geometry. But there are classical skeptical arguments challenging the objectivity of both these aspects of theoretical structure: Leibniz's argument against the reality of absolute motion and Poincaré's argument against the determinateness of spatial geometry. Leibniz's argument appeared above in the context of Newton's three-dimensional Euclidean embedding space. For the present, it suffices to point out that virtually the same argument applies to Einstein's four-dimensional Minkowskian space-time if one simply changes the "positions" of all physical *events* while preserving their spatio-temporal relations.

Poincaré developed his skeptical argument in 1902. Imagine the interior of an apparently Euclidean three-dimensional sphere. Now suppose that this space is filled by a nonuniform field of force—call it "temperature"—causing all objects to contract as they move from the center of the sphere toward its circumference. Suppose also that this same field of force affects the refractive index of light in such a way that light rays no longer follow Euclidean straight lines. Finally, suppose that there are sentient beings living in the space who are ignorant of the existence of this force. According to these beings: the space is infinite not finite; its geometry is Lobachevskian or hyperbolic not Euclidean.[7] In other words, we have two apparently distinct states of affairs. In one, we are living in a finite, Euclidean space filled by a nonuniform "universal" force; in the other, we are living in an infinite, non-Euclidean space with no "universal" distorting force. But, by hypothesis, these two states of affairs are completely indistinguishable. Therefore, according to Poincaré, the geometrical structure of a given space cannot be an objective or determinate matter:

> What then are we to think of the question: Is Euclidean geometry true? It has no meaning. We might as well ask if the metric system is true, and if the old weights and measures are false; if Cartesian coordinates are true and polar coordinates false. One geometry cannot be more true than another; it can only be more convenient. ([79], 50)

A choice of geometry, like a choice of coordinate system, is a matter of arbitrary convention.

The early positivists were strongly attracted to both these skeptical arguments. Thus, Schlick criticized special relativity from a Leibnizean point of view in 1917: "From the philosophic standpoint it is desirable to be able to affirm that *every* motion is relative, i.e., not the particular class of uniform translations only. According to the special theory, irregular motions would still be absolute in

---

[7] Poincaré [79], 65–68. See Tuller [111], 165–173, for a good discussion of Poincaré's model of a hyperbolic space and the related model due to Klein.

character; in discussing them we could not avoid speaking of Space and Time 'without reference to an object'" ([97], 4). Reichenbach echoed these sentiments in 1920: "There is no reason for the philosopher to single out the uniform translation. As soon as space is characterized as a scheme of order and not as a physical entity, all arbitrarily moving coordinate systems become equivalent for the description of events" ([84], 8). Schlick also endorsed Poincaré in 1917: "The reflections of Poincaré ... teach us beyond doubt that we can imagine the world transformed by means of far-reaching geometrical-physical changes into a new one, which is completely indistinguishable from the first, and which is completely identical with it physically, so that the transformation would not actually signify a real happening" ([97], 26). He concluded that space is "metrically amorphous": "Space itself in no wise has a form of its own; it is neither Euclidean nor non-Euclidean in constitution, just as it is not a peculiarity of distance to be measured in kilometers and not in miles" ([97], 35). Moreover, Schlick thought that both Leibniz's conception of a thoroughgoing relativity of motion and Poincaré's conception of a thoroughgoing relativity of space were realized in the general theory of relativity.

The idea here is disarmingly simple. We know that the class of objective or factual elements postulated by a given theory is inversely proportional to the class of admissible coordinate systems: as we widen the latter, we narrow the former. In general relativity, however, the class of admissible coordinate systems includes all coordinate systems whatsoever. The laws of motion of our theory are generally covariant: they are preserved by arbitrary (one-one and sufficiently continuous) transformations, and not merely by a restricted group of transformations like the Lorentz group. But what elements of structure are invariant under arbitrary (bicontinuous) transformations (7)? Only the topological features of events are preserved by this group: that is, the notion of the *coincidence* of two events and the notion of two events being *near* one another in space-time. Therefore, the argument goes, only these latter notions are objective or factual in general relativity; everything else is arbitrary or conventional.

This is a remarkable result. In Schlick's words, it is "the realization of an eminently desirable point of contact between physical theory and the theory of knowledge" ([97], 82). For the features of the world that are now objective according to general relativity are precisely the observable features of the world:

If we suppose a complete change of this sort to take place, by which every physical point is transferred to another space-time point in such a way that its new coordinates x'_1, x'_2, x'_3, x'_4 are quite arbitrary (but continuous and single-valued) functions of its previous coordinates x_1, x_2, x_3, x_4: then the new world is, as in previous cases, not in the slightest degree different from the old one physically, and the whole change is only a transformation to other coordinates. For that which we can alone observe by means of our instruments, viz. space-time coincidences, remains unaltered. Hence points which coincided at the world-point x_1, x_2, x_3, x_4 in the one universe would again coincide in the other world-point x'_1, x'_2, x'_3, x'_4. Their coincidence—and this is all that we can observe—takes place in the second world precisely as in the first ([97], 52–53)

The "previous cases" to which Schlick refers are the skeptical "indistinguishability" arguments of Leibniz and Poincaré. The coordinate transformations (7) are supposed to give precise expression to these arguments.

In particular, then, the general theory of relativity is supposed to eliminate the two aspects of theoretical structure that were found objectionable in the special theory: absolute motion (absolute acceleration and rotation) and the determinate Euclidean spatial geometry in any single inertial frame. The conventional/factual distinction is no longer drawn within the theoretical realm. According to Schlick, it has, in effect, become the theoretical/observational distinction:

The desire to include, in our expression for physical laws, only what we physically observe leads to the postulate that the equations of physics do not alter their form in the above arbitrary transformation, i.e., that they are valid for any space-time coordinate systems *whatever*. In short, expressed mathematically, they are "covariant" for *all* substitutions. This postulate contains our general postulate of relativity; for, of course, the term "*all* substitutions" includes those which represent transformations of entirely arbitrary three-dimensional systems in motion. But it goes further than this, inasmuch as it allows the relativity of space, in the most general sense discussed above, to be valid even *within* these coordinate systems. In this way Space and Time are deprived of the "last vestige of physical objectivity," to use Einstein's words. ([97], 53)

Schlick is referring to a passage from Einstein's fundamental 1916 paper, which reads as follows:

> That this requirement of general covariance, which takes away from space and time the last vestige of physical objectivity, is a natural one, will be seen from the following reflection. All our space-time verifications invariably amount to a determination of space-time coincidences. If, for example, events consisted merely in the motion of material points, then ultimately nothing would be observable but the meeting of the material points of our measuring instruments with other material points, coincidences between the hands of a clock and points on the clock dial, and observed point-events happening at the same place at the same time.
>
> The introduction of a system of reference serves no other purpose than to facilitate the description of the totality of such coincidences. ([28], 117)

I think it is fair to say that this passage represents the birth of the modern observational/theoretical distinction. It also contains the beginnings of the empiricist and verificationist interpretations of science characteristic of later positivism.[8]

One might object to my story in the following way. Are not the skeptical arguments of Leibniz and Poincaré themselves empiricist or verificationist arguments? After all, both proceed by arguing that states of affairs indistinguishable to us should be identical in reality. Since these arguments play an important role in the development of general relativity, is it not more correct to say—following the usual interpretation—that empiricism gave birth to general relativity rather than the other way around?

It is true, of course, that the arguments of Leibniz and Poincaré are broadly empiricist in spirit. It is also true that such broadly empiricist currents of thought had a significant influence on the development of relativity theory. Nevertheless, empiricist arguments and empiricist currents of thought are not the same as an empiricist philosophy. In an important sense, these kinds of arguments are always available—and available to many different philosophical

---

[8] Of course, these ideas are only the beginnings of the modern observational/ theoretical distinction. For one thing, not all "space-time coincidences" are literally observable: consider the collision of two elementary particles. Schlick himself was quite clear on this point ([97], 83–85). Perhaps what we should say is this: with respect to geometrical structure, the observable = the totality of space-time coincidences.

24

positions (note that neither Leibniz nor Poincaré was an empiricist). Rather, the question is: Why did these arguments and tendencies come together at just this time and with just this result? How did they contribute to the evolution of the philosophical movement we now call logical positivism or logical empiricism in the decade 1925–1935?

My claim is that developments in mathematics and physics—especially the idea of general covariance—were absolutely decisive here. If we ignore these developments, we cannot make sense of the evolution of logical positivism into an empiricist philosophical movement, for these mathematical-physical developments—not any independently grounded empiricist position—give both clear content and philosophical force to the observational/theoretical distinction. The transformations (7) provide us for the first time with a precise idea of what features of the world are observable, accessible, or distinguishable by us: namely, space-time coincidences—pointer readings, matchings of end-points of measuring rods, and the like. Further, these same transformations, in conjunction with the idea that the objective = the invariant, provide us an interesting reason for thinking that the objective = the observable. Thus, any serious empiricist position must supply a link connecting *indistinguishable to us* and *identical in reality*. There is a *prima facie* gap here that needs to be bridged by an appropriate theory of meaning, metaphysics, epistemology, or some combination of the three. In 1920 this gap was not bridged by an empiricist or verificationist theory of meaning but by a neo-Kantian analysis of objectivity applied to the mathematical physics of the day.

Both the earlier, neo-Kantian positivists and the later, empiricist positivists saw the development of relativity theory as a progressive elimination of more and more aspects of theoretical structure. Special relativity eliminates some aspects of the theoretical structure taken for granted in classical physics (absolute rest, absolute simultaneity); general relativity eliminates all such aspects of theoretical structure. As we shall see in considerable detail in the following chapters, this conception of the development of relativity theory is seriously confused and misleading. Perhaps the biggest villain of the piece is a fundamental misunderstanding about the "principle of general covariance" and the role of this principle in general relativity. In particular, the use of general (arbitrary) coordinate systems and the transformations (7) does not, contrary to the above story, lead to the elimination of all theoretical structure. In fact, the principle of general covariance does not involve an elimination of absolute motion nor an elimination of determinate spatial or spatio-temporal geometry.

The most basic confusion here is just the confusion between intrinsic and extrinsic features of a space. Any space, regardless of its geometrical structure (or lack thereof), can be described by general (arbitrary) coordinates. In this sense, any theory of space-time is generally covariant and, hence, can be formulated in such a way that its laws are preserved by the transformations (7). What distinguishes general relativity from previous theories is the use of a *nonflat* space-time in which (for that very reason) inertial or "Cartesian" coordinates do not exist. Thus, for example, we can just as well formulate special relativity in a generally covariant way by using the more general form (13) for the line element. But, since Minkowski space-time is flat, or semi-Euclidean, we are able to prove that there exists a privileged subclass of coordinate systems in which the line element takes the special form (12). So we can adequately describe Minkowski space-time using only such special (inertial) coordinates. By contrast, in the space-time of general relativity there will be no coordinate systems satisfying (12). So we have no choice but to use the generally covariant formulation (13).

In other words, the use of the full group of admissible transformations (7) in general relativity does not imply that we are working in the context of a structureless, "metrically amorphous" manifold, but, rather, that we are working in the context of a highly structured, non-Euclidean manifold. The space-time of general relativity is endowed with a perfectly definite metric (13), which is related in a definite way to the distribution of mass-energy by Einstein's field equations. There is no sense in which this metric is determined by arbitrary choice or convention. Moreover, the metric (13) induces a definite (usually non-Euclidean) three-dimensional geometry on any spacelike hypersurface in our manifold. So any three-dimensional reference frame has a perfectly determinate spatial geometry as well (though, of course, different reference frames will have different spatial geometries). In this respect, the space-time of general relativity has precisely as much structure as the space-time of special relativity; the difference is just that this structure is curved rather than flat.

Similarly, although there are no inertial frames in general relativity, there are inertial (geodesic) *trajectories*, and these trajectories

26

give rise to an absolute distinction between inertial and noninertial (accelerating or rotating) motion, just as in previous theories. It is not true, therefore, that all frames of reference are "equivalent" or "indistinguishable." We still have a privileged subclass of frames, the *local inertial* frames, that are neither accelerating nor rotating according to the geodesics of the metric (13).[9] The existence of such a privileged subclass of frames clearly shows that the general theory does not institute a thoroughgoing relativity of motion. Because acceleration and rotation have essentially the same status as in special relativity, the principle of general covariance does not amount to a general principle of relativity. A more subtle analysis of reference frames and coordinate transformations, an analysis distinguishing the notion of *covariance* from the notions of *equivalence* and *indistinguishability* (and from the related notions of *symmetry* and *invariance*), is required.

Nor does general relativity implement Mach's idea that we can explain the distorting effects of absolute rotation in terms of relative rotation with respect to the total distribution of matter. It is true that the geodesics of the metric (13) are affected by the distribution of matter; *a fortiori*, absolute rotation is so affected as well. Nevertheless, neither the metric (13) nor the notions of absolute acceleration and rotation are determined by the distribution of matter. In fact, the effects of the surrounding distribution are actually very slight. In particular, absolute rotation is still quite possible even if there is no surrounding distribution of matter. Once again, absolute motion has much the same status as in previous theories. The difference is that some trajectories previously classified as absolutely accelerating—namely, freely falling or gravitationally affected trajectories—are now classified as inertial or nonaccelerating, for these trajectories are the geodesics of the nonflat metric (13).

But what about Einstein's "elevator" and the *principle of equivalence*? Does this latter principle not imply that accelerating frames are indistinguishable from nonaccelerating frames in the presence of gravity? Are we not led to a thoroughgoing relativization of motion after all? This line of thought also rests on a misunderstanding. The principle of equivalence is better understood as recom-

---

[9] Local inertial frames can be characterized by the condition that the metric take the form (12) at the origin. However, since we cannot obtain the form (12) on any finite neighborhood of the origin, due to the nonvanishing curvature, the existence of local inertial frames does not alleviate the need for a generally covariant formulation.

27

mending a choice between *two* ways of describing gravitation: in terms of a flat space-time in which gravitational trajectories deviate from the geodesics via a *gravitational force* (as in traditional Newtonian gravitation theory); and in terms of a nonflat space-time in which gravitational trajectories follow the geodesics and there is no gravitational force (as in general relativity). In the first kind of description nonaccelerating frames are indistinguishable from accelerating frames. The class of inertial frames cannot be picked out from the wider class of Galilean (arbitrarily accelerating) frames by the laws of our theory. In fact, the situation here is precisely analogous to the situation in Newtonian kinematics, where the class of rest frames cannot be distinguished from the wider class of inertial frames by the laws of our theory. So the first way of describing gravitation is defective in just the same way as Newtonian kinematics, and the principle of equivalence recommends the second. When we move to the second kind of theory, however, we do not simply eliminate the distinction between inertial and noninertial motion; rather, we replace the old, flat space-time distinction with a new, curved space-time distinction: the distinction between local inertial (freely falling) frames and non-local-inertial (arbitrarily accelerating and rotating) frames. Moreover, the class of local inertial frames can be distinguished from the wider class of arbitrarily moving frames. Therefore, although the transition from the first kind of theory to the second certainly leads to a change in our concept of acceleration, it does not lead to a true relativization of that concept.

For these and other reasons I conclude that the positivists were mistaken in drawing generalized verificationist and conventionalist conclusions from the development of relativistic physics. Relativistic physics is not based on a rejection of all theoretical entities and properties. Nevertheless, there is something undeniably important in the positivist interpretation of our theories. For the development of relativity theory does involve the elimination of some theoretical (unobservable, "metaphysical") entities and properties, most notably, the elimination of absolute space and absolute velocity in the transition from classical physics to special relativity (in accordance with the special principle of relativity) and the elimination of the class of inertial frames in the transition from classical physics to general relativity (in accordance with the principle of equivalence). Furthermore, both these transitions are partly motivated by conceptual or methodological arguments—arguments that appeal to some version of the principle of parsimony and some version of the identity of indiscernibles. We can best represent this conceptual or methodological motivation by formulating *Newtonian counterparts* of the above two theoretical transitions: a four-dimensional version of Newtonian kinematics that dispenses with absolute space (as in special relativity) and a curved space-time version of Newtonian gravitation theory that dispenses with inertial frames (as in general relativity). The point is that these reformulations are empirically equivalent, but methodologically superior, to the usual formulations.

It is clear, then, that the positivistic picture of relativity does not rest on thin air. Important elements of relativistic theory and practice do correspond to empiricist and verificationist methodology. In traditional Newtonian kinematics we find that inertial reference frames are observationally indistinguishable yet theoretically distinct (having different absolute velocities); by eliminating absolute space, we obtain a new, more parsimonious theory. In traditional Newtonian gravitation theory we find that Galilean reference frames are observationally indistinguishable yet theoretically distinct (having different absolute accelerations); by eliminating the flat affine structure, we again obtain a new, more parsimonious theory. So both the special principle of relativity and the principle of equivalence involve the identity of indiscernibles and the principle of parsimony. In both cases we restore the former by applying the latter: we eliminate "excess" theoretical structure that "makes no observable difference."

It is equally clear, however, that these parsimonious methodological moves do not extend to all theoretical structure. As we have seen, general relativity does not carry empiricist and verificationist tendencies to their "ultimate consequences." In essence, it merely replaces a flat affine-metrical structure with a nonflat one. In this sense, the general principle of relativity is not at all analogous to the special principle of relativity. We need an explanation for this asymmetry. We want to know the difference between those theoretical entities rejected in the development of relativity theory (Newton's three-dimensional embedding space, the flat affine structure) and those accepted (Minkowski's four-dimensional space-time, the nonflat affine structure). We want to know especially why the verificationist arguments that apply so well to the former do not apply equally well to the latter. Hence, it is not enough to point to mistakes and confusions in the positivists'—and Einstein's own—interpretations of relativity theory; we need an alternative account of the evolution of relativity that allows us to draw a distinction between "good" theoretical entities and properties and "bad" theoretical entities and properties. We need to articulate both the methodological limitations and the considerable force of the principle of parsimony.

The actual development of relativity theory has a much finer structure than the simple diagram we drew above. On the one hand, aspects of this development express an empiricist attitude toward theoretical entities; these aspects show us what is right about the identity of indiscernibles and the principle of parsimony. On the other hand, the development of relativity theory itself commits us to higher-level theoretical structure—the variably curved space-time of general relativity—and this fact shows us what is wrong with the identity of indiscernibles and the principle of parsimony. In particular, the general theory of relativity realizes neither of the classical indistinguishability arguments of Leibniz and Poincaré. Contrary to the relationalism of Leibniz, general relativity retains absolute motion (that is, absolute acceleration and rotation) and conceives of this motion in terms of a four-dimensional "container" or embedding space. Contrary to the conventionalism of Poincaré, general relativity uses a perfectly determinate and objective spatial and spatio-temporal physical geometry. Careful attention to the fine structure of our theoretical evolution will reveal how these cases differ from the cases of absolute velocity and gravitational acceleration. We shall see in detail why the indistinguishability arguments of Leibniz and Poincaré do not have the same methodological import as the special principle of relativity and the principle of equivalence.

Here we can use the development of relativity theory to advance beyond positivism. A central problem facing postpositivist philosophy of science is *theoretical underdetermination*: the problem of elucidating, and perhaps justifying, methodological criteria for choosing between incompatible theories that are empirically equivalent or agree on all observations. Such criteria will *ipso facto* provide a distinction between "good" and "bad" theoretical entities: an entity is "good"—its postulation is legitimate—just in case it is postulated by a methodologically preferred theory. I shall show how the evolution of relativity reveals these methodological criteria in action, allowing us to drive a wedge between "good" theoretical entities postulated by methodologically well-behaved theories

30

(special relativistic kinematics, general relativistic gravitation theory) and "bad" theoretical entities postulated by methodologically ill-behaved theories (traditional Newtonian kinematics, traditional Newtonian gravitation theory). Most important, we shall see what is "good" about the former entities and theories—what point or function they serve in the process of theoretical inference, explanation, and confirmation—and what is "bad" about the latter entities and theories—how they are pointless or superfluous in this process.

Note, however, that we cannot even begin to develop such an account unless we accept the distinction between observational and theoretical. Otherwise, we miss the problem of theoretical under-determination and, *a fortiori*, remain unaware of the methodological criteria actually used for choosing between empirically equivalent theories. If, on the other hand, we accept the positivistic conception of the observable—roughly, that the observable = the totality of space-time coincidences of material processes—we can formulate precise notions of observational indistinguishability and equivalence, show in detail how the traditional relativity principles function methodologically, and see how "good" theoretical entities contribute to well-confirmed and explanatory theories about the observable. This provides additional support for my central moral: namely, that we cannot develop a more satisfactory philosophy of science unless we grant the positivists their due and consider their arguments on their own terms. If we simply reject all their basic distinctions out of hand, we shall never learn from their mistakes.

31

II
Space-Time Theories

1. General Properties of Space-Time Theories

All the physical theories considered in this book will be treated as theories about space-time: the set of all places-at-a-time or all actual and possible events. Our theories postulate various types of geometrical structure on this set and picture the material universe—the set of all *actual* events—as embedded within in it. According to the present point of view, then, the basic or primitive elements of our theories are of two kinds: space-time and its geometrical structure; and matter fields—distributions of mass, charge, and so on—which represent the physical processes and events occurring within space-time. Our theories seek to explain and predict the properties of material processes and events by relating them to the geometrical structure within which they are "contained." This point of view contrasts with standard philosophical formulations, such as Reichenbach's [85], which characteristically take more observational entities—reference frames, light rays, particle trajectories, material rods and clocks—as primitive and attempt to define geometrical structure in terms of the behavior of such relatively observational entities. In the present treatment we explicitly take the more abstract geometrical entities as primitive and define the more observational entities in terms of them. Reference frames are treated as particular kinds of coordinate systems, light rays and particle trajectories as particular kinds of curves in space-time, material rods and clocks as particular configurations of the fundamental matter fields.

Our theories all agree that space-time is a four-dimensional differentiable manifold. What does this mean? First, space-time has a *topology*: given any point p in space-time, we have the notion of a *neighborhood* of p—a set of points all of which are "close" to p. Second, space-time is *coordinatizable* by R^4—the set of quadruples of real numbers. That is, given any point p in space-time, there exists a neighborhood A of p and a one-one map ϕ from A into R^4 that is sufficiently *continuous* (ϕ maps "nearby" points in A onto "nearby" points in R^4 and vice versa). ϕ is called a coordinate system, or

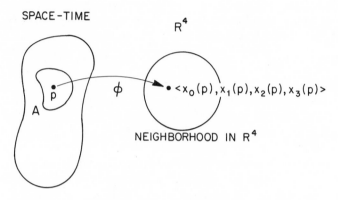

Coordinatization of space-time.

chart, around p. Such a coordinate system enables us to translate statements about geometrical entities in space-time into statements about real numbers: for example, we can describe curves in space-time by numerical equations. We are also given a natural notion of *differentiability*: if f is a real-valued function (representing mass-density, say) defined on a neighborhood in space-time, we say that f is differentiable just in case $f \circ \phi^{-1}$ (the result of applying f to the result of applying the inverse of ϕ—a function from R^4 into R) is differentiable for every chart ϕ.

The assertion that space-time is a four-dimensional differentiable manifold is a purely *local* assertion. It says that space-time is well-behaved in the small—on a neighborhood of each point p. This leaves quite undetermined what space-time is like *globally*, or in the large. In particular, we do not know that space-time as a whole can be mapped continuously onto R^4. Space-time in the large can be finite or infinite, closed like a sphere or open like a plane, connected (no holes or missing pieces) or disconnected (with arbitrary deletions), and so on. Moreover, it is extremely convenient to view all the assertions of our space-time theories as purely local in this sense. Each of our theories will have a large and interesting class of

33

cosmological models, since there are many different global topologies compatible with a given local structure; and all of our theories will look more like general relativity, which is standardly formulated in just this local fashion.

Viewing space-time as a four-dimensional manifold does not prejudge the question in favor of the unification of space and time characteristic of relativity theory. It simply expresses the need for four coordinates to specify any event uniquely. The separate space and time of common sense (and some formulations of Newtonian theory) can also be described in terms of a four-dimensional manifold, one with the simple product structure $E^3 \times R$ (the Cartesian product of three-dimensional Euclidean space and one-dimensional Newtonian time). As we shall see, we effect a relativistic unification of space and time only if we view space-time as a four-dimensional *semi-Riemannian* manifold. On the other hand, since our basic object is four-dimensional, we are not automatically committed to a meaningful notion of "absolute" motion. For an arbitrary four-dimensional manifold does not necessarily have a relation of being-at-the-same-three-dimensional-spatial-position defined between pairs of space-time points. To get a notion of absolute motion we need additional geometrical structure, as would be provided, for example, by the simple product structure $E^3 \times R$.

The most important aim of our space-time theories is to describe the trajectories or histories of certain classes of physical particles: free particles, or particles affected only by gravitational forces, or charged particles subject to an external electromagnetic field, and

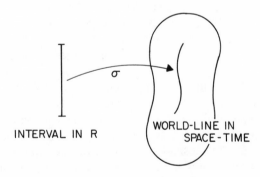

INTERVAL IN R

WORLD-LINE IN
SPACE-TIME

so on. We represent such a trajectory or history by a *curve* in space-time, a continuous and differentiable map σ from an interval

34

of the real line into our manifold. Intuitively, we can think of the real numbers in the domain of σ as *times*, although, as we shall see, "time" refers to very different things in the different theories. Following tradition, let us call such curves *world-lines* in space-time.

We can describe the configuration of a world-line by tracing its twists and turns as it winds its way through our manifold. We want to be able to say, for each point p along the curve, how much and in what direction it is "curved" or "bent" at p. How do we do this? First, we need the idea of a *tangent vector* to the curve at p—intuitively, an indicator of the direction in which the curve is going at p. By tracing the changes in the tangent vector, we can trace the path of our curve. But what is a tangent vector? We know that for curves in ordinary Euclidean three-space R^3 the tangent vector is given by the triple $\langle dx/dt, dy/dt, dz/dt \rangle$, where the coordinates of our curve are given by the functions $x(t), y(t), z(t)$:

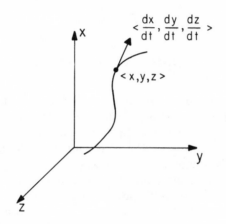

Tangent vector in R^3.

Thus, in R^3 the tangent vector to a curve at a given point p is just a triple of real numbers $\langle dx/dt(t_0), dy/dt(t_0), dz/dt(t_0) \rangle$, where $p = \langle x(t_0), y(t_0), z(t_0) \rangle$.

Relative to a given coordinatization of our manifold, we have an analogous tangent vector for curves in space-time. Let σ be a curve through p (that is, for some t in the domain of σ, $\sigma(t) = p$), and let the x_i ($i = 0,1,2,3$) be the coordinates of a chart ϕ around p. Then, relative to ϕ, the tangent vector to σ at p is the quadruple of real

35

numbers

$$\left\langle \frac{d(x_0 \circ \sigma)}{du}\bigg|_t, \frac{d(x_1 \circ \sigma)}{du}\bigg|_t, \frac{d(x_2 \circ \sigma)}{du}\bigg|_t, \frac{d(x_3 \circ \sigma)}{du}\bigg|_t \right\rangle$$

where for each $i = 0,1,2,3$, $(x_i \circ \sigma)$ is a continuous and differentiable

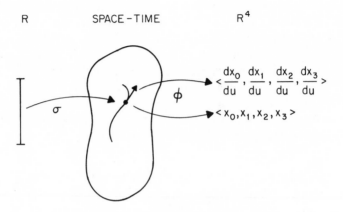

Tangent vector in space-time.

function from R into R. Of course, this quadruple of real numbers will change if we change our coordinate system: it only represents the tangent vector relative to the particular chart ϕ. So it is better to identify the tangent vector itself with a slightly more abstract object: namely, the function $T_\sigma(t)$, which, given any coordinate x_i as input, results in the number $d(x_i \circ \sigma)/du(t)$ as output. We call the numbers $d(x_i \circ \sigma)/du(t)$ the *components* of the vector $T_\sigma(t)$ in the coordinate system $\langle x_i \rangle$. Moreover, since x_i is arbitrary, $T_\sigma(t)$ actually operates on any continuous and differentiable real-valued function f defined on a neighborhood of p:

$$T_\sigma(t)[f] = \frac{d(f \circ \sigma)}{du}\bigg|_t .$$

The collection of all such operators $T_{\sigma'}(t)$ for all curves σ' with $\sigma'(t) = p$ is called T_p: the *tangent space* at p.

We now have tangent vectors attached to each point along each curve in space-time. To describe the shape of a curve we must be able to describe how the tangent vector changes as we move along the curve, to record the variations in direction that it experiences

36

successively. We therefore need to be able to say, for vectors at "nearby" points, whether they point in the same direction or not. However, given the machinery we have so far, there is simply no answer to this question on an arbitrary manifold. In ordinary Euclidean three-space this question is easy to answer: we can simply "read off" changes in direction because Euclidean three-space is itself one big vector space within which the tangent spaces at different points fit together in a natural way (a). In general, however, a

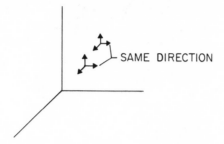

(a) Tangent spaces at nearby points in R^3.

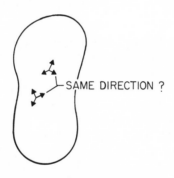

(b) Tangent spaces at nearby points in a differentiable manifold.

differentiable manifold is not itself a vector space. We simply do not know how the tangent spaces at different points fit together (b). Thus, to answer the question about changes in direction we need additional geometrical structure.

This additional structure is called *affine structure*. It coordinates the vector spaces at "nearby" points in our manifold so that, given

37

two different vectors at two "nearby" points, we are able to say whether they point in the same direction or not. There are many ways to introduce such structure onto our manifold. Perhaps the simplest is by means of a *derivative operator*, or *affine connection*. Suppose we are given a curve σ and a *vector field* $X(p)$ defined on points along σ, where $X(p)$ is a continuous and differentiable selection of vectors from the tangent space at each point along σ: for each p, $X(p)$ is in T_p (although in general $X(p)$ need not be the tangent to σ at p). Given T_σ (the tangent vector field to σ), $X(p)$, and a point q along σ, a derivative operator D gives us a vector $(D_{T_\sigma}X)(q)$ in T_q that records the rate and direction of the change

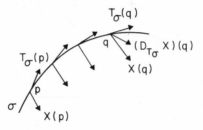

Derivative operator acting on vector field $X(p)$ along curve σ.

in $X(p)$ at q. Such an assignment of vectors counts as a derivative operator or affine connection if it has the properties that a *derivative* should have (see (A.10)).

Let us suppose, then, as all our space-time theories do, that we have a derivative operator $D_{T_\sigma}X$ defined on our manifold for all curves σ and vector fields X. We can now coordinate the tangent spaces at different points. Thus, for example, if $D_{T_\sigma}X = 0$ on a neighborhood A of q, then we know that the field X is constant on A and, therefore, that all vectors $X(p)$ with p in A point in the same direction. (Of course, this notion of "same direction" is relative to the choice of derivative operator D: given a different derivative operator, we have a different notion of "same direction.") Moreover, we can apply the derivative operator to the tangent vector field of our initial curve, thereby recording the changes in direction of the curve itself. So if $D_{T_\sigma}T_\sigma = 0$, the curve does not change direction: it is a "straight" world-line (such a world-line is called a space-time *geodesic*). If $D_{T_\sigma}T_\sigma \neq 0$ at q, then our curve "bends" at q, and $(D_{T_\sigma}T_\sigma)(q)$ provides us with a natural measure of the rate and

38

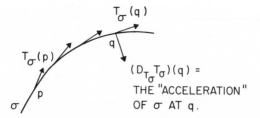

$(D_{T_\sigma}T_\sigma)(q) =$
THE "ACCELERATION"
OF σ AT q.

Derivative operator acting on tangent vector field $T_\sigma(p)$ along curve σ.

direction of the "bend." $(D_{T_\sigma}T_\sigma)(q)$ measures the curvature or "acceleration" of our world-line at q.

An *affine manifold*, a manifold equipped with a derivative operator, or affine connection, has a *geometry*. There is a privileged class of "straight lines," or geodesics (curves σ such that $D_{T_\sigma}T_\sigma = 0$), and we can ask whether this class of "straight lines" satisfies a Euclidean or non-Euclidean geometry. How can we tell? We know that $D_{T_\sigma}T_\sigma$ is a vector field. So, given a coordinate system $\langle x_i \rangle$, $D_{T_\sigma}T_\sigma$ becomes a quadruple of real-valued functions $\langle D_{T_\sigma}T_\sigma^i \rangle$. It can be shown (see (A.11)) that these functions are explicitly given by[1]

$$D_{T_\sigma}T_\sigma^i = \frac{d^2 x_i}{du^2} + \Gamma_{jk}^i \frac{dx_j}{du}\frac{dx_k}{du}$$

where $x_i = x_i \circ \sigma(u)$ and the Γ_{jk}^i ($i,j,k = 0,1,2,3$) are real-valued functions (called the *components* of the affine connection D) depending on the coordinate system $\langle x_i \rangle$. An affine manifold is said to be *flat*, or *Euclidean*, at a point p just in case we can find a coordinate system around p in which the functions Γ_{jk}^i all vanish. Such a coordinate system is called a *Cartesian*, or *inertial*, system. For, if $\langle x_i \rangle$ satisfies this condition, then

$$D_{T_\sigma}T_\sigma^i = \frac{d^2 x_i}{du^2} = \text{the ordinary acceleration,}$$

and geodesics satisfy

$$\frac{d^2 x_i}{du^2} = 0, \qquad x_i = a_i u + b_i$$

---

[1] In this equation and all others, we use the *summation convention*: repeated indices are summed over. So $\Gamma_{jk}^i (dx_j/du)(dx_k/du) =$

$$\sum_{j,k=0}^{3} \Gamma_{jk}^i \frac{dx_j}{du}\frac{dx_k}{du} = \Gamma_{00}^i \left(\frac{dx_0}{du}\right)^2 + \Gamma_{01}^i \frac{dx_0}{du}\frac{dx_1}{du} + \cdots + \Gamma_{23}^i \frac{dx_2}{du}\frac{dx_3}{du} + \Gamma_{33}^i \left(\frac{dx_3}{du}\right)^2.$$

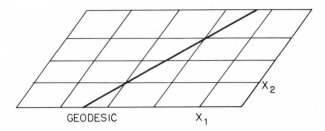

Flat, Euclidean manifold. Cartesian coordinates exist and geodesics satisfy the Euclidean conditions $x_i = a_i u + b_i$.

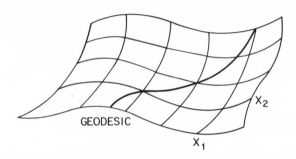

Nonflat, non-Euclidean manifold. Cartesian coordinates do not exist and geodesics are not Euclidean straight lines.

FIGURE 1. Flat versus Nonflat Affine Manifolds

with the a_i and b_i constants. So in Cartesian coordinate systems geodesics are ordinary Euclidean straight lines (see Fig. 1).

Although an affine manifold has some geometrical structure—a privileged class of straight lines or geodesics—it still lacks metrical structure: there is no notion of length or distance. How can we introduce this additional structure? The most intuitive way is by means of a *distance function* $d(p,q)$, which assigns a real number to every pair of points in our manifold and satisfies the conditions

$$d(p,q) \geq 0 \qquad \text{(positive-definiteness)}$$
$$d(p,q) = d(q,p) \qquad \text{(symmetry)}$$
$$d(p,q) = 0 \quad \text{iff} \quad p = q \qquad \text{(nonsingularity)}$$
$$d(p,q) \leq d(p,r) + d(r,q) \qquad \text{(triangle inequality)}.$$

40

However, even though a distance function is the most intuitive way to introduce metrical structure, it is not the most general way. (For example, the metric of a relativistic space-time satisfies only the symmetry condition.) Moreover, a distance function is not very easy to work with analytically, since it is not a local or field-theoretic object. The distance function $d(p,q)$ is defined for arbitrary pairs of points, no matter how far apart, whereas a true field-theoretic object is defined "infinitesimally" or one point at a time. It would be desirable, then, if we could reconstruct $d(p,q)$ from some "infinitesimal" or field-theoretic object.

Think of a curve in ordinary Euclidean space. We can define the length of the curve by dividing it up into small segments and considering the distances $d(p_i,p_j)$ between nearby points in our division:

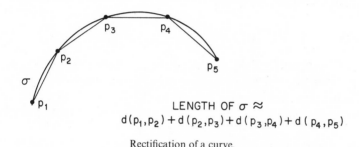

LENGTH OF $\sigma \approx$
$$d(p_1,p_2) + d(p_2,p_3) + d(p_3,p_4) + d(p_4,p_5)$$

Rectification of a curve.

Adding up these distances for all the points in our division results in an approximation to the length of the curve. If we take the limit of this sum as our division becomes finer and finer, as our points become closer and closer together, we get the true length of the curve. This process is called rectification of the curve. It views the length of a curve as the infinite sum of a set of "infinitesimal" distances defined at each point along the curve. Now let $|\sigma|$ be the length of the curve $\sigma(t)$ defined by the rectification process. It is straightforward to show that

$$|\sigma| = \int_\sigma |T_\sigma(t)| dt$$

where

$$|T_\sigma(t)| = \sqrt{\left(\frac{dx}{dt}\right)^2 + \left(\frac{dy}{dt}\right)^2 + \left(\frac{dz}{dt}\right)^2}$$

In other words, the length of a curve in Euclidean three-space is

41

obtained by integrating the "infinitesimal" length of the tangent vector $T_\sigma(t) = \langle dx/dt, dy/dt, dz/dt \rangle$ over the whole curve.[2]

Look at the quantity under the square root sign in our formula for $|T_\sigma(t)|$. This quantity is the Euclidean *inner product*, or *dot* product, of the tangent vector $T_\sigma(t)$ with itself. In general, the inner product of two vectors $X = \langle a_1, a_2, a_3 \rangle$ and $Y = \langle b_1, b_2, b_3 \rangle$ in R^3 is given by

$$X \cdot Y = a_1 b_1 + a_2 b_2 + a_3 b_3$$

and this product is easily seen to have the properties

$X \cdot X > 0$ if $X \neq 0$ (positive-definiteness)

$X \cdot Y = Y \cdot X$ (symmetry)

$X \cdot Y = 0$ for all Y iff $X = 0$ (nonsingularity).

$X \cdot Y$ is also linear; that is,

$$X \cdot (aY + bZ) = a(X \cdot Y) + b(X \cdot Z)$$

where a and b are real numbers (by symmetry, the product is linear in both arguments). This inner product, a bilinear real-valued function defined on the tangent space at each point on a curve, is the "infinitesimal" metric we are looking for. It gives us a length for each tangent vector $T_\sigma(t)$; a length for any suitably well-behaved curve by integration; and a distance function $d(p,q) =$ the length of the shortest curve between p and q.

It is easy to generalize all this to arbitrary, and not necessarily Euclidean, manifolds. A *Riemannian metric tensor* is any bilinear, real-valued function $g(X,Y)$ defined on the tangent space at each point satisfying the above three conditions. (If $g(X,Y)$ satisfies only the second and third conditions, it is called a *semi*-Riemannian metric tensor.) Given such a metric tensor, we can define the length $|X|$ of any vector in T_p by $|X|^2 = g_p(X,X)$; the angle α between two vectors X, Y in T_p by $\cos(\alpha) = g_p(X,Y)/|X||Y|$, and the length of any well-behaved curve $\sigma(u)$ by

$$|\sigma| = \int_\sigma |T_\sigma(u)| \, du = \int_\sigma \sqrt{g_{\sigma(u)}(T_\sigma(u), T_\sigma(u))} \, du.$$

We also have a notion of *metric geodesic*: a curve of shortest (extremal) length. Moreover, if our manifold has both affine structure (a derivative operator) and metric structure (a Riemannian metric tensor), we require that the two notions of geodesic coincide: a curve is an affine geodesic (straightest curve) just in case it is a

---

[2] Compare, e.g., Taylor [106], 361–362.

metric geodesic (shortest curve). A metric g that satisfies this condition for a given affine connection D is said to be *compatible* with D.
A manifold equipped with both g and D, where g is compatible with
D, is called a *Riemannian manifold*. (If g is only semi-Riemannian,
the situation is a bit more complicated. In general, affine geodesics
will not be extremal. However, all affine geodesics will either be *time-
like* $(g(T_\sigma,T_\sigma) > 0)$, *spacelike* $(g(T_\sigma,T_\sigma) < 0)$, or *null* $(g(T_\sigma,T_\sigma) = 0)$.
Timelike geodesics are extremal in the class of timelike curves;
spacelike geodesics are extremal in the class of spacelike curves—
where $|T_\sigma|^2 = -g(T_\sigma,T_\sigma)$; etc. See §A.6.)

How can we tell whether a given metric tensor g is Euclidean or
non-Euclidean? Given a coordinate system $\langle x_i \rangle$, $T_\sigma(u)$ becomes a
quadruple of real-valued functions $\langle d(x_i \circ \sigma)/du \rangle$, and it can be
shown (see §A.6) that the function $g(T_\sigma(u),T_\sigma(u))$ is explicitly given by

$$g(T_\sigma(u),T_\sigma(u)) = g_{ij} \frac{dx_i}{du} \frac{dx_j}{du}$$

where $x_i = x_i \circ \sigma(u)$ and the g_{ij} $(i,j = 0,1,2,3)$ are real-valued functions—the components of the metric tensor—depending on the
coordinate system $\langle x_i \rangle$. Thus, the components g_{ij} form a 4×4
matrix (g_{ij}). For each point p, we can always find a coordinate
system such that this matrix is diagonalized with all diagonal elements equal to ± 1 *at p*. That is,

$$(g_{ij}) = \begin{pmatrix} \pm 1 & 0 & 0 & 0 \\ 0 & \pm 1 & 0 & 0 \\ 0 & 0 & \pm 1 & 0 \\ 0 & 0 & 0 & \pm 1 \end{pmatrix} = \mathrm{diag}(\pm 1, \pm 1, \pm 1, \pm 1)$$

at p. (The pair $\langle n_+, n_- \rangle$, where n_+ is the number of positive diagonal
elements and n_- is the number of negative diagonal elements, is
called the *signature* of g.) Now, a (semi-) Riemannian metric g is
said to be flat or (semi-) Euclidean at p just in case we can find a
coordinate system (a (semi-) Cartesian system) around p in which
$(g_{ij}) = \mathrm{diag}(\pm 1, \pm 1, \pm 1, \pm 1)$ on a *neighborhood* of p. So on a flat
Riemannian manifold the formula for $|T_\sigma(u)|$ reduces to

$$|T_\sigma(u)|^2 = \left(\frac{dx_0}{du}\right)^2 + \left(\frac{dx_1}{du}\right)^2 + \left(\frac{dx_2}{du}\right)^2 + \left(\frac{dx_3}{du}\right)^2$$

which is just the Euclidean formula (in four dimensions) we encountered above. Of course, a (semi-) Riemannian manifold is flat

43

in the metrical sense just in case it is flat in the affine sense (see §A.6).

We now have enough machinery to say, in general terms, how our space-time theories work. They all agree that space-time is a four-dimensional, differentiable, affine manifold. Where they differ is, first, over whether space-time is flat or nonflat, Euclidean or non-Euclidean, and, second, over what additional geometrical structure—especially metrical structure—they attribute to space-time. Further, in connection with metrical structure, the various theories differ over the individual natures of space and time, how the underlying space-time "splits up" into three-dimensional space and one-dimensional time. Regardless of these differences, however, our theories all attempt to describe the trajectories, or world-lines, of physical particles by means of a distinguished affine connection. Whether space-time is taken to be flat or nonflat, the theories all agree that *free particles* (particles acted on by no external forces) follow geodesics of our postulated affine structure. So free particles have the *equation of motion*

$$D_{T_\sigma} T_\sigma = 0$$

or, in terms of a coordinate system $\langle x_i \rangle$:

$$\frac{d^2 x_i}{du^2} + \Gamma^i_{jk} \frac{dx_j}{du} \frac{dx_k}{du} = 0.$$

If our manifold happens to be flat, we can find a coordinate system in which free particles satisfy

$$\frac{d^2 x_i}{du^2} = 0$$

which the reader should recognize as (almost) Newton's Law of Inertia. Thus, a coordinate system in which $\Gamma^i_{jk} = 0$ is called an *inertial frame*.

What about particles that are acted on by external forces? Here, our space-time theories provide theories of *interaction*—gravitational interaction, electromagnetic interaction, and so on. A theory of interaction postulates additional geometrical objects on the underlying manifold: *source variables* (mass density, charge density, and the like) represent the sources of the interaction in question, and *field variables* (gravitational field, electromagnetic field, and so on) represent the forces arising from the interaction. The theory provides *field equations*, which relate the source variables to the field variables, and *equations of motion*, which pick out a privileged class of trajectories with the help of the field variables.

There are two basic ways to formulate equations of motion. The first way, exemplified by electrodynamics and traditional formulations of Newtonian gravitation theory, is to construct a vector field F out of the field variables (representing a particular type of force) and to describe the class of trajectories in question by an equation of motion of the form

$$D_{T_\sigma} T_\sigma = cF$$

where c is a constant (charge/mass ratio, for example) characterizing the relevant physical properties of the particle. Once again, in terms of a coordinate system $\langle x_i \rangle$ this becomes

$$\frac{d^2 x_i}{du^2} + \Gamma^i_{jk} \frac{dx_j}{du} \frac{dx_k}{du} = cF^i.$$

Therefore, if our manifold is flat, we can find inertial coordinate systems in which the equation of motion is just

$$\frac{d^2 x_i}{du^2} = cF^i$$

which the reader should recognize as (almost) Newton's Second Law.

The second way of describing an interaction, exemplified by general relativity, is to "geometrize away" the forces in question by incorporating the field variables into the affine structure of space-time. In this kind of theory the field equations relate the affine properties of space-time directly to the source variables (thus, the only field variable is the affine connection itself), and the equation of motion takes the simple geodesic form

$$D_{T_\sigma} T_\sigma = 0.$$

In the first kind of theory the trajectory of a particle is explained by variations in F due to the distribution of source variables; in the second kind of theory the trajectory of a particle is explained by variations in the affine structure itself, again due to the distribution of source variables (see Fig. 2). Hence, in the second kind of theory space-time is in general not flat; rather, it has *variable curvature*. So in this kind of theory there are in general no inertial coordinate systems. Note that a necessary condition for our being able to "geometrize away" the forces arising from a particular interaction is that the forces in question affect all particles in the same way, independently of their differing physical constitutions. That is, the *coupling factor c* must be the same for all particles. Although only

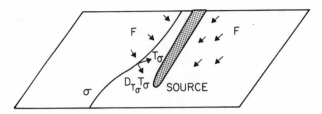

Field of force and nongeodesic equation of motion in flat space-time.

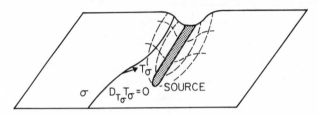

Variable curvature and geodesic equation of motion in nonflat space-time.

FIGURE 2. Two Ways of Describing an Interaction

general relativity has been traditionally formulated in this second way, we shall see that Newtonian gravitation theory can be given such a formulation as well.

2. Covariance, Invariance, and Relativity Principles

One of the most confusing chapters in the history of traditional discussions of our space-time theories centers around the notions of *covariance* and *invariance*. These notions arise in the context of certain groups of transformations that are traditionally associated with our theories. Associated with Newtonian mechanics is the Galilean group; associated with special relativity is the Lorentz group; associated with general relativity is a much larger group consisting of all admissible (one-one and sufficiently continuous) transformations. It is often thought that our theories can be somehow defined in terms of their associated groups: a theory is special relativisitic just in case its associated group is the Lorentz group, and so on. Unfortunately, it has been quite unclear exactly how these groups are associated with their respective theories. Are they

46

covariance groups or invariance groups? For that matter, is invariance the same as covariance? If not, how are the two related?

These questions typically elicit different answers when different theories are considered. In the case of general relativity, the associated group is almost invariably taken to be a group of coordinate transformations, and this group is said to be the covariance group of the theory. In the case of special relativity, on the other hand, the Lorentz group is sometimes treated as a group of coordinate transformations and sometimes as a group of *automorphisms* (mappings of the space-time manifold onto itself). Sometimes it is said to be a covariance group (for example, in discussions of relativistic electrodynamics);[3] but sometimes it is said to be an invariance group, where the notion of invariance appealed to appears to be quite different from the notion of covariance. However, this notion of invariance is seldom precisely defined. It is still more difficult to see how Lorentz invariance is supposed to relate to the general covariance attributed to general relativity—how the latter is supposed to be a generalization of the former, for example.

Of course, this problem is not merely of technical interest. The groups in question are intimately connected with the traditional *relativity principles*, and these, in turn, are supposed to reflect an evolving *relativization* of motion. Thus, both the Galilean principle of relativity and the special principle of relativity are supposed to express the relativity of inertial, or nonaccelerated, motion: the fact that different states of inertial motion are "equivalent" or "indistinguishable" (the special principle goes beyond the Galilean principle in extending this "indistinguishability" from mechanical to electromagnetic phenomena). The general principle of relativity is supposed to express the relativity of all motion: the (alleged) fact that all states of motion are "equivalent" or "indistinguishable." A clarification of the technical notions of covariance and invariance is therefore a prerequisite for understanding the implications of our space-time theories for the philosophical problem about the relativity of motion. In this section I shall concentrate on the technical problem of precisely defining covariance and invariance. In the next section, and in the remainder of the book, I shall discuss the relativity of motion.

What, then, is a covariance group? To answer this question we must first realize that a space-time theory can be formulated in two

---

[3] Compare Bergmann [4], 111–113.

very different ways. In the coordinate-independent, or *intrinsic*, style of formulation adopted here, we view the objects of the theory as various kinds of abstract mappings. Tangent vectors are mappings from real-valued functions to real numbers; affine connections are mappings from vector fields to vector fields; metric tensors are mappings from pairs of vectors to real numbers. (We shall encounter other types of tensors—(linear) mappings from (sequences of) vectors to real numbers (see §A.3)—in what follows.) This way of formulating a theory makes no reference to coordinate systems. The equations of a space-time theory T pick out a class of *dynamically possible models* $\langle M, \Phi_1, \ldots, \Phi_n, T_{\hat{\sigma}} \rangle$—where M is a four-dimensional manifold; $\Phi_1, \ldots, \Phi_n$ are the geometrical objects postulated by T; $T_{\hat{\sigma}}$ is the tangent vector field to a class of curves $\hat{\sigma}$ on M; $\Phi_1, \ldots, \Phi_n$ satisfy the field equations of T; and $T_{\hat{\sigma}}$ satisfies the laws of motion of T—and the class of models picked out by the theory is independent of the choice of any particular coordinatization of M. On the other hand, in the coordinate-dependent style of formulation found in most physics books one does not deal directly with the geometrical objects postulated by T; instead, one works with the components of these objects in some particular coordinate system.

Relative to a coordinate system $\langle x_i \rangle$, the components of a geometrical object form a k-tuple of real-valued functions. The components of a tangent vector T_σ form a quadruple $\langle d(x_i \circ \sigma)/du \rangle$; the components of an affine connection D form a 64-tuple $\langle \Gamma^i_{jk} \rangle$; the components of a metric tensor g form a 16-tuple $\langle g_{ij} \rangle$; and similarly for other types of geometrical objects. Thus, one can view a geometrical object Θ as a mapping that takes a coordinate system $\langle x_i \rangle$ as input and yields a set of real-valued functions $\{\Theta^\alpha_{\langle x_i \rangle}\}$ defined on the domain of $\langle x_i \rangle$ as output. Moreover, in terms of any coordinate system $\langle x_i \rangle$, the equations of a space-time theory T become differential equations in R^4, that is, differential equations for the components of the geometrical objects of T in the coordinate system $\langle x_i \rangle$. For example, the intrinsic, or coordinate-independent, form of a geodesic equation of motion is

$$D_{T_\sigma} T_\sigma = 0$$

while in terms of a coordinate system $\langle x_i \rangle$ we have the system of equations

$$\frac{da^i}{du} + \Gamma^i_{jk} a^j a^k = 0$$

where the a^i are the components of the tangent vector T_σ in $\langle x_i \rangle$—$a^i = d(x_i \circ \sigma)/du$—and the Γ^i_{jk} are the components of the connection D in $\langle x_i \rangle$. In general, a coordinate-independent equation

$$\mathscr{F}(\Phi, \Theta) = 0$$

on M becomes a system of differential equations

$$\mathscr{D}^{\mathscr{F}}_{\langle x_i \rangle}(\Phi^\alpha_{\langle x_i \rangle}, \Theta^\beta_{\langle x_i \rangle}) = 0$$

on R^4 relative to a choice of coordinate system $\langle x_i \rangle$.

As we change from one coordinate system $\langle x_i \rangle$ to a second coordinate system $\langle y_j \rangle$, the components of a geometrical object Θ change from one set of real-valued functions $\{\Theta^\alpha_{\langle x_i \rangle}\}$ to a second set of real-valued functions $\{\Theta^\alpha_{\langle y_j \rangle}\}$. The rule that describes this change is called the *transformation law* for the type of geometrical object in question. Thus, suppose the components of a tangent vector are given by $a^i = d(x_i \circ \sigma)/du$ in $\langle x_i \rangle$. In $\langle y_j \rangle$ the components of the same vector become

$$\bar{a}^j = \frac{d(y_j \circ \sigma)}{du} = \frac{\partial y_j}{\partial x_i} a^i$$

by the "chain rule" from elementary calculus. This is the transformation law for tangent vectors. It can be shown (see (A.12)) that the transformation law for affine connections is

$$\bar{\Gamma}^k_{ij} = \left[\frac{\partial^2 x_l}{\partial y_j \partial y_i} + \frac{\partial x_m}{\partial y_j} \frac{\partial x_n}{\partial y_i} \Gamma^l_{nm} \right] \frac{\partial y_k}{\partial x_l}$$

where the $\bar{\Gamma}^k_{ij}$ are the components of D in $\langle y_j \rangle$ and the Γ^l_{nm} are the components of D in $\langle x_i \rangle$. Similarly, the transformation law for metric tensors is given by

$$\bar{g}_{ij} = \frac{\partial y_i}{\partial x_k} \frac{\partial y_j}{\partial x_l} g_{kl},$$

where the $\bar{g}_{ij}$ are the components of g in $\langle y_j \rangle$ and the g_{kl} are the components of g in $\langle x_i \rangle$. Analogous transformation laws hold for other types of geometrical objects (see (A.7)).

Moreover, as we change from one coordinate system $\langle x_i \rangle$ to a second coordinate system $\langle y_j \rangle$, the equations of our theory change from one system of differential equations to a second system of differential equations. Consider again the geodesic equation

$$D_{T_\sigma} T_\sigma = 0. \tag{1}$$

Suppose our manifold happens to be flat, and let $\langle x_i \rangle$ be an *inertial* system in which $\Gamma^i_{jk} = 0$. In $\langle x_i \rangle$ our geodesic equation therefore becomes

$$\frac{da^i}{du} = 0. \tag{2}$$

On the other hand, in a *noninertial* coordinate system $\langle y_j \rangle$ the very same geodesic equation becomes

$$\frac{d\bar{a}^j}{du} + \frac{\partial y_j}{\partial x_i} \frac{\partial^2 x_i}{\partial y_k \, \partial y_l} \bar{a}^k \bar{a}^l = 0 \tag{3}$$

where (3) is obtained from

$$\frac{d\bar{a}^j}{du} + \bar{\Gamma}^j_{kl} \bar{a}^k \bar{a}^l = 0 \tag{4}$$

and the transformation law for affine connections. What is happening is this. The true, intrinsic geodesic equation has two arguments: the tangent vector T_σ and the affine connection D. In the special co-ordinate system $\langle x_i \rangle$, however, one of these arguments drops out: the components of D vanish. As a result, our system of differential equations in $\langle x_i \rangle$ has only the components a^i of the tangent vector T_σ as arguments. But when we move to $\langle y_j \rangle$, the components $\bar{\Gamma}^i_{jk}$ of D reappear, so we obtain a differential equation for the arguments $\bar{a}^j$ with a very different form.

So far our discussion has proceeded from the intrinsic to the extrinsic. We started with a coordinate-independent equation (1), and we then saw what forms this equation took in certain particular coordinate systems (2), (3). Suppose, however, we start with the extrinsic point of view; this, of course, is how our space-time theories were actually developed. We start with a particular coordinate system $\langle x_i \rangle$ and a system of differential equations in R^4 for the components of some geometrical object(s): for example, the equations (2) for the components a^i of a tangent vector field. Any such system of equations will pick out a class of geometrical objects on M, but the class picked out will depend on the particular coordinate system with which we start. If we start with $\langle y_j \rangle$ and the same differential equations

$$\frac{d\bar{a}^j}{du} = 0 \tag{5}$$

we will pick out a class of tangent vector fields different from the class picked out by (2). For any field T_σ satisfying (2) will satisfy (3), not (5), in $\langle y_j \rangle$. Thus, the extrinsic style of formulation specifies no well-defined class of models: relative to different choices of co-ordinate system, the same equations pick out distinct classes of models.

Suppose we are given a system of differential equations

$$\mathscr{D}(\Phi^\alpha, \Theta^\beta) = 0$$

in R^4. Relative to $\langle x_i \rangle$ this system of equations picks out a class of models of the form $\langle M, \Phi, \Theta \rangle$. Our system of equations is said to be *covariant* under the *coordinate transformation* to a second coordi-nate system $\langle y_j \rangle$ just in case the same class of models is picked out relative to $\langle y_j \rangle$: that is, just in case

$$\mathscr{D}_{\langle x_i \rangle}(\Phi^\alpha_{\langle x_i \rangle}, \Theta^\beta_{\langle x_i \rangle}) = 0 \quad \text{iff} \quad \mathscr{D}_{\langle y_j \rangle}(\Phi^\alpha_{\langle y_j \rangle}, \Theta^\beta_{\langle y_j \rangle}) = 0$$

for all Φ, Θ on M. The *covariance group* of a system of equations is

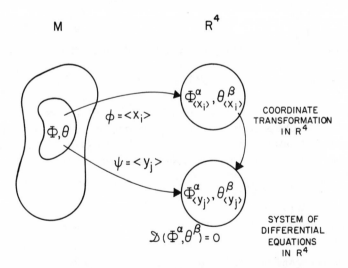

Covariance under a coordinate transformation. Does the system of equations $\mathscr{D}(\Phi^\alpha, \Theta^\beta) = 0$ in R^4 pick out the same geometrical objects Φ, Θ on M relative to the two charts ϕ and ψ?

the largest group of coordinate transformations such that each transformation in the group satisfies the above condition. Thus, for example, the covariance group of the system of equations (2) is just the group of all *linear* transformations: that is, transformations such that $\partial^2 x_i / \partial y_k \partial y_l = 0$. For if $\langle y_j \rangle$ is linearly related to $\langle x_i \rangle$, (5) picks out the very same fields T_σ as (2).

There is a second, equivalent way of approaching the notion of covariance.[4] Instead of viewing the transformations in our group as coordinate transformations, we view them as manifold transformations, or *automorphisms*: one-one, suitably continuous and differentiable mappings of a neighborhood of M into M. For any coordinate transformation $\langle x_i \rangle \to \langle y_j \rangle$ induces a manifold transformation h such that $x_i(hp) = y_i(p)$. Conversely, any manifold transformation h induces a coordinate transformation $y_i = x_i \circ h$.[5] We know that under a coordinate transformation the components of a geometrical object Θ change from $\Theta^\alpha_{\langle x_i \rangle}$ to $\Theta^\alpha_{\langle y_j \rangle}$. Under a manifold transformation h the object itself changes from Θ to $h\Theta$, where $h\Theta$ is a second geometrical object whose components relative to $\langle x_i \rangle$ at hp are equal to the components of Θ relative to $\langle y_j \rangle$ at p. That is,

$$h\Theta^\alpha_{\langle x_i \rangle}(hp) = \Theta^\alpha_{\langle y_j \rangle}(p)$$

where $x_i(hp) = y_i(p)$.

Suppose we are again given a system of differential equations

$$\mathscr{D}(\Phi^\alpha, \Theta^\beta) = 0$$

in R^4. Relative to $\langle x_i \rangle$ this equation picks out a class of models $\langle M, \Phi, \Theta \rangle$. Our system of equations is said to be *covariant* under the *manifold transformation* h just in case $\langle M, h\Phi, h\Theta \rangle$ is also a model (relative to $\langle x_i \rangle$) for each $\langle M, \Phi, \Theta \rangle$, that is, just in case

$$\mathscr{D}_{\langle x_i \rangle}(\Phi^\alpha_{\langle x_i \rangle}, \Theta^\beta_{\langle x_i \rangle}) = 0 \quad \text{iff} \quad \mathscr{D}_{\langle x_i \rangle}(h\Phi^\alpha_{\langle x_i \rangle}, h\Theta^\beta_{\langle x_i \rangle}) = 0$$

for all Φ, Θ on M.

---

[4] Compare Anderson [3], 75–83.

[5] Actually, these statements are true only for suitably well-behaved manifold transformations and coordinate transformations. Thus, h(p) is well-defined only if $\langle y_j(p) \rangle \in$ range $\langle x_i \rangle$; $y_j(p)$ is well-defined only if h(p) $\in$ domain $\langle x_i \rangle$. See (A.34).

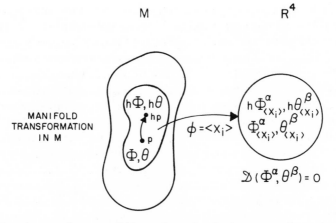

Covariance under a manifold transformation. Is the system of equations $\mathscr{D}(\Phi^\alpha,\Theta^\beta) = 0$ in R^4 satisfied (relative to ϕ) by both $\langle\Phi,\Theta\rangle$ and $\langle h\Phi,h\Theta\rangle$?

Since

$$\mathscr{D}_{\langle y_j\rangle}(\Phi^\alpha_{\langle y_j\rangle},\Theta^\beta_{\langle y_j\rangle})\big|_p = \mathscr{D}_{\langle x_i\rangle}(h\Phi^\alpha_{\langle x_i\rangle},h\Theta^\beta_{\langle x_i\rangle})\big|_{hp},$$

where $x_i(hp) = y_i(p)$, our two notions of covariance give the same results. Thus, for example, the manifold transformations under which (2) is covariant—which take a solution T_σ of (2) to a second solution hT_σ of (2)—are again the linear transformations:

$$x_i(hp) = a_{ij}x_j(?) + b_i,$$

with the a_{ij} and b_i constants.

It may be helpful to think of the relationship between our two approaches in picturesque terms. Think of a coordinate system as a window through which one can view space-time and its geometrical contents; when we change to a second coordinate system, we view space-time through a different window. Think of a system of equations in R^4 as a frame or filter that can be moved from window to window and through which only those geometrical objects are visible that satisfy the equations relative to the coordinate system in question. When we ask whether a system of equations is covariant under a coordinate transformation, we are asking whether the same objects are visible when we move our frame from one window to another. On the other hand, when we are dealing with a manifold transformation h, we keep both the frame and the window fixed. Instead, we change the geometrical objects themselves from Φ,Θ to $h\Phi,h\Theta$. When we ask whether a system of equations (a frame) is covariant

53

under h, we are asking whether both $\langle\Phi,\Theta\rangle$ and $\langle h\Phi,h\Theta\rangle$ are visible through the same window. Because of the way $h\Phi$ is defined, a frame is covariant under the manifold transformation h just in case it is covariant under the coordinate transformation $y_i = x_i \circ h$.

We now know what the covariance group of a system of equations is. What is the significance of such a group? In particular, what is the significance of *general* covariance, that is, covariance under all admissible coordinate transformations? Consider the systems of equations (2) and (4). We know that (2) is not generally covariant; rather, it is covariant only under linear transformations. On the other hand, (4) is generally covariant: it is satisfied by the very same geometrical objects in every coordinate system. Let us note two points. First, ordinary differentiation of a vector field (2) is not a generally covariant operation, whereas application of a derivative operator to a vector field (4) is. In fact, we can view the components of a derivative operator as "correction terms" that make differentiation generally covariant (historically, this is how the affine connection was discovered).[6] Thus, the affine connection is also called the *covariant derivative*.[7] We can often turn a non-generally-covariant equation into a generally covariant equation simply by replacing ordinary derivatives by covariant derivatives. Second, (4) is the general coordinate representation of an intrinsic or coordinate-independent equation: namely, the geodesic equation (1). Equation (2) is not the general coordinate representation for any intrinsic equation. At best (if our manifold happens to be flat), it represents the geodesic equation (1) in certain particular coordinate systems (inertial or Cartesian systems). Thus, general covariance is best understood in terms of complete coordinate-independence: *a theory will be generally covariant just in case it can be given an intrinsic, or coordinate-independent, formulation.*

Only generally covariant theories, then, give us an intrinsic description of space-time. Only generally covariant theories pick out a well-defined class of models $\langle M,\Phi_1, \ldots ,\Phi_n\rangle$ independently of any choice of coordinate system. Non-generally-covariant theories—for example, the traditional formulations of Newtonian physics and special relativity—arise in the following way. One implicitly assumes that space-time is flat and, therefore, that there are special coordinate systems (inertial or Cartesian systems) in which the components

---

[6] See Ricci and Levi-Civita [91].

[7] Compare, e.g., Einstein [29], 68–74, and (A.22).

of certain geometrical objects (the affine connection and some spatio-temporal metrical objects) become constants. In inertial coordinate systems the components of these objects do not appear explicitly as arguments in the differential equations governing the remaining objects. The system of equations (2), for example, contains only the components of T_σ explicitly. Such a system of equations will not be generally covariant; it will be covariant only under transformations preserving the special form of the objects used to define the inertial systems: for example, the form $\Gamma^i_{jk} = 0$ for a connection D or the form $(g_{ij}) = \mathrm{diag}(\pm 1, \pm 1, \pm 1, \pm 1)$ for a metric g. In general, the covariance group will be a subgroup of the linear group. Thus, traditional formulations fall short of general covariance by failing to include explicitly all geometrical objects that are actually assumed. We can turn such a formulation into a generally covariant theory by putting these objects back in, that is, by using them to pick out the special coordinate systems in which the traditional formulation is actually valid. For example, we can restore (2) to general covariance by saying that there exist inertial coordinate systems in which $\Gamma^i_{jk} = 0$ and (2) is valid. This statement, unlike (2) itself, makes a genuinely intrinsic assertion about space-time: namely, that the trajectories σ are geodesics of a flat affine connection.

It should now be clear that the traditional relativity principles cannot be understood simply as covariance requirements, as constraints on the covariance groups of their respective theories. Consider, for example, the general principle of relativity. This principle is almost always equated with the principle of general covariance—the requirement that the covariance group of our theory be the group of all admissible transformations. However, since this is just the requirement that our theory be formulable in the intrinsic, coordinate-independent style, it will be satisfied by any theory that can be formulated in this way. In particular, it is satisfied by all the space-time theories we shall consider! It cannot, therefore, be used to classify or single out general relativity. As Kretschmann first pointed out in 1917 [64], the principle of general covariance has no physical content whatever: it specifies no particular physical theory; rather, it merely expresses our commitment to a certain style of formulating physical theoreis. For the same reason, general covariance has nothing at all to do with the relativity of motion. The simple space-time $E^3 \times R$, in which there are meaningful notions of "absolute" motion in every possible sense (rest, velocity, acceleration, and so on), can be given a coordinate-independent (and

55

therefore generally covariant) description just as easily as the space-time of general relativity.

Hence, the notion of a covariance group is not the appropriate notion for interpreting any of the traditional relativity principles. Since all our space-time theories are covariant under the group of all admissible coordinate transformations, they are all equally, and trivially, covariant under any subgroup of coordinate transformations (Galilean transformations, Lorentz transformations, or whatever). In order to give content to the traditional group-theoretic classification of our space-time theories, we need to find a different way of associating the groups in question with their respective theories. We need a notion of *invariance* group that is quite distinct from the notion of covariance group. Moreover, in developing such a notion, we should try as much as possible to stay away from coordinate systems and coordinate transformations. We should try to formulate a notion of invariance group in an intrinsic or coordinate-independent way.

The most promising suggestion along these lines has been made by the physicist J. L. Anderson [3]. Anderson treats all groups under consideration uniformly as groups of manifold transformations or automorphisms. As we saw above, any manifold transformation h maps a geometrical object Φ onto a second geometrical object $h\Phi$ having the "same values" at hp that Φ has at p. Such a mapping is said to leave Φ *invariant*, or to be a *symmetry* of Φ, just in case it maps Φ onto itself, that is, just in case $h\Phi = \Phi$. Anderson defines the *symmetry group*, or *invariance group*, of a theory as the largest subgroup of $\mathcal{M}$ (the group of all automorphisms)[8], leaving certain of the geometrical objects of the theory—those he calls the *absolute objects* of the theory—invariant. On the basis of this definition, Anderson proceeds to argue that the symmetry group of Newtonian physics is the Galilean group, that of special relativity the Lorentz group, and that of general relativity just $\mathcal{M}$ itself (since general relativity has no absolute objects).

The heart of Anderson's construction is a distinction between two types of geometrical objects: between *absolute* objects and what he calls *dynamical* objects. The absolute objects of a theory are thought to be those objects not affected by the interactions described by the theory. They are independent of the dynamical objects, part of the

---

[8] To be perfectly precise, this notion of symmetry group must be relativized to certain (topologically) well-behaved models of T. See §A.7.

fixed "background framework" within which interaction takes place. Examples of absolute objects are the metric of special relativity and the absolute time of Newtonian mechanics. Examples of dynamical objects are the metric of general relativity (which is affected by the mass-energy distribution) and the electromagnetic field (which is affected by the charge-current distribution). However, although the intuitive distinction between absolute and dynamical objects is clear from these examples, Anderson's actual formal definition of *absolute object* ([3]), 83–84) is rather obscure, at least to me. It is particularly difficult to see how one can use his definition to prove that any given geometrical object is absolute or dynamical. In what follows I shall try to formulate a more perspicuous definition of *absolute object*, one that will, I hope, capture Anderson's intentions. In succeeding chapters I shall apply this definition to our space-time theories and show that it divides up the objects of our theories in the intuitively correct way, yielding the intuitively correct groups as the symmetry (or invariance) groups of our theories.

The absolute objects of a space-time theory are intended to be those objects (such as the metric of special relativity) that are intuitively unaffected by interaction with the other objects of the theory. The dynamical objects are those objects (such as the metric of general relativity) that are affected by interaction with other objects (in this case the mass-energy tensor). Since the absolute objects are supposed to be fixed independently of the other objects, the theory's field equations should determine them in a way in which they do not determine the other objects. For example, the metric of special relativity is determined to be flat or semi-Euclidean: there are always coordinate systems in which $(g_{ij}) = \text{diag}(1, -1, -1, -1)$. The metric of general relativity, on the other hand, is not necessarily flat: although we can always find coordinate systems in which $(g_{ij}) = \text{diag}(1, -1, -1, -1)$ at a single point, the g_{ij} will not in general be constant on any finite neighborhood. Metrics with wildly different curvatures are compatible with the field equations of general relativity. Therefore, the metric of general relativity is not determined by the field equations alone, but only by the field equations plus a fixed distribution of mass-energy.

Our problem, then, is to give a precise sense to the idea that an absolute object is determined by the field equations of a theory T. We want to find a natural equivalence relation $\simeq$ such that the absolute objects, but not the dynamical objects, are determined up

57

to equivalence under $\simeq$. That is, an object Φ_i should be absolute just in case $\Phi_i \simeq \Psi_i$ for any pair of models $\langle M,\Phi_1, \ldots ,\Phi_n\rangle$, $\langle M,\Psi_1, \ldots ,\Psi_n\rangle$ of the field equations of T.

There is such a natural equivalence relation. Let $\langle M,\Phi_1, \ldots ,\Phi_n\rangle$ be a model for the field equations of T. Since T is generally covariant, we know that $\langle M,h\Phi_1, \ldots ,h\Phi_n\rangle$ is also a model, where h is any automorphism. More precisely, if T is generally covariant then

(GC)   If $\langle M,\Phi_1, \ldots ,\Phi_n\rangle$ is a model for T, then for every $p \in M$, neighborhood A of p, and transformation $h:A \rightarrow B$, where B is also a neighborhood of p, $\langle h\Phi_1, \ldots ,h\Phi_n\rangle$ satisfies T on $A \cap B$.

(This more complicated formulation is necessary because an automorphism h is in general defined only on a neighborhood $A \subset M$. Thus, $h\Phi_1, \ldots ,h\Phi_n$ are defined only on the neighborhood $B = hA$ and not on all of M.) On the other hand, the converse of (GC) does not hold. That is, if $\langle M,\Phi_1, \ldots ,\Phi_n\rangle$ and $\langle M,\Psi_1, \ldots ,\Psi_n\rangle$ are two models for T, then it is not necessarily the case that $\Psi_i = h\Phi_i$ for some automorphism h. More precisely, we do not in general have

(DE)   If $\langle M,\Phi_1, \ldots ,\Phi_n\rangle$ and $\langle M,\Psi_1, \ldots ,\Psi_n\rangle$ are both models for T, then for every $p \in M$, there are neighborhoods A, B of p, and a transformation $h:A \rightarrow B$, such that $\Psi_i = h\Phi_i$ on $A \cap B$.

To see this, consider two models $\langle M,g\rangle$ and $\langle M,g'\rangle$ for general relativity. Suppose g is everywhere flat and g' is nowhere flat (such solutions obviously exist). Since g is flat, we know that there is a coordinate system $\langle x_i\rangle$ around p in which the components of g are constants; in fact, $(g_{ij}) = \mathrm{diag}(1,-1,-1,-1)$. We know that the components of hg in $\langle hx_i\rangle$ at hq are equal to the components of g in $\langle x_i\rangle$ at q. So the components of hg are constants—in fact, $(hg_{ij}) = \mathrm{diag}(1,-1,-1,-1)$—in $\langle hx_i\rangle$. Thus, hg is flat at hp if g is flat at p; and since g' by hypothesis is nowhere flat, $g' \neq hg$ on any neighborhood A. An analogous argument shows that if Θ is any tensor field that vanishes everywhere in some models of T but not in others, then Θ does not satisfy (DE). So, for example, the electromagnetic field does not satisfy (DE).

However, (DE) does hold in some cases. Consider two models $\langle M,g\rangle$ and $\langle M,g'\rangle$ for special relativity. We know that both g and g' are flat and, therefore, that the components of both g and g' can be put in the form $\mathrm{diag}(1,-1,-1,-1)$ around every point $p \in M$. Let $\phi(\langle x_i\rangle)$ be a chart around p in which $(g_{ij}) = \mathrm{diag}(1,-1,-1,-1)$ and

58

let $\psi\,(\langle y_j\rangle)$ be a chart around p in which $(g'_{ij}) = \mathrm{diag}(1,-1,-1,-1)$. ϕ and ψ can be so chosen that p has coordinates $(0,0,0,0)$ in both: that is, $x_i(p) = 0 = y_i(p)$. Therefore the set $C = \mathrm{range}(\phi) \cap \mathrm{range}(\psi)$ is non-empty, and $\phi^{-1}(C)$, $\psi^{-1}(C)$ are both neighborhoods of p. Now consider the map $\mathrm{h}\!:\!\phi^{-1}(C) \to \psi^{-1}(C)$ such that $\mathrm{h}(q) = \psi^{-1} \circ \phi(q)$ for $q \in \phi^{-1}(C)$. h is an automorphism, since it is the com-

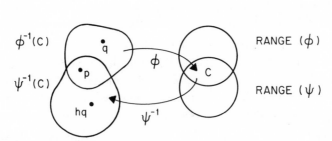

M   R^4

$\phi^{-1}(C)$ RANGE (ϕ)

$\psi^{-1}(C)$ RANGE (ψ)

position of two diffeomorphisms ϕ and ψ^{-1}. Furthermore, $y_i = \mathrm{h}x_i$. But the components of g' in $\langle y_j\rangle$ at $\mathrm{h}q$ equal the components of g in $\langle x_i\rangle$ at q (both take the form $\mathrm{diag}(1,-1,-1,-1)$). It follows that $g' = \mathrm{h}g$ on $\psi^{-1}(C)$. (Note that this argument is essentially the reverse of the argument we just used to show that the general relativistic metric does not satisfy (DE).)

What these examples suggest, of course, is that (DE) is the natural equivalence relation we are looking for (note that (DE) in fact is an equivalence relation). Let us call the relation between Φ_i and Ψ_i described in (DE) *d-equivalence*, and let us require that the absolute objects of a space-time theory be determined up to *d*-equivalence:[9]

---

[9] I have received much help in formulating this definition of absolute object, especially from Clark Glymour, Clifton McIntosh, and Dan Friedan. I should point out that the definition still has some problems. For, as Robert Geroch has observed, since any two timelike, nowhere-vanishing vector fields defined on a relativistic space-time are *d*-equivalent, it follows that any such vector field counts as an absolute object according to (A); and this is surely counter-intuitive. Fortunately, however, this problem does not arise in the context of any of the space-time theories I discuss. It could arise in the general relativistic theory of "dust" if we formulate the theory in terms of a quintuple $\langle M,D,g,\rho,U\rangle$, where ρ is the density of the "dust" and U is its velocity field. U is nonvanishing and thus would count as an absolute object by my definition. But here it seems more natural to formulate the theory as a quadruple $\langle M,D,g,\rho U\rangle$, where ρU is the momentum field of the "dust." Since ρU does vanish in some models, it will not be absolute. (Geroch's observation was conveyed to me by Roger Jones, who also suggested the example of the general relativistic theory of "dust." For further discussion see Jones [62].)

    (A) A geometrical object Φ_i is an *absolute object* of a space-time theory T just in case for any two models $\langle M,\Phi_1,\ldots,\Phi_n\rangle$ and $\langle M,\Psi_1,\ldots,\Psi_n\rangle$ of T, Φ_i and Ψ_i are d-equivalent.

How can we tell in general whether an object is absolute or dynamical? Consider the argument we just used to show that the special relativistic metric g is determined up to d-equivalence. Nothing in that argument depended on the particular set of constants $(g_{ij}) = \text{diag}(1,-1,-1,-1)$; the same argument works whenever the components of a geometrical object can always be made to take on any fixed set of constants. In other words

    (*) Let Φ be a geometrical object of a space-time theory T. Suppose that there is a fixed k-tuple of *constants* $\langle A_1,\ldots,A_k\rangle$ such that around every point in any model of T there is a chart in which the components of Φ are $\langle A_1,\ldots,A_k\rangle$. Then Φ is an absolute object of T.

(*) gives us a simple test for an object's being absolute. Moreover, it shows that there is a close connection between absolute objects on the one hand and theories that describe flat space-times in which there exist inertial or Cartesian coordinate systems on the other. Inertial or Cartesian systems are defined by the components of some object(s) becoming constants—by the connection taking the form $\Gamma^i_{jk} = 0$, by a metric taking the form $(g_{ij}) = \text{diag}(\pm1,\pm1,\pm1,\pm1)$, and so forth.

    This last point allows us to connect Anderson's symmetry or invariance groups with our earlier discussion of covariance groups. The symmetry group of a theory, in Anderson's sense, is a group of manifold transformations preserving the absolute objects of the theory. These objects, in turn, can be used to specify a class of privileged coordinate systems in which the equations of the theory become differential equations for the dynamical objects alone—the components of the absolute objects become constants and drop out. Call the system of differential equations we obtain in this way the *standard formulation* of the theory. Now, the symmetry group of the theory is just the covariance group of the standard formulation of the theory. To see this, consider a theory T whose models contain two objects, Φ and Θ. Suppose that Φ is used to define the inertial coordinate systems of T—and is therefore absolute—and that Θ is dynamical. We know (since T is generally covariant) that if $\langle M,\Phi,\Theta\rangle$ is a model for T, then so is $\langle M,h\Phi,h\Theta\rangle$. Moreover, if $\langle M,\Phi,\Theta\rangle$ is a model for T, then $\langle M,\Theta\rangle$ is a model for the standard formulation of T relative to the inertial systems of Φ. However, if h is an element of the symmetry group of T, then $h\Phi = \Phi$, the inertial systems of $h\Phi =$ the inertial systems of Φ, and $\langle M, h\Theta \rangle$ is a model for the standard formulation of T relative to the inertial systems of Φ. Therefore, h is an element of the covariance group of the standard formulation of T. For the other direction, let $\langle M, \Phi, \Theta \rangle$ again be a model for T. $\langle M, \Theta \rangle$ will satisfy the standard formulation of T precisely in the inertial systems of Φ. Therefore, h will be in the covariance group of the standard formulation only if the coordinate transformation $\langle x_i \rangle \rightarrow \langle y_j \rangle$, where $x_i(hp) = y_i(p)$, takes inertial systems onto inertial systems. It follows that

$$
\begin{aligned}
\Phi^{\alpha}_{\langle y_j \rangle}(p) &= \Phi^{\alpha}_{\langle x_i \rangle}(p) && \text{(since } \langle x_i \rangle \text{ and } \langle y_j \rangle \text{ are both inertial)} \\
&= \Phi^{\alpha}_{\langle x_i \rangle}(hp) && \text{(since the } \Phi^{\alpha} \text{ are constants)} \\
&= h\Phi^{\alpha}_{\langle x_i \rangle}(hp) && \text{(by definition of } h\Phi \text{).}
\end{aligned}
$$

So $h\Phi = \Phi$, and h is a symmetry of Φ.

Thus, for example, consider a simple theory T with a single field variable—a flat affine connection D—and the geodesic law of motion (1). The standard formulation of T will be the system of equations (2), whose covariance group is just the linear group, that is, the group of all transformations preserving the condition $\Gamma^i_{jk} = 0$. Therefore, the symmetry group of D is the linear group. We can also turn this process around to find the covariance group of a given system of equations $\mathscr{D}(\Theta^{\beta})$. We look for a generally covariant theory T in which $\mathscr{D}(\Theta^{\beta})$ is valid in precisely the inertial systems of T. Suppose these systems are defined by the components of a set of objects $\{\Phi_1, \ldots, \Phi_m\}$ becoming constants; $\Phi_1, \ldots, \Phi_m$ will then be absolute objects of T, and the covariance group of $\mathscr{D}(\Theta^{\beta})$ will be just the symmetry group of $\{\Phi_1, \ldots, \Phi_m\}$. Finally, suppose T has no absolute objects and, therefore, no privileged inertial coordinate systems. There will be no standard formulation of T in which some of the objects—the absolute objects—drop out; the standard formulation will be just the generally covariant formulation. In this kind of theory both the symmetry group and the covariance group are degenerate: both contain all admissible transformations. As we shall see, this is exactly what happens in general relativity. General relativity, unlike Newtonian physics or special relativity, describes a nonflat space-time in which there are no inertial coordinate systems. This is the only sense in which general relativity is "more covariant" than the other two theories.

3. *"Absoluteness" and Space-Time Structure*

One of the main purposes of this book is to clarify and evaluate traditional disputes about the "absoluteness" of spatio-temporal structures. I shall treat various aspects of this controversy—for example, the role of "absolute" space and time in Newtonian mechanics—when I discuss the individual space-time theories in detail. I shall also devote a later chapter to an overall assessment of the problem in the light of the physical theories considered. Here, I shall simply try to clarify the issue in a preliminary way by distinguishing the different senses that "absolute" has had in the literature. As we shall see, distinguishing these different senses is especially important for evaluating the bearing that relativity theory has on our traditional disputes. I think that there are three basic senses of "absolute" relevant to our discussion. We can indicate these by three contrasts: (i) absolute-relational, (ii) absolute-relative, and (iii) absolute-dynamical.

(i) ABSOLUTE-RELATIONAL   The absolute-relational contrast represents a debate about the ontological status of space-time structures: whether theories ostensibly about space-time structure are merely theories about the spatio-temporal relations between physical objects, or whether they describe independently existing entities—space, time, or space-time—in which physical objects are located; whether spatio-temporal relations and properties are reducible to or definable from other relations and properties—for example, "causal" relations and properties. Note that there really are two separable issues here. The first concerns the ontology or domain of our space-time theories. Is our domain the set of physical objects, or, in keeping with the four-dimensional point of view, is it the set of *actual* physical events? Or, rather, is it necessary to view the set of actual physical events as located or embedded in a more inclusive (since there may be unoccupied points) space-time manifold? The second issue concerns the status of the spatio-temporal properties and relations defined on our given domain. Can these properties and relations be defined in terms of other, somehow more fundamental properties and relations or not?

Let us call the first kind of relationalism *Leibnizean relationalism* and the second kind *Reichenbachian relationalism*. The Leibnizean relationalist is perfectly happy to take spatio-temporal properties and relations as primitive, so long as the domain over which they are defined does not exceed the set of actual physical events. There can be primitive relations of distance, simultaneity, or temporal precedence between actual physical events, but such relations never hold between unoccupied space-time points. The Reichenbachian relationalist, on the other hand, is not particularly concerned with the domain of definition of spatio-temporal relations; he may even be willing to admit unoccupied positions. But he insists that only relations defined in the "proper" way (for example, in terms of "causal" connectibility) are admissible. These two kinds of relationalism can be combined in various ways to yield constraints of varying strength on the interpretation of our space-time theories. The strictest view requires both an ontology limited to actual physical events and the definability of all relations in terms of preferred vocabulary. Intermediate views require only limited ontology or only limited vocabulary. The most tolerant view admits an inclusive ontology: the entire space-time manifold, containing, if need be, unoccupied space-time points; and any spatio-temporal properties and relations that are explanatorily useful,[10] regardless of their definability. This, of course, is the full absolutist point of view, which will be adopted and defended in this book.

(ii) ABSOLUTE-RELATIVE In the sense of this contrast, an absolute element of spatio-temporal structure is one that is well-defined independently of reference frame or coordinate system. For example, in the space-time $E^3 \times R$ of common sense and some versions of Newtonian mechanics there are well-defined relations of being-at-the-same-spatial-position and being-at-the-same-temporal-position (simultaneity) that hold between points of space-time independently of reference frame: both space and time are absolute in this sense in such a space-time. In other formulations of Newtonian mechanics sameness-of-temporal-position remains well-defined, but sameness-of-spatial-position is defined only relative to a given inertial reference frame: time is absolute, but space is not. In relativity theory both relations are well-defined only relative to a given inertial (or local inertial) reference frame: neither time nor space is absolute in this sense. Of course, sameness-of-*space-time*-position is frame independent, and therefore absolute in the present sense, in all space-time theories.

[10] This is far too easy to say, of course. The hard problem for such a view is to give content to the notion of *explanatorily useful*. I shall address this problem in the last two chapters.

(iii) ABSOLUTE-DYNAMICAL   This contrast is precisely the distinction we examined in detail in §II.2 between geometrical structure that is fixed independently of the processes and events occurring within space-time and geometrical structure that is not so fixed. In this sense, an absolute geometrical structure is one that affects the material content of space-time (through laws of motion, for example) but is not affected in turn. Einstein found this kind of absoluteness especially disturbing, and he endeavored to eliminate it in the general theory of relativity. He felt such independence of geometrical structure to be "contrary to the mode of thinking in science," since it violates a general methodological (ontological?) principle of action and reaction ([29], 55–56).

According to Anderson's program for interpreting symmetry groups, this is the sense of absoluteness most directly reflected in the symmetry group of a space-time theory. In general, the larger the symmetry group, the fewer the absolute elements postulated by a given theory. Thus, the space-time $E^3 \times R$ of common sense (and some versions of Newtonian mechanics), whose symmetry group is $O^3 \times T$—the Euclidean group (see (A.36)) plus time translations—has more absolute elements than the Galilean space-time of Newtonian mechanics (§III.2), whose symmetry group is the Galilean group: $O^3 \times T \times$ Galilean transformations. The former theory postulates an absolute enduring three-space that the latter lacks. Similarly, the general theory of relativity, whose symmetry group is just $\mathscr{M}$, has no absolute elements whatever.

What are the logical relations between these different senses of "absolute"? At first sight, it may appear that the third sense implies the first, in that if a geometrical structure is not even affected by physical events and processes, it cannot be reducible to them. However, this would be a misunderstanding of the technical sense in which absolute objects are independent of physical processes. All it means for an object to be independent or absolute in this technical sense is that it is determined (up to d-equivalence) by the field equations of the theory in question. So it is the "same" (up to d-equivalence) in every dynamically possible model. This, by itself, does not rule out the possibility that the object may be reducible to physical processes; it merely implies that any such reducing physical processes must themselves be the "same" (up to d-equivalence) in every model of the theory. What is true is that if an object is dynamical and depends explicitly on some matter field

64

or fields, then one at least knows where to look for reducing physical processes: namely, to the matter field(s) on which the object depends. Of course, this is far from a guarantee of reduction; for an object may depend on matter fields without being even uniquely determined by them, much less explicitly definable from them. An example of this situation is the metric of general relativity, which depends on the distribution of mass-energy but is not uniquely determined by it. This shows that the first sense of "absolute" does not imply the third either.

The second and third senses of "absolute" are also independent of each other. A feature can be frame independent and still be dynamical or matter-field dependent. An example is proper time in general relativity. Proper time is a function of the general relativistic metric (it measures "distance" along timelike curves) and therefore dynamical, but it is absolute in the sense of being frame independent. In special relativity, on the other hand, proper time is absolute in both the second and the third senses. Similarly, a feature can be frame dependent yet not dynamical. Of course, if a feature is frame dependent it is not really a geometrical object, so the dichotomy absolute-dynamical does not strictly apply. What can happen, however, is that a feature may be a function of some of the components of a geometrical object that is itself absolute in the third sense. An example is coordinate time in special relativity. Such a feature will be frame dependent but matter-field independent.

Although the first and second senses of "absolute" are logically independent as well, they are nonetheless intimately related due to the problem of motion. On an absolutist view—according to which space or space-time is an independently existing entity—the path of a moving object is represented as a curve or trajectory in an inclusive space or space-time. One attempts to explain physical properties of motion in terms of the geometrical properties of such curves. On a Leibnizean relationalist view there is no inclusive space or space-time. Motion is represented in terms of the changing spatial relations of a given physical object to some other physical object: all motion is relative to some actual physical reference object.[11] On this kind of view, therefore, a single object alone in an

---

[11] At any rate, this is what the traditional relationalist thinks. An alternative view, which we shall explore in detail in Chapter VI, conceives of motion in terms of primitive *one-place* properties of physical objects. Motion is not relative to *anything*— neither an inclusive space-time nor other physical objects.

otherwise "empty" universe can have no assignable state of motion. Similarly, if the world contains just two objects, they cannot differ with respect to their motion-related properties. For, as far as relative motion goes, the two are perfectly symmetrical.

How could we implement such a relativistic conception of motion? One strategy is to invoke a *relativity principle*: we relativize a certain class of motions by declaring that all objects from the class, whether alone in the universe or not, are physically "equivalent" or "indistinguishable." As a result, any pair from such a class, whether alone in the universe or not, will be perfectly symmetrical; there will be no need to invoke anything more than relative motion. For a limited class of motions—inertial or non-accelerated motions— this kind of relativity principle does in fact hold. As far as we can tell, all such motions are indeed physically "equivalent"; so variations within this class—being at rest, having velocity 50 mph, and so on—can be interpreted purely relativistically. An object is at rest or has velocity 50 mph only relative to this or that (inertial) reference frame. There is no need to single out any privileged states of motion within this class and to assert, for example, that an object is "really" at rest just in case it is at rest relative to such a privileged state of motion. There is no need, that is, to introduce a frame-independent notion of rest or velocity.

However, as Newton pointed out in his rotating bucket thought-experiment ([75], 10–11), this kind of relativity principle fails if we widen our class of motions to include accelerated or rotational motions. In general, two mutually accelerating or rotating objects are not physically "equivalent." Acceleration and rotation produce distorting forces—the so-called pseudo forces (like centrifugal and Coriolis forces)—that can affect one member of a pair of mutually accelerating or rotating objects but not the other. Consider, for example, Einstein's pair of rotating globes ([28], 112–112). Only

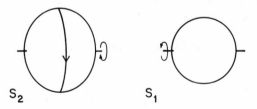

Einstein's rotating globes.

the globe that is "really" rotating (namely, S_2) will experience distorting centrifugal forces causing it to bulge at the equator. These kinds of distortions cannot be explained in terms of the acceleration or rotation of the two bodies relative to each other.

Our space-time theories all solve this problem by introducing an affine connection on the space-time manifold, which immediately divides all motions into two classes: motions whose trajectories in space-time are geodesics of the postulated connection, and motions whose trajectories are not geodesics. The first class comprises the inertial motions; the second class comprises the noninertial or accelerated motions. Furthermore, the laws of motion of our theories all entail the observed distortions that affect noninertially moving objects. They imply that in noninertial reference frames additional (pseudo force) terms appear in the equation $F = ma$. Only in inertially moving reference frames is the equation $F = ma$ (where F is the actual "impressed" force and a is the relative acceleration) valid. (In general relativity and one version of Newtonian gravitation theory $F = ma$ is valid only "locally"; see Chaps. III and V.)

The way in which our space-time theories explain the observed differences between mutually accelerating and rotating objects is *prima facie* antirelationalist. It involves the attribution of abstract geometrical structure—affine structure—to the space-time manifold and the explanation of physical properties of motion in terms of the geometrical properties of curves in the manifold. Moreover, this explanation involves the postulation of a privileged class of trajectories, the inertial trajectories, which may not be—and, in fact, are highly unlikely to be—occupied by any actual physical objects (since it is highly unlikely that any actual physical object has exactly zero net external force acting on it). We therefore appear to be committed to unoccupied space-time points. But, of course, relationalism is not ruled out in principle. For it may still be possible to give a relationalist account of the affine properties of space-time: that is, a reduction of the affine properties of space-time to some physical properties of the total distribution of matter. It may be possible thereby to exhibit the distorting effects experienced by an accelerating or rotating object as depending directly on the relations of that object to the rest of the matter in the universe, bypassing any appeal to a privileged class of (probably unoccupied) inertial trajectories. This is precisely the program envisioned by Mach. According to Mach, the distorting pseudo forces affecting Einstein's rotating globe S_2, for example, are to be attributed to rotation

67

relative to the surrounding mass of the "fixed stars," not to rotation relative to some unobservable state of inertial motion. If the "fixed stars" were removed, no differential effects would occur: if S_1 and S_2 were alone in the universe, S_2 would not (could not) bulge.

This second strategy for relativizing motion, which I shall call *Machianization*, should be sharply distinguished from the first strategy I discussed. If we relativize by means of a relativity principle, we declare that any pair of moving objects is symmetrical or "indistinguishable": there are no differential effects to be explained. Following the strategy of Machianization, on the other hand, we explicitly recognize that there are pairs of nonsymmetrical, relatively moving objects. We admit differential effects and attempt to explain them in terms of relations to some third object (like the "fixed stars"). Thus, the two strategies agree only when the pair of objects in question is the only pair in the universe: on both strategies this case exhibits no differential effects. But when there are more than two objects in the universe, the two strategies are actually inconsistent. Moreover, only the second strategy has any chance at all for success. A general relativity principle in the sense of our first strategy is simply false. In fact, it is so obvious that there are pairs of non-symmetrical, relatively moving objects that it is sometimes difficult to understand how anyone could have imagined that a general relativity principle in this sense is a live option. Part of the answer, I think, is that our two strategies for relativization were not clearly distinguished.

General relativity (and one version of Newtonian gravitation theory) goes some of the way toward implementing the strategy of Machianization. In general relativity affine structure is not absolute in our third sense. The affine connection is a dynamical object, depending on the distribution of mass-energy. Thus, the total distribution of mass-energy influences the affine structure of space-time and, therefore, the differential effects experienced by accelerating and rotating objects. However, since, as noted above, absoluteness in the first sense does not imply absoluteness in the third, the Machian program remains unfulfilled. Although the distribution of mass-energy influences the affine structure of space-time, it does not uniquely determine this affine structure; *a fortiori* the latter cannot be defined in terms of the former. In particular, according to general relativity, a world that contained only two mutually rotating objects (such as Einstein's rotating globes) would still possess affine structure. In such a world it would still be possible

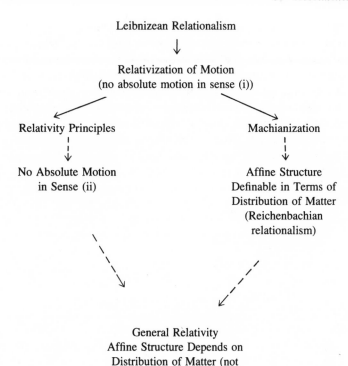

Leibnizean Relationalism

↓

Relativization of Motion
(no absolute motion in sense (i))

Relativity Principles

No Absolute Motion
in Sense (ii)

Machianization

Affine Structure
Definable in Terms of
Distribution of Matter
(Reichenbachian
relationalism)

General Relativity
Affine Structure Depends on
Distribution of Matter (not
absolute in sense (iii))

FIGURE 3. The Different Senses of "Absolute" and the Problem of Motion
(Dotted lines indicate unsuccessful programs.)

for one and only one of these objects—the one whose world-lines are nongeodesic trajectories according to the given affine structure—to experience differential effects.

In brief, then, Leibnizean relationalism leads traditionally to attempts to relativize motion—to interpret all motion in terms of changing spatial relations between actual physical objects. There are two traditional strategies for relativization. On the first, implementation of a relativity principle, we declare that all pairs of relatively moving objects are symmetrical or "indistinguishable" (we try to eliminate absolute motion in the second sense). This strategy applies smoothly to the class of inertial motions but runs into a dead end when applied to accelerated and rotational motion. A general relativity principle in this sense is simply false. The second strategy, Machianization, fares a bit better. On this strategy, we attempt to explain the nonsymmetrical, differential effects that actually occur in nature in terms of relations to the "fixed stars": we attempt to reduce the affine properties of space-time to the total distribution of matter (Reichenbachian relationalism). It turns out, however, that general relativity does not really implement this strategy either. The affine structure depends on the distribution of matter (it is not absolute in the third sense), but the former is not reducible to the latter. Acceleration and rotation remain unregenerately absolute in senses one and two (see Fig. 3).

III
Newtonian Physics

In this chapter I intend to formulate Newtonian physics from the space-time point of view. The basic object of Newtonian physics will be the space-time manifold M; various geometrical structures will be defined on M; and these geometrical structures will be related to the processes and events that occur within space-time by field equations and laws of motion. Looking at our Newtonian theories in this way leads to several surprising results. First, in an important sense Newtonian physics is as four-dimensional as special relativity and general relativity. Second, it is as generally covariant as general relativity. Third, Newtonian gravitation theory, like general relativity, can be formulated in such a way that gravitational forces are incorporated into the geometry of space-time and are thus "geometrized away."

From the present point of view the different Newtonian theories turn out to be the most complicated space-time theories we shall see. This is because four-dimensional formulations of Newtonian physics require the introduction of a relatively large number of geometrical objects on M and because there are alternative formulations that result in essentially the same laws of motion. For clarity, I shall treat Newtonian kinematics and Newtonian gravitation theory separately and give two versions of each: kinematics with absolute space and without it, and gravitation theory with gravitational forces and with gravitational forces "geometrized away." I shall also present a formulation of classical electrodynamics. Finally, I shall discuss the relevance of these formulations to traditional philosophical questions about the structure of Newtonian theory.

1. Kinematics: Absolute Space

Newtonian space-time, like all the space-times considered here, is a four-dimensional differentiable manifold. Many authors (e.g.,

71

Trautman [109]) place global restrictions on the topology of Newtonian space-time: for example, it is often required to be homeomorphic (topologically equivalent) to R^4. As I mentioned in the last chapter, however, I prefer to use only *local* conditions that can be expressed in ordinary field equations. The first local condition we need is that Newtonian space-time is *flat*: that is, there is a flat affine connection D defined on M. As we know, this means that around every point p in M there is a chart ϕ such that the components Γ^i_{jk} of D all vanish: $\Gamma^i_{jk} = 0$. It can be shown that there is a geometrical object K, called the *curvature tensor* of D, such that flatness at p is equivalent to the condition that $K = 0$ at p (see (A.19)). So our first field equation is just $K = 0$ (for all p in M). Our equation of motion for free particles is just the geodesic law

$$D_{T_\sigma} T_\sigma = 0$$

and, since our derivative operator is flat, there are always coordinate systems $\langle x_i \rangle$ in which

$$\frac{d^2 x_i}{du^2} = 0.$$

This last is almost Newton's Law of Inertia. To reconstruct Newton's actual law we have to say more about space and time individually; that is, we have to explain how M "splits up" into three-dimensional space and one-dimensional time.

Newtonian space-time, in contrast to all relativistic space-times, possesses an *absolute time*. Given any pair of points p and q in M, there is a notion of the temporal interval or duration between p and q—the elapsed time between p and q—that is well-defined independently of coordinate system or reference frame (that is, it is absolute in sense (ii) of §II.3). As a consequence, given any single point p in M, there is a notion of the set of all points *simultaneous* with p (the set of all points whose temporal interval from p is zero), and this notion is also well-defined independently of coordinate system or reference frame. The "simultaneity sets" so defined are called *planes of absolute simultaneity*. Hence, Newtonian space-time can be pictured as follows on any well-behaved neighborhood (in this diagram, and all others like it, we have to suppress one spatial dimension). The four-dimensional space-time M is "stratified" into a succession of three-dimensional instantaneous spaces, each of

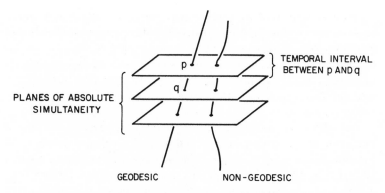

Stratification of Newtonian space-time.

which represents the set of all events simultaneous with a given event. The "distance" between any two such planes of simultaneity represents temporal duration.

Each three-dimensional plane of absolute simultaneity is a Euclidean three-space. There is a Euclidean distance function, or metric, $d_s(p,q)$ defined on pairs of points in any single plane of simultaneity S ($d_s(p,q)$ is undefined if p and q lie on different planes), and, since $d_s(p,q)$ is Euclidean, we can find coordinates x_1, x_2, x_3 on S such that

$$d_S(p,q)^2 = (x_1 - x_1')^2 + (x_2 - x_2')^2 + (x_3 - x_3')^2$$

where $p = \langle x_1, x_2, x_3 \rangle$ and $q = \langle x_1', x_2', x_3' \rangle$. This, of course, is the Pythagorean Theorem. Therefore, we can find coordinate systems

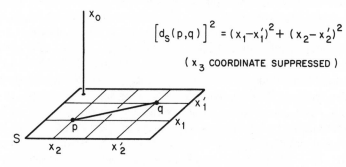

Euclidean geometry on a plane of simultaneity.

73

$\langle x_i \rangle$ on M such that x_1, x_2, x_3 range over S, x_0 ranges over times, and the above condition is satisfied. We have a separate Euclidean geometry on each instantaneous three-space, but we have so far defined no spatial relations at all between points on different planes of simultaneity. To combine our different instantaneous three-spaces into one big "enduring" three-space we need an additional geo-metrical structure.

The additional structure we need is *absolute space*: a relation of occurring-at-the-same-place defined between arbitrary points in M (not just between pairs of points in the same plane of simultaneity). We can introduce such a relation by means of a *rigging* of space-time: a family of non-intersecting geodesics that "penetrates" each plane

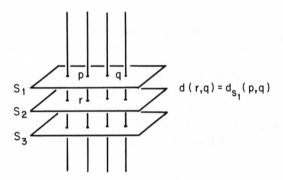

$$d(r,q) = d_{S_1}(p,q)$$

Rigging of Newtonian space-time.

of simultaneity. Thus, two points p and r occur at the same spatial position just in case they lie on the same line of our rigging. More-over, we can extend our family of Euclidean metrics $\{d_{S_1}, d_{S_2}, \ldots\}$ defined on the planes of simultaneity $S_1, S_2, \ldots$ to a single distance function d defined on all of M by "projecting" any pair of points q and r onto the same plane. Note, however, that d remains an es-sentially three-dimensional object. It is not a true four-dimensional metric because it is *singular*: $d(p,r) = 0$ for some distinct points p and r. Note also that our choice of rigging is arbitrary: any family of nonintersecting geodesics, regardless of their "angle" with the planes of simultaneity, will serve equally well. In other words, there simply is no notion of "angle" between lines in a rigging and a plane of simultaneity antecedent to a choice of rigging. The rigging is not chosen to be "orthogonal" to the planes of simultaneity; rather, the choice of rigging itself defines such "orthogonality."

We now know what the geometrical structure of Newtonian space-time looks like. Our next problem is to describe this structure by means of geometrical objects and local field equations. This will enable us to connect our geometrical structure with traditional formulations of the laws of Newtonian physics and to compare Newtonian theory with other field theories, notably, with general relativity.

First of all, how do we describe absolute time and the planes of absolute simultaneity? Perhaps the simplest way is by means of a continuous and differentiable *time-function t* mapping points in M onto real numbers. We interpret $t(p)$ as the time of the event p, $|t(p) - t(q)|$ as the temporal interval between p and q, and we say that p and q are simultaneous just in case $t(p) = t(q)$. In practice, however, this method is a bit awkward, because it assigns absolute significance not only to the temporal interval between p and q but also to the actual real number that any event is assigned. Thus, if t is one such function, and if $t' = t + b$, with b an arbitrary constant, is a second such function, then t and t' agree on temporal intervals and simultaneity but disagree on the specific number assigned to any event. But this last disagreement is no real disagreement at all.

To get around this problem, we observe that associated with any function t from *points* in M to real numbers is a second function dt from *tangent vectors* on M to real numbers such that for any tangent vector X

$$dt(X) = Xt.$$

(Recall that a tangent vector X is itself a mapping from real-valued functions on M to real numbers.) Moreover, t and t' define the very same function dt, since $X(t + b) = Xt$. The function dt is also a *linear* operator on the tangent space T_p:

$$dt(aX + bY) = a\,dt(X) + b\,dt(Y)$$

for all vectors X and Y in T_p and constants a and b. The collection of all such linear operators on T_p is called T_p^*: the *cotangent* space at p. Elements of T_p^* are called *co vectors*, and tangent vectors in T_p are often called *contra* vectors. In general, associated with any co vector w in T_p^* is a family of real-valued functions $\{f_w\}$ such that

75

$w(X) = X(f_w)$ for all contra vectors X. The real numbers $w_i = \partial(f_w)/\partial x_i$—which are well-defined independently of the particular f_w chosen—are called the *components* of w in $\langle x_i \rangle$ (see (A.4)). So the components of dt are given by $t_i = \partial t/\partial x_i = \partial(t + b)/\partial x_i$.

We represent absolute time, therefore, not by a function t but by a co vector field (a continuous and differentiable selection of co vectors from T_p^* for each point p) dt. Associated with dt is not a single time-function t but a family of time-functions $\{t + b\}$ for arbitrary constants b (all of which agree on duration and simultaneity). The co vector field dt induces a division of the tangent space at p as follows: $X \in T_p$ is *timelike* just in case $dt_p(X) \neq 0$; $X \in T_p$ is *spacelike* just in case $dt_p(X) = 0$. Thus, the spacelike vectors are "orthogonal" to the "direction" of absolute time. Furthermore, the spacelike vectors form a three-dimensional *subspace* of T_p (itself four-dimensional, of course) that is "tangent" at p to the plane of absolute simultaneity through p. This latter is defined by the condition $t = \text{constant} = t(p)$ for any associated time-function t. Now, we want our absolute time dt to be compatible with the affine structure that already exists on M. In particular, we want our planes of simultaneity to be flat (Euclidean in the affine sense). To ensure this we require that the covariant derivative of dt with respect to our affine connection D vanish: for every vector field X, $D_X(dt) = 0$ (see (A.20)). We write this condition as

$$\bar{D}(dt) = 0.$$

This is our second local field equation.

Next, how do we describe the Euclidean (metrical) geometry of the planes of simultaneity? Our Euclidean distance functions $d_S(p,q)$ work perfectly well, but, as we observed in the last chapter, $d_S(p,q)$ is not a local or field-theoretic object. The local or field-theoretic counterpart of a distance function $d_S(p,q)$ is a *metric tensor* g_S. We do not want a four-dimensional metric tensor defined on all of M; we just want a family of three-dimensional metric tensors $\{g_S\}$, each of which is defined on a single three-dimensional plane of simultaneity S. Nevertheless, it would be nice if we could capture this family of three-dimensional metrices with a single four-dimensional geometrical object.

We proceed in a slightly devious way: we introduce a symmetric tensor h_p on the *cotangent* space at each point p (so h_p is a bilinear, real-valued function on pairs of co vectors from T_p^*). Since h_p is symmetric, it is almost a semi-Riemannian metric on T_p^*. However,

76

h_p is singular: in fact, we require that $h_p(dt_p,w) = 0$ for all w in T_p^*. Despite its singularity, h_p has properties closely analogous to a "true" metric tensor on T_p^*. In particular, h_p has a 4×4 matrix of components (h^{ij}) in every coordinate system $\langle x_i \rangle$, and we can always find systems in which (h^{ij}) is diagonalized with all diagonal elements equal to $+1$, -1, or zero (see §A.6). The triple $\langle n_+,n_-,n_0 \rangle$, where n_0 is the number of vanishing diagonal elements, is the *signature* of h_p.

A metric tensor is induced by h_p on the three-dimensional subspace of spacelike vectors in T_p in the following way. Consider the map H_* from T_p^* to T_p such that for all w, v in T_p^*, $w(H_*(v)) = h_p(w,v)$ (see §A.6). H_* takes only spacelike vectors in T_p as values: $dt(H_*(v)) = 0$ for all v in T_p^*. Moreover, H_* is not one-one: $H_*(v) = H_*(v + k\, dt_p)$ for any constant k and v in T_p^*. Nevertheless, the metric h_p^* defined on the spacelike subspace of T_p by

$$h_p^*(X,Y) = h_p(w,v), \qquad H_*(w) = X, \qquad H_*(v) = Y$$

is well-defined, since $h_p(dt,v) = h_p(dt,w) = h_p(dt,dt) = 0$. Note that h_p^* is defined only on the three-dimensional spacelike subspace of T_p. On the other hand, h_p is defined on all of T_p^*. This is why we do not start with a symmetric (but singular) tensor on T_p: we come through the back door via T_p^*.[1]

We now have our desired family of three-dimensional metrics. If $S(p)$ is the plane of simultaneity through p, then $g_{S(p)} = h_q^*$ for all q in $S(p)$. As we have seen, $g_{S(p)}$ gives us a notion of length, distance, and angle—a metrical geometry—on $S(p)$. How do we ensure that the tensors $g_{S(p)}$ are both Riemannian and flat—that the geometry on each $S(p)$ is Euclidean? We require h to have signature $\langle 3,0,1 \rangle$, and we require h to be compatible with our flat connection $D: \bar{D}(h) = 0$. This equation and our previous condition $h(dt,w) = 0$ are our two local field equations for h.

Finally, how do we describe absolute space, our rigging of Newtonian space-time? This is easy: we introduce a tangent vector field V. The curves σ that fit V—$T_\sigma = V$ along σ—are the lines in our rigging. We want these lines to be timelike, so we require $dt(V) = 1$;

---

[1] Our complete version of Newtonian space-time will involve two distinguished structures on T_p: (i) the Euclidean metric on the spacelike subspace; (ii) a distinguished "null" timelike vector tangent to the rigging at p. If we started with a singular tensor h_p' on T_p, we would thereby have picked out a "null" vector V_p by the condition $h_p'(V_p,Y) = 0$ for all Y. Our "metric" h' would have absolute space built into it. Thus, it serves clarity to separate (i) and (ii): hence, the singular tensor h_p on T_p^*.

and we want them to be geodesics, so we require $\bar{D}(V) = 0$ (as a consequence, $D_V V = 0$). These conditions are our local field equations for V. Now, for each point p, we already have a notion of length for spacelike vectors in T_p: if X is spacelike, $|X|^2 = h_p^*(X,X)$. V_p allows us to extend this notion of length to all vectors in T_p. For every vector X in T_p, consider the vector $S_X = X - dt_p(X)V_p$. S_X is spacelike, since $dt(S_X) = 0$, and $X = dt_p(X)V_p + S_X$. Thus, S_X is the "projection" of X onto the three-dimensional spacelike subspace of T_p. Define the length of X to be $|S_X|$. (This is the "projective" extension of the metrics d_S to a single function d that we mentioned above.) A vector X in T_p such that $|X| = 0$ is called *null*. The null vectors at p form a one-dimensional subspace of T_p "orthogonal" to the spacelike subspace of T_p, and they have the form $X = bV_p$. A trajectory σ is said to be at *absolute rest* if its tangent vector field T_σ is null, that is, if $T_\sigma = bV$. Thus, the trajectories at absolute rest are just the elements of our rigging.

This, then, is our first version of Newtonian kinematics. We postulate the following geometrical objects on the space-time manifold M: an affine connection D, a co vector field dt, a symmetric tensor field h of type (2,0) (that is, h_p is defined on $T_p^* \times T_p^*$) and signature $\langle 3,0,1 \rangle$, and a contra vector field V. These objects are required to satisfy the following field equations:

$$K = 0 \tag{1}$$
$$\bar{D}(dt) = 0 \tag{2}$$
$$\bar{D}(h) = 0 \tag{3}$$
$$h(dt,w) = 0 \tag{4}$$
$$\bar{D}(V) = 0 \tag{5}$$
$$dt(V) = 1 \tag{6}$$

where K is the curvature tensor of D. Our equations of motion require that particles follow timelike curves and that free particles follow timelike geodesics. That is, the trajectories of free particles satisfy

$$D_{T_\sigma} T_\sigma = 0 \tag{7}$$

and we have $dt(T_\sigma) \neq 0$ for any trajectory σ.

If $\langle x_i \rangle$ is a coordinate system around any point p in M, the field equations take the following form in terms of the components of

78

our objects with respect to $\langle x_i \rangle$:

$$R^i_{jkl} = 0 \tag{1a}$$

$$t_{i;j} = 0 \tag{2a}$$

$$h^{ij}_{;k} = 0 \tag{3a}$$

$$h^{ij}t_i = 0 \tag{4a}$$

$$v^i_{;j} = 0 \tag{5a}$$

$$t_i v^i = 1 \tag{6a}$$

where the v^i are the components of V, the R^i_{jkl} are the components of K (see (A.19)), and a semicolon preceding an index i indicates covariant differentiation with respect to the coordinate x_i (see (A.22)). The equations of motion of course take the form

$$\frac{d^2x_i}{du^2} + \Gamma^i_{jk} \frac{dx_j}{du}\frac{dx_k}{du} = 0 \tag{7a}$$

for free particles and $t_i(dx_i/du) \neq 0$ for arbitrary particles.

I would now like to show that this theory has the desired consequences, that it really can serve as a foundation for Newtonian kinematics with absolute space. First of all, since $R^i_{jkl} = 0$, we know that around every point p in M there is a coordinate system $\langle y_j \rangle$ in which the components of D vanish: $\Gamma^i_{jk} = 0$. In such a coordinate system covariant differentiation reduces to ordinary differentiation (A.22), so (2a) and (3a) become

$$t_{i,j} = \frac{\partial t_i}{\partial y_j} = 0; \qquad h^{ij}_{,k} = \frac{\partial h^{ij}}{\partial y_k} = 0.$$

(Thus, a comma preceding an index i indicates ordinary differentiation with respect to the coordinate y_i.) In $\langle y_j \rangle$, therefore, the t_i and the h^{ij} are *constants*. Now consider any time-function t such that $t_i = \partial t/\partial y_i$. Since the t_i are constants, t is a linear function of the coordinates: $t = a_iy_i + b$. Hence, by a linear transformation we can obtain new coordinates $\langle z_k \rangle$ such that $t = z_0$; that is, the coordinate z_0 is itself a suitable time-function. Furthermore, by a second linear transformation that does not disturb the preceding conditions we can find a coordinate system $\langle x_i \rangle$ in which $(h^{ij}) =$ diag(0,1,1,1) (because of the signature of h). A coordinate system in which $\Gamma^i_{jk} = 0$, $(t_i) = (1,0,0,0)$, and $(h^{ij}) = $ diag(0,1,1,1) is called an *inertial* coordinate system. We have just seen that there are inertial coordinate systems around every point in Newtonian space-time.

What does an inertial coordinate system look like? Consider the *coordinate curves* with the equations $x_i = u$, $x_j = a_j$ ($j \neq i$), where the a_j are constants and u ranges over an interval in R. The tangent vector T_{x_i} to a coordinate curve satisfies

$$T_{x_i}(f) = \frac{df}{du} = \frac{\partial f}{\partial x_j} \frac{dx_j}{du} = \frac{\partial f}{\partial x_i}.$$

In other words, T_{x_i} is just the operator $\partial/\partial x_i$. In our coordinate system T_{x_i} has components (δ_{ij}) ($j = 0,1,2,3$), where $\delta_{ij} = 1$ if $i = j$ and $\delta_{ij} = 0$ if $i \neq j$. Moreover, it is easy to see that each coordinate curve is a geodesic—$D_{T_{x_i}} T_{x_i} = 0$; that T_{x_0} is timelike—$t_i T^i_{x_0} = 1$; and that the T_{x_j} ($j = 1,2,3$) are spacelike—$t_i T^i_{x_j} = 0$. For each

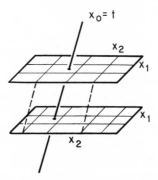

Inertial coordinate system on a neighborhood of Newtonian space-time.

value t_0 of t the hypersurface $x_0 = t_0$ is a plane of simultaneity over which the coordinates x_1, x_2, x_3 range. Since $(h^{ij}) = \mathrm{diag}(0,1,1,1)$ in our coordinate system, the induced metric h^* is Euclidean on any such plane of simultaneity. So our distance functions $d_S(p,q)$ have the Cartesian, or Pythagorean, form as desired.

A timelike geodesic $\sigma(u)$ satisfies the equations

$$\frac{d^2(x_i \circ \sigma)}{du^2} = 0 \tag{8}$$

in an inertial coordinate system. The solutions of (8) are $x_i = a_i u + b_i$, with the a_i and b_i constants. Therefore, $t = x_0 = a_0 u + b_0$ is an affine parameter on timelike geodesics (see (A.15)). In general, then,

80

timelike geodesics satisfy

$$\frac{d^2 y_i}{dt^2} + \Gamma^i_{jk} \frac{dy_j}{dt} \frac{dy_k}{dt} = 0 \tag{9}$$

which in inertial coordinate systems takes the simple form

$$\frac{d^2 x_i}{dt^2} = 0. \tag{10}$$

This last is of course Newton's Law of Inertia. As we shall see, the terms involving Γ^i_{jk} in (9) represent the so-called pseudo forces (such as centrifugal and Coriolis forces) of classical mechanics, which arise from adopting noninertial coordinate systems.

More generally, we can use a time-function t as a parameter for any timelike curve σ and write its equation as $y_i(t) = y_i \circ \sigma(t)$ in an arbitrary coordinate system. Such a curve has a well-defined *velocity* four-vector $\mathbf{u}$ with components $u^i = dy_i/dt$. In an inertial coordinate system (u^i) takes the form $(1, dx_1/dt, dx_2/dt, dx_3/dt)$, and consequently $dt(\mathbf{u}) = 1$. We also have a well-defined *acceleration* four-vector $\mathbf{a}$ with components

$$a^i = \frac{d^2 y_i}{dt^2} + \Gamma^i_{jk} \frac{dy_j}{dt} \frac{dy_k}{dt}.$$

In an inertial coordinate system (a^i) takes the form $(0, d^2x_1/dt^2, d^2x_2/dt^2, d^2x_3/dt^2)$, and consequently $dt(\mathbf{a}) = 0$: $\mathbf{a}$ is spacelike. If $\sigma(t)$ is a timelike geodesic, (10) implies that in an inertial coordinate system $\sigma(t)$ satisfies the equations

$$\begin{aligned} x_0 &= t + b_0 \\ x_i &= v_i t + b_i \qquad (i = 1,2,3). \end{aligned} \tag{11}$$

Hence, the v_i are the components of the ordinary three-velocity of the particle represented by $\sigma(t)$.

So far we have talked about inertial coordinate systems but not about inertial frames. An *inertial frame* is an inertial coordinate system that is adapted to a moving body (particle): the trajectory $\sigma(t)$ of the particle satisfies $x_0 = t, x_i = 0 \, (i = 1,2,3)$ in the associated coordinate system. Thus, a particle is at rest at the origin of its associated frame. There exists an inertial coordinate system adapted to the trajectory $\sigma(t)$ just in case $\sigma(t)$ is a timelike geodesic. To see this from left to right, suppose there is such an associated coordinate system. Then $\sigma(t)$ satisfies $x_0 = t$, $x_i = 0$ ($i = 1,2,3$) and $d^2x_i/dt^2 = 0$. Hence, $\sigma(t)$ is a geodesic. The tangent to $\sigma(t)$ has the form $(1,0,0,0)$; so $t_i u^i = 1$, and $\sigma(t)$ is timelike. For the other direction, suppose $\sigma(t)$ is a timelike geodesic. Then in an arbitrary inertial coordinate system $\sigma(t)$ satisfies $d^2x_i/dt^2 = 0$; so $\sigma(t)$ satisfies $x_0 = t + b_0$, $x_i = v_i t + b_i$ ($i = 1,2,3$). A linear transformation (in fact, a Galilean transformation) gives us an inertial system $\langle y_j \rangle$ in which $\sigma(t)$ satisfies $y_0 = t$, $y_j = 0$ ($j = 1,2,3$). Therefore, inertial frames can be associated only with particles performing inertial motion, that is, only with particles satisfying the law of motion (7).

Although we cannot associate inertial frames with particles undergoing noninertial, accelerated motion, we can associate *rigid Euclidean* frames with them. Call a coordinate system $\langle y_j \rangle$ rigid Euclidean if $y_0 = t$ (for some time-function t) and $(h^{ij}) = \mathrm{diag}(0,1,1,1)$. A rigid Euclidean frame is a rigid Euclidean coordinate system together with an arbitrary (not necessarily geodesic) trajectory $\sigma(t)$ to which it is adapted: $\sigma(t)$ satisfies $y_j = 0$ ($j = 1,2,3$) in $\langle y_j \rangle$. The

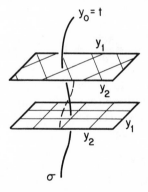

Rigid Euclidean frame adapted to (arbitrary) trajectory $\sigma(t)$.

coordinates y_1, y_2, y_3 are Cartesian on any single plane of simultaneity, but successive planes of simultaneity are not related to one another the way they are in inertial frames. In particular, the coordinates y_1, y_2, y_3 may be "rotated" on successive planes of simultaneity.

Since our rigid Euclidean coordinate system is not inertial, the Γ^i_{jk} will not vanish in such a frame. Instead, they give important information about the frame's acceleration and rotation. First, what do the Γ^i_{jk} look like along σ itself? Since $y_0 = t$, $(t_i) = (1,0,0,0)$ and σ has four-velocity $(u^i) = (1,0,0,0)$ and four-acceleration $(a^i) = (0,a^1,a^2,a^3)$. Hence, $\mathbf{u} = T_\sigma = \partial/\partial y_0$. Now, $D_{\mathbf{u}}(\partial/\partial y_j)$ is always space-like, so $\Gamma^0_{\,0j} = 0$. Moreover, if we consider a purely *spatial* geodesic $\rho(u)$ that intersects σ, $\rho(u)$ must satisfy $d^2 y_j/du^2 = 0$ (since y_1, y_2, y_3 are Cartesian on each hypersurface $y_0 = t_0 = $ constant). Consequently, $\Gamma^i_{jk} = 0$ $(j,k = 1,2,3)$. Therefore, the only nonvanishing components of D are Γ^i_{00} and $\Gamma^i_{0j} = \Gamma^i_{j0}$ $(i,j = 1,2,3)$. To see what these represent note that

$$(D_{\mathbf{u}}\mathbf{u})^i = a^i = \frac{d^2 y_i}{dt^2} + \Gamma^i_{jk} \frac{dy_j}{dt} \frac{dy_k}{dt}.$$

So

$$\Gamma^i_{00} = a^i \qquad (i = 1,2,3). \tag{12}$$

Furthermore, $D_{\mathbf{u}} h = 0$, so $\Gamma^j_{0i} = -\Gamma^i_{0j}$ $(i,j = 1,2,3)$, and

$$\Gamma^i_{0j} = \Gamma^i_{j0} = \Omega^i_j \qquad (i,j = 1,2,3) \tag{13}$$

where (Ω^i_j) is an antisymmetric rotation matrix. Along σ we have the equation of motion

$$\frac{d^2 y_i}{dt^2} + a^i + 2\Omega^i_j \frac{dy_j}{dt} = 0 \qquad (i = 1,2,3). \tag{14}$$

Hence, a^i is the inertial force due to the frame's acceleration, and $2\Omega^i_j(dy_j/dt)$ is the Coriolis force due to its rotation (of course, no centrifugal force is felt at the origin).

So far we have only found the Γ^i_{jk} at the origin of $\langle y_j \rangle$, that is, along σ itself. However, we can easily calculate them at an arbitrary point of $\langle y_j \rangle$ in the following way. If $\langle x_i \rangle$ is a given inertial system, then for each time t, $\langle y_j \rangle$ is related to $\langle x_i \rangle$ by the composition of a rotation and a translation. For each t,

$$\begin{aligned} y_0 &= x_0 = t \\ y_i &= a_{ij}(t) x_j - b_i(t) \qquad (i,j = 1,2,3) \end{aligned} \tag{15}$$

where $a_{ij}(t)$ is orthogonal: $a_{ij}(t) a_{kj}(t) = \delta_{ik}$. And since the components of D vanish in $\langle x_i \rangle$, our transformation law (A.12) reduces to

$$\Gamma^i_{jk} = \frac{\partial^2 x_m}{\partial y_k \, \partial y_j} \frac{\partial y_i}{\partial x_m}. \tag{16}$$

83

Substituting (15) into (16) we find

$$\Gamma^i_{00} = a_{ij}\ddot{a}_{kj}y_k + a_{ij}\ddot{b}_j$$
$$\Gamma^i_{j0} = \Gamma^i_{0j} = a_{ik}\dot{a}_{jk} \qquad (i,j,k = 1,2,3) \qquad (17)$$

where $b_j = a_{ij}b_i$ (and the dots indicate time derivatives). All the other Γ^i_{jk} vanish. Comparing (17) with (12) and (13), we must have

$$a^i = a_{ij}\ddot{b}_j$$
$$\Omega^i_j = a_{ik}\dot{a}_{jk}$$

along σ. The general law of motion becomes

$$\frac{d^2y_i}{dt^2} + a^i + 2a_{ik}\dot{a}_{jk}\frac{dy_j}{dt} + a_{ij}\ddot{a}_{kj}y_k = 0. \qquad (18)$$

Hence, $a_{ij}\ddot{a}_{kj}y_k$ is the centrifugal force.

It is now time to consider the remaining object of our theory, the vector field v^i. From (5a) and (6a) it follows that (v^i) takes the form $(1,v^1,v^2,v^3)$, with v^1,v^2,v^3 constants in any inertial coordinate system. A linear transformation gives us an inertial system in which $(v^i) = (1,0,0,0)$. Such a coordinate system is called a *rest system*, which is characterized by the conditions $\Gamma^i_{jk} = 0$, $(t_i) = (1,0,0,0)$, $(h^{ij}) = \text{diag}(0,1,1,1)$, and $(v^i) = (1,0,0,0)$. Recall that a particle is said to be in a state of absolute rest if its trajectory σ is a null geodesic, that is, if $T_\sigma = bV$. Such a trajectory satisfies the equations $x_0 = bt$, $x_i = b_i$ ($i = 1,2,3$) in any rest system. Its velocity vector therefore takes the form $(b,0,0,0)$. An arbitrary timelike (but not necessarily null) geodesic has the equations $x_0 = t$, $x_i = v_it + b_i$ ($i = 1,2,3$) in a rest system. Thus, null geodesics have vanishing three-velocity in a rest system, while arbitrary timelike geodesics have constant three-velocity. Analogously, we can define a *rest frame* as a rest system in which a given particle is at rest at the origin. Similarly, it is easy to show that there exists an associated rest frame for a particle with trajectory σ just in case σ is a null geodesic. Rest frames can be associated only with particles in a state of absolute rest.

What is the symmetry group of this theory? It is clear from (*) in §II.2 that all objects of our theory are absolute objects: rest systems are privileged coordinate systems for all objects. Consequently, the symmetry group must leave the connection D, the co vector field dt, the "metric" h, and the vector field V invariant.

I shall derive a coordinate representation for the required symmetry group in a rest system $\langle x_i \rangle$ using the technique of §A.7. First of all, we know that the transformations preserving the flat connection D are just the linear transformations. In particular, for any 1-parameter subgroup $\{h_t\}$ we must have

$$\frac{\partial^2 (x_i \circ h_t)}{\partial x_k \, \partial x_j} = 0.$$

Since this is true for every t,

$$\frac{\partial}{\partial t} \left(\frac{\partial^2 (x_i \circ h_t)}{\partial x_k \, \partial x_j} \right) = 0.$$

Interchanging the order of differentiation, we get

$$\frac{\partial^2}{\partial x_k \, \partial x_j} \left[\frac{\partial (x_i \circ h_t)}{\partial t} \right] = 0.$$

So it follows from (A.45) that

$$\frac{\partial^2 a^i}{\partial x_k \, \partial x_j} = 0.$$

That is, if the 1-parameter subgroup $\{h_t\}$ is linear, then its generator (a^i) must be linear as well.

What about the other objects of our theory? Invariance of h means that the required generators must satisfy $L_X h = 0$, which in rest systems takes the simple form

$$\frac{\partial a^i}{\partial x_k} h^{kj} + \frac{\partial a^j}{\partial x_k} h^{ik} = 0. \tag{19}$$

So the a^i are given by

$$a^i = b_{ij} x_j + c_i. \tag{20}$$

Since $(h^{ij}) = \mathrm{diag}(0,1,1,1)$, (19) implies $b_{ij} = -b_{ji}$ $(i,j = 1,2,3)$. Moreover, we must also have $L_X \, dt = 0$ and $L_X V = 0$, which in our coordinate system reduce to

$$\frac{\partial a^k}{\partial x_i} t_k = 0, \qquad \frac{\partial a^i}{\partial x_k} v^k = 0. \tag{21}$$

Since $(t_i) = (v^i) = (1,0,0,0)$, (21) implies $b_{0j} = b_{i0} = 0$. We end up, therefore, with three independent b_{ij} and four independent c_i. The

85

independent b_{ij} form the matrix

$$(b_{ij}) = \begin{pmatrix} 0 & 0 & 0 & 0 \\ 0 & 0 & 1 & 1 \\ 0 & -1 & 0 & 1 \\ 0 & -1 & -1 & 0 \end{pmatrix}.$$

It follows that the desired symmetry group is a 7-parameter Lie group with seven independent generators:

Translations	Rotations
$(1,0,0,0)$	$(0,x_2,-x_1,0)$
$(0,1,0,0)$	$(0,0,x_3,-x_2)$
$(0,0,1,0)$	$(0,x_3,0,-x_1)$
$(0,0,0,1)$	

Hence, the symmetry group is just the symmetry group of Euclidean three-space of §A.7 plus the additional generator $(1,0,0,0)$. This generator yields the equations

$$\frac{\partial x_0^*}{\partial s} = 1, \qquad \frac{\partial x_1^*}{\partial s} = 0, \qquad \frac{\partial x_2^*}{\partial s} = 0, \qquad \frac{\partial x_3^*}{\partial s} = 0$$

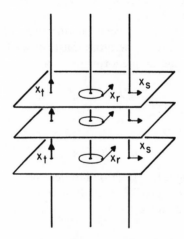

FIGURE 4. The Symmetry Group of Newtonian Space-Time
The planes of simultaneity S, the Euclidean metrics d_S, and the rigging are invariant. So the transformations in our group are just spatial translations (generated by X_s), temporal translations (generated by X_t), and spatial rotations (generated by X_r).

which have the solution

$$x_0^* = x_0 + s$$
$$x_1^* = x_1$$
$$x_2^* = x_2$$
$$x_3^* = x_3.$$

This group is just the *time-translation* group T. Since the symmetry group of E^3 is often denoted by "O^3", the symmetry group of Newtonian space-time with absolute space can be written as $O^3 \times T$ (see Fig. 4).

2. *Kinematics: Galilean Relativity*

As we observed above, our choice of a rigging for Newtonian space-time is essentially arbitrary: any class of nonintersecting timelike geodesics would do equally well. Our choice of a rigging, via our choice of the vector field V, can be viewed as the arbitrary selection of one particular inertial system over all others as a system at rest with respect to absolute space. Our different inertial systems all agree on simultaneity—on the relation of occurring-at-the-same-time—but they disagree on the relation of occurring-at-the-same-place. Our arbitrary choice of one such system fixes this latter relation once and for all, therby giving us notions of absolute rest and absolute velocity.

To formulate Newtonian kinematics in a way that dispenses with absolute space and treats all inertial systems on a par we simply drop the rigging—the vector field V—while retaining the flat connection D and the notion of inertial frame. Consequently, we retain an absolute distinction between inertial motion (constant velocity) and noninertial motion (acceleration and rotation). But we no longer have notions of absolute rest and absolute velocity. Similarly, we retain the absolute time dt and its planes of absolute simultaneity, and we also retain the tensor field h, thereby retaining the Euclidean geometry on each plane of simultaneity. However, we can no longer extend h to all vectors in T_p, nor, consequently, can we define the spatial distance between any two points in space-time. In other words, our new space-time is simply Newtonian space-time without the rigging. This new space-time is called *Galilean* space-time.[2]

[2] See Weyl [116], 156.

Thus, our new theory postulates the following geometrical objects on M: an affine connection D, a co vector field dt, and a symmetric tensor field h of type $(2,0)$ and signature $\langle 3,0,1 \rangle$. These objects satisfy the field equations:

$$K = 0 \tag{22}$$
$$\bar{D}(dt) = 0 \tag{23}$$
$$\bar{D}(h) = 0 \tag{24}$$
$$h(dt,w) = 0. \tag{25}$$

For the equations of motion we again require that particles follow timelike curves and that free particles follow timelike geodesics. In terms of components our field equations are:

$$R^i_{jkl} = 0 \tag{22a}$$
$$t_{i;j} = 0 \tag{23a}$$
$$h^{ij}_{\;;k} = 0 \tag{24a}$$
$$h^{ij}t_i = 0. \tag{25a}$$

The equations of motion are again

$$\frac{d^2 x_i}{du^2} + \Gamma^i_{jk}\frac{dx_j}{du}\frac{dx_k}{du} = 0. \tag{26}$$

Within this theory we can define the notion of inertial frame and formulate Newton's Law of Inertia, but we cannot define the notion of rest frame.

By the reasoning of §III.1, it follows that there are inertial co-ordinate systems—characterized by $\Gamma^i_{jk} = 0$, $(t_i) = (1,0,0,0)$, and $(h^{ij}) = \mathrm{diag}(0,1,1,1)$—around every point p in M. Similarly, any time-function t is an affine parameter on timelike geodesics, which consequently satisfy

$$\frac{d^2 x_i}{dt^2} + \Gamma^i_{jk}\frac{dx_j}{dt}\frac{dx_k}{dt} = 0. \tag{27}$$

So in inertial coordinate systems we have Newton's Law of Inertia

$$\frac{d^2 x_i}{dt^2} = 0. \tag{28}$$

We can again use any time-function t as a parameter for arbitrary timelike curves. We have a well-defined velocity four-vector $u^i = dx_i/dt$ and acceleration four-vector $a^i = (D_\mathbf{u}\mathbf{u})^i$, which in inertial coordinate systems take the form $(1,\vec{v})$ and $(0,\vec{a})$ respectively, with $\vec{v}$ the ordinary three-velocity and $\vec{a}$ the ordinary three-acceleration. Thus, in inertial coordinate systems free particles satisfy the equations

$$x_0 = t + b_0$$
$$x_i = v_i t + b_i \qquad (i = 1,2,3) \tag{29}$$

where the v_i are the components of $\vec{v}$. Finally, we again have the notion of inertial frame and the notion of rigid Euclidean frame, but not, of course, the notion of rest frame.

The symmetry group of Galilean space-time is larger than the symmetry group of §III.1. It is again clear that all objects of our theory are absolute objects, so the symmetry group must preserve D, dt, and h. By the reasoning of §III.1 we find that the components of the generators of the 1-parameter subgroups of our symmetry group take the form

$$a^i = b_{ij}x_j + c_i \tag{30}$$

in an inertial coordinate system. The requirement $L_X \, dt = 0$ again implies $b_{0j} = 0$, while the requirement $L_X h = 0$ implies in addition $b_{ij} = -b_{ji}$ $(i,j = 1,2,3)$. Due to the elimination of V, however, we no longer have $b_{i0} = 0$. We obtain, therefore, additional independent b_{ij}: six independent b_{ij} and four independent c_i. The independent b_{ij} form the matrix

$$(b_{ij}) = \begin{pmatrix} 0 & 0 & 0 & 0 \\ -1 & 0 & 1 & 1 \\ -1 & -1 & 0 & 1 \\ -1 & -1 & -1 & 0 \end{pmatrix}$$

The symmetry group is thus a 10-parameter Lie group with ten independent generators.

		Galilean
Translations	*Rotations*	*Transformations*
$(1,0,0,0)$	$(0,x_2,-x_1,0)$	$(0,-x_0,0,0)$
$(0,1,0,0)$	$(0,0,x_3,-x_2)$	$(0,0,-x_0,0)$
$(0,0,1,0)$	$(0,x_3,0,-x_1)$	$(0,0,0,-x_0)$
$(0,0,0,1)$		

This symmetry group is just $O^3 \times T$ plus the additional generators of the Galilean transformations. To see that these latter really do generate the Galilean transformations consider, for example, the generator $(0, -x_0, 0, 0)$, which yields the equations

$$\frac{\partial x_0^*}{\partial v} = 0, \qquad \frac{\partial x_1^*}{\partial v} = -x_0^*, \qquad \frac{\partial x_2^*}{\partial v} = 0, \qquad \frac{\partial x_3^*}{\partial v} = 0$$

having the solution

$$\begin{aligned}
x_0^* &= x_0 \\
x_1^* &= x_1 - vx_0 \\
x_2^* &= x_2 \\
x_3^* &= x_3.
\end{aligned} \tag{31}$$

This represents a Galilean transformation along the x_1-axis. The full 10-parameter group—$O^3 \times T \times$ Galilean transformations—is called the *Galilean group* (see Fig. 5).

What does all this have to do with traditional ways of talking about the Galilean group? First of all, we can reconstrue our sym-

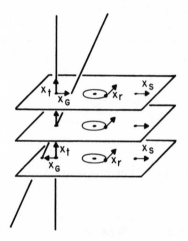

FIGURE 5. The Symmetry Group of Galilean Space-Time
The inertial trajectories, the planes of simultaneity S, and the Euclidean metrics d_S are invariant. So the transformations in our group are just spatial translations (generated by X_s), temporal translations (generated by X_t), spatial rotations (generated by X_r), and Galilean transformations (generated by X_G). A Galilean transformation "tilts" the inertial trajectories relative to the planes of simultaneity. In Newtonian space-time (Fig. 4) invariance of the rigging forbids "tilting."

metry group as a group of coordinate transformations. Every such coordinate transformation preserves the special form of the components of D, dt, and h, that is, it preserves the conditions $\Gamma^i_{jk} = 0$, $(t_i) = (1,0,0,0)$, and $(h^{ij}) = \mathrm{diag}(0,1,1,1)$. These coordinate transformations take inertial coordinate systems onto inertial coordinate systems, and they preserve the special form (28) of the Law of Inertia. But it is important to realize that our transformations preserve more than (28). The form of the Law of Inertia is actually preserved under a wider group of transformations, transformations of the form

$$y_0 = x_0 + c_0$$
$$y_i = a_{ij}x_j + c_i \qquad (i = 1,2,3) \tag{32}$$

where the a_{ij} and c_i are arbitrary constants. Thus, (32) is the covariance group of (28). These are just the transformations preserving only D and dt. The Galilean group, on the other hand, is a subgroup of (32). It is further restricted by the requirement that the a_{ik} ($i,k = 1,2,3$) form an orthogonal matrix (this is the import of the condition $b_{ij} = -b_{ji}$ ($i,j = 1,2,3$) on (30)). This requirement preserves the Euclidean form of the three-dimensional spatial metrics induced by h.

Second, the Galilean group has an obvious connection with the relativity of motion. Let there be given two inertial frames, I and II, associated with two timelike geodesics, $\sigma(t)$ and $\rho(t)$. Suppose that the trajectory $\rho(t)$ represents a state of uniform motion with velocity v along the x_1-axis of frame I. Let $\langle x_i \rangle$ be the coordinates of frame I, $\langle y_j \rangle$ the coordinates of frame II; and suppose for simplicity that $\sigma(t)$ and $\rho(t)$ intersect at $(0,0,0,0)$ in both frames and that $x_2 = y_2$,

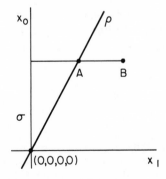

Frame I.

$x_3 = y_3$. Hence, in frame I $\sigma(t)$ has the equations $x_i = 0$ ($i = 1,2,3$) and $\rho(t)$ has the equations $x_1 = vt$, $x_2 = x_3 = 0$. Let B be an arbitrary point having coordinates (y_0,y_1,y_2,y_3) in frame II. In frame II point A has coordinates $(y_0,0,y_2,y_3)$, while in frame I A has coordinates (x_0,vx_0,x_2,x_3). Therefore, in frame I point B (by Euclidean geometry) has coordinates $(x_0,vx_0 + y_1,x_2,x_3)$. Since $x_0 = t = y_0$ we have the relations

$$
\begin{aligned}
y_0 &= x_0 \\
y_1 &= x_1 - vx_0 \\
y_2 &= x_2 \\
y_3 &= x_3
\end{aligned}
\tag{33}
$$

This is just the classical transformation connecting two inertial frames in relative motion with constant velocity v. Thus, the use of the letter "v" for the group parameter in (31) is justified.

3. Gravitation Theory: Gravitational Force

We saw above that the law of motion in Newtonian kinematics can be expressed in the form

$$
\frac{d^2x_i}{dt^2} + \Gamma^i_{jk} \frac{dx_j}{dt} \frac{dx_k}{dt} = 0.
$$

A natural way of expressing the law of motion for Newtonian *dynamics* is

$$
m\left(\frac{d^2x_i}{dt^2} + \Gamma^i_{jk} \frac{dx_j}{dt} \frac{dx_k}{dt}\right) = F^i
\tag{34}
$$

where m is the mass of the particle in question and F^i is a spacelike vector field. In inertial coordinate systems (F^i) has the form $(0,\vec{F})$, with $\vec{F}$ the ordinary three-dimensional force, and (34) takes the familiar form

$$
m\frac{d^2x_i}{dt^2} = F^i
\tag{35}
$$

of Newton's Second Law.

In the case of gravitation theory, we know that the three-dimensional force appearing on the right-hand side of (35) is given by $\vec{F} = m\vec{G}$, where $\vec{G}$ is the gravitational field. The gravitational field $\vec{G}$ is the negative gradient of the gravitational potential Φ:

92

$\vec{G} = -\vec{\nabla}\Phi$, where $\vec{\nabla}$ is the ordinary three-dimensional gradient of a scalar field (in R^3, $\vec{\nabla}$ maps the function Φ onto the vector $(\partial\Phi/\partial x, \partial\Phi/\partial y, \partial\Phi/\partial z)$). Finally, the potential Φ is in turn related to the mass density ρ by Poisson's equation:

$$\vec{\nabla}^2\Phi = 4\pi k\rho \tag{36}$$

where k is the Newtonian gravitational constant and $\vec{\nabla}^2$ is the three-dimensional Laplacian (in R^3, $\vec{\nabla}^2$ maps the function Φ onto the function $\partial^2\Phi/\partial x^2 + \partial^2\Phi/\partial y^2 + \partial^2\Phi/\partial z^2$). A four-dimensional formulation of Newtonian gravitation theory should therefore have the consequence that (35) and (36) are satisfied in inertial frames, where the three-force $\vec{F}$ in (35) has the form $\vec{F} = -m\vec{\nabla}\Phi$.

Newtonian gravitation theory can be formulated within the framework of either of our two versions of Newtonian kinematics. For simplicity I shall use the Galilean space-time of §III.2. The tensor h induces the map $H_*: T_p^* \to T_p$, which allows us to define an operation of "raising" indices (see §A.6). However, since H_* does not possess an inverse, the operation of "lowering" indices is not well-defined. We need the map H_* to define the generalized four-dimensional gradient and Laplacian necessary for the formulation of (35) and (36): if Φ is a function on M, we define grad(Φ) as $H_*(d\Phi)$, and del(Φ) as div(grad(Φ)).[3] Thus, if Φ is the gravitational potential (on M), we can define the gravitational force as $F = -m(\text{grad}(\Phi)) = -mH_*(d\Phi)$; and, since the range of H_* consists only of spacelike vectors, F has the desired properties.

Our theory will have five geometrical objects: an affine connection D, a co vector field dt, a symmetric tensor field h of type (2,0) and signature $\langle 3,0,1 \rangle$, and two scalar fields—the gravitational potential Φ and the mass density ρ. These objects satisfy the field equations:

$$K = 0 \tag{37}$$
$$\bar{D}(dt) = 0 \tag{38}$$
$$\bar{D}(h) = 0 \tag{39}$$
$$h(dt,w) = 0 \tag{40}$$
$$\text{del}(\Phi) = 4\pi k\rho. \tag{41}$$

For our equation of motion we take the covariant form of Newton's Second Law:

$$mD_{T_\sigma}T_\sigma = -m(\text{grad}(\Phi)). \tag{42}$$

---

[3] For div, our generalized *divergence*, see (A.23).

In terms of components we have

$$R^i_{jkl} = 0 \tag{37a}$$

$$t_{i;j} = 0 \tag{38a}$$

$$h^{ij}_{\;;k} = 0 \tag{39a}$$

$$h^{ij}t_i = 0 \tag{40a}$$

$$h^{ij}\Phi_{;i;j} = 4\pi k\rho \tag{41a}$$

for the field equations and

$$m\left(\frac{d^2x_i}{dt^2} + \Gamma^i_{jk}\frac{dx_j}{dt}\frac{dx_k}{dt}\right) = -mh^{ir}\Phi_{;r} \tag{42a}$$

(where t is any time-function) for the laws of motion.

To see that this really is a formulation of Newtonian gravitation theory, consider an inertial coordinate system $\langle x_i \rangle$. In this system we have $\Gamma^i_{jk} = 0$, $(t_i) = (1,0,0,0)$, and $(h^{ij}) = \text{diag}(0,1,1,1)$. So (41a) becomes

$$\frac{\partial^2\Phi}{\partial x_1^2} + \frac{\partial^2\Phi}{\partial x_2^2} + \frac{\partial^2\Phi}{\partial x_3^2} = 4\pi k\rho \tag{43}$$

which is, of course, identical to (36). In free space we have $\rho = 0$ and del(Φ) = 0, which in inertial systems reduces to

$$\vec{\nabla}^2\Phi = 0. \tag{44}$$

Equations (43) and (44) are the usual field equations for the gravitational field. Further, the equations of motion (42a) become

$$m\frac{d^2x_i}{dt^2} = -m\frac{\partial\Phi}{\partial x_i} \tag{45}$$

in an inertial coordinate system. Since the right-hand side of (45) is just the gravitational force acting on the given particle of mass m, we have recovered the familiar form of Newton's Second Law (35).

From our field equations it is clear that Φ and ρ are dynamical objects. There can be everywhere zero and not everywhere zero solutions to (41) for both Φ and ρ. So the absolute objects of this formulation of gravitation theory are D, dt, and h: the fixed background space-time structure of our theory is just Galilean space-time. The symmetry group of this formulation, consequently, is simply the Galilean group. Moreover, the Galilean group is the covariance

94

group of (43) and (45). The law of motion (42a), unlike the law of motion (26), contains the "metric" h^{ij} explicitly. So the covariance group of (45) is narrower than the covariance group (32) of (28).

4. Gravitation Theory: Curved Space-Time

In this section we shall see that it is possible to "geometrize away" gravitational forces in the context of Newtonian theory by incorporating the gravitational potential into the affine connection. That is, we shall construct a new, nonflat derivative operator D and assert that freely falling particles, particles affected only by gravity, follow the geodesics of D. To avoid confusion, let us write $\overset{\circ}{D}$ for the flat connection of §III.3 and $\overset{\circ}{\Gamma}{}^i_{jk}$ for its components in any coordinate system.

To motivate our new theory of gravitation, consider the following problem that arises for the theory of §III.3. Let there be given an inertial system $\langle x_i \rangle$, and let $\sigma(t)$ be an arbitrary accelerating trajectory. Let $\langle y_j \rangle$ be a rigid Euclidean system adapted to $\sigma(t)$, and suppose that $\langle y_j \rangle$ is *not* rotating relative to $\langle x_i \rangle$. Hence, $\langle y_j \rangle$ is

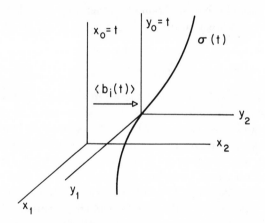

Accelerated reference frame in Galilean space-time. Is $\langle y_j \rangle$ really *accelerating* by an amount $\ddot{b}_i$, or have we merely "miscalculated" the gravitational potential by an amount $-\ddot{b}_j x_j$?

related to $\langle x_i \rangle$ by a transformation of the form (15). In this case

$$
\begin{aligned}
y_0 &= x_0 = t \\
y_i &= x_i - b_i(t) \qquad (i = 1,2,3).
\end{aligned}
\tag{46}
$$

For each time t, $\langle y_j \rangle$ is displaced by an amount $b_i(t)$ along the x_i-axis of $\langle x_i \rangle$. Clearly, $\ddot{b}_i(t) = d^2 b_i / dt^2$ is the acceleration of σ relative to $\langle x_i \rangle$.

But how do we know that $\langle y_j \rangle$ is not an inertial frame? We know that in inertial frames $\overset{\circ}{\Gamma}{}^i_{jk} = 0$, whereas in $\langle y_j \rangle$ we have $\overset{\circ}{\Gamma}{}^i_{00} = \ddot{b}_i$ ($i = 1,2,3$) and all other $\overset{\circ}{\Gamma}{}^i_{jk}$ vanish (17). So our law of motion in $\langle x_i \rangle$ is

$$\frac{d^2 x_i}{dt^2} + \frac{\partial \Phi}{\partial x_i} = 0.$$

In $\langle y_j \rangle$ we have

$$\frac{d^2 y_i}{dt^2} + \frac{\partial \Phi}{\partial y_i} + \ddot{b}_i = 0.$$

This is not an inertial law of motion, so $\langle y_j \rangle$ is not an inertial frame.

But how do we know that the true gravitational potential is Φ? Perhaps the gravitational potential is really

$$\Psi = \Phi + \ddot{b}_j x_j. \tag{47}$$

If Ψ is the true gravitational potential, then $\langle y_j \rangle$ is an inertial frame after all. For the law of motion in $\langle y_j \rangle$ is just

$$\frac{d^2 y_i}{dt^2} + \frac{\partial \Psi}{\partial y_i} = 0$$

where we have replaced the acceleration $\ddot{b}_i$ by an "extra" gravitational potential $\ddot{b}_j x_j$. So which is the true potential, Φ or Ψ? The answer, of course, is that by local conditions alone we cannot determine the true potential! To see this, let $\overset{\circ}{D}$ be our original flat connection whose components vanish in $\langle x_i \rangle$, and let $\overset{\circ}{D}'$ be a second flat connection whose components vanish in $\langle y_j \rangle$; thus, $\langle x_i \rangle$ is an inertial frame according to $\overset{\circ}{D}$, and $\langle y_j \rangle$ is an inertial frame according to $\overset{\circ}{D}'$. It is easy to show that if $\langle M, \overset{\circ}{D}, dt, h, \Phi, \rho \rangle$ satisfies the theory of §III.3, then so does $\langle M, \overset{\circ}{D}', dt, h, \Psi, \rho \rangle$. Our local field equations do not enable us to pick out a unique class of inertial frames and a unique gravitational potential.

Note, however, that we can single out a unique potential, and thus a unique flat connection, by imposing sufficiently strong global boundary conditions. Let us call a coordinate system *Galilean* if it is both rigid Euclidean and nonrotating (both $\langle x_i \rangle$ and $\langle y_j \rangle$ are

Galilean systems). In Galilean systems we have the law of motion

$$\frac{d^2 x_i}{dt^2} + \frac{\partial \Phi}{\partial x_i} = 0 \qquad \text{for } some \text{ potential } \Phi. \tag{48}$$

We can single out a unique potential Φ by requiring that there exists a global (everywhere defined) Galilean system $\langle x_i \rangle$; that all the matter in the universe is concentrated in a finite region ("island universe"); and, finally, that $\Phi \to 0$ as $x_i \to \pm\infty$ ($i = 1,2,3$) in any global Galilean system. It now follows from (47) that $\ddot{b}_j = 0$ and $\Phi = \Psi$. That is, two global Galilean systems must agree on the gravitational potential and must be unaccelerated relative to each other; hence, (46) becomes a Galilean transformation: $b_i = v_i t$ for constant v_i. On the other hand, if our universe happens to be everywhere filled with matter (which appears actually to be the case), even such global conditions will not help us to single out a unique Φ and $\overset{\circ}{D}$.

These considerations suggest that the gravitational potential Φ and the flat connection $\overset{\circ}{D}$ may not have absolute (in sense (ii) of §II.3) existence: they exist only relative to an arbitrary choice of Galilean coordinate system. There is no privileged class of inertial systems and no distinguished gravitational potential. All we can say is that there are Galilean systems in which the law of motion takes the form (48). But what does this mean in terms of geometrical structure? It means that we have replaced the flat connection $\overset{\circ}{D}$ (whose components vanish in inertial systems) with a nonflat connection D whose components satisfy the conditions $\Gamma^i_{00} = \Phi_{,i}$ for some potential Φ and all other Γ^i_{jk} vanish in an arbitrary Galilean system. Our general law of motion is now the geodesic law

$$\frac{d^2 x_i}{dt^2} + \Gamma^i_{jk} \frac{dx_j}{dt} \frac{dx_k}{dt} = 0$$

which in Galilean systems reduces to (48) as desired. Thus, the geodesics of D are just the freely falling trajectories: trajectories of particles affected only (at most) by gravitational forces.

It is important to note here that we are not eliminating (or relativizing) the notion of acceleration. It is true that we are eliminating (relativizing) the notion of acceleration with respect to the flat connection $\overset{\circ}{D}$; but *we retain an absolute (in sense (ii) of §II.3) notion of acceleration with respect to the new, nonflat connection D*. Every

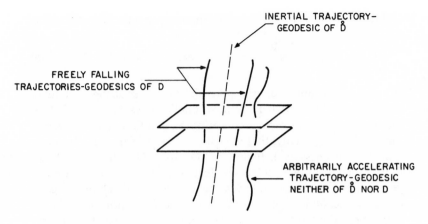

INERTIAL TRAJECTORY–
GEODESIC OF $\mathring{D}$

FREELY FALLING
TRAJECTORIES-GEODESICS OF D

ARBITRARILY ACCELERATING
TRAJECTORY–GEODESIC
NEITHER OF $\mathring{D}$ NOR D

Flat versus nonflat Newtonian gravitation theory. Note that the geodesics of $\mathring{D}$ (inertial trajectories) are *not* in general geodesics of D (freely falling trajectories).

trajectory has a well-defined four-acceleration

$$a^i = \frac{d^2 x_i}{dt^2} + \Gamma^i_{jk} \frac{dx_j}{dt} \frac{dx_k}{dt}$$

and we can pick out nonaccelerating (relative to D) trajectories by the condition $a^i = 0$. This enables us to distinguish a subclass of Galilean frames by the condition $\Gamma^i_{jk} = 0$ along the origin. Such a Galilean frame is called a *local inertial* frame: its defining trajectory $\sigma(t)$ is a geodesic of D. Consider, for example, our trajectory $\sigma(t)$ and frame $\langle y_j \rangle$ above. The four-acceleration of $\sigma(t)$ relative to D is given by

$$a^i = \frac{\partial \Phi}{\partial y_i} + \ddot{b}_i$$

where $\ddot{b}_i$ is the three-acceleration relative to the "inertial" frame $\langle x_i \rangle$. Therefore, $\sigma(t)$ is a geodesic of D just in case $a^i = 0$; just in case $\ddot{b}_i = -\Phi_{,i}$; just in case $\sigma(t)$ is falling freely in our original gravitational field. If $\sigma(t)$ were experiencing a nongravitational acceleration, our frame would be Galilean but not local inertial, and $\sigma(t)$ would no longer be a geodesic of D. It is in no sense true, then, that all accelerations are "equivalent" or "indistinguishable"; at most, it is true that all *gravitational* accelerations are "equivalent." We can pick out local inertial (freely falling) frames from the wider class of

98

Rigid Euclidean Frames (arbitrary acceleration and rotation)

$(t_i) = (1,0,0,0)$

$(h^{ij}) = \text{diag}(0,1,1,1)$

$\Gamma^i_{00} = a_{ij}\dot{a}_{kj}y_k + \Phi_{,i}$ for some potential Φ

$\Gamma^i_{j0} = \Gamma^i_{0j} = a_{ik}\dot{a}_{jk}$ $(i,j,k = 1,2,3)$

all other Γ^i_{jk} vanish

Galilean Frames (arbitrary acceleration; no rotation)

$(t_i) = (1,0,0,0)$

$(h^{ij}) = \text{diag}(0,1,1,1)$

$\Gamma^i_{00} = \Phi_{,i}$ for some potential Φ

all other Γ^i_{jk} vanish

Local Inertial Frames (freely falling; no rotation)

$(t_i) = (1,0,0,0)$

$(h^{ij}) = \text{diag}(0,1,1,1)$

$\Gamma^i_{00} = \Phi_{,i}$ for some potential Φ

all other Γ^i_{jk} vanish

all Γ^i_{jk} vanish along the *origin*

FIGURE 6. The Different Classes of Reference Frames in Newtonian Gravitation Theory
Each class is wider than the class below it.

Galilean (arbitrarily accelerating) frames, and, of course, we can pick out Galilean frames from the wider class of rigid Euclidean (arbitrarily rotating) frames (see Fig. 6).

Let us now proceed to the formal development of our new theory. The key is to find an explicit formula for our new connection D. Recall that the equations of motion for the theory of §III.3 are

$$m\left(\frac{d^2x_i}{dt^2} + \overset{\circ}{\Gamma}^i_{jk}\frac{dx_j}{dt}\frac{dx_k}{dt}\right) = -mh^{ir}\Phi_{;r}. \tag{49}$$

Now notice that the mass m can be eliminated from both sides of (49) to yield

$$\frac{d^2x_i}{dt^2} + \overset{\circ}{\Gamma}^i_{jk}\frac{dx_j}{dt}\frac{dx_k}{dt} = -h^{ir}\Phi_{;r}. \tag{50}$$

If we introduce a new connection D with components

$$\Gamma^i_{jk} = \overset{\circ}{\Gamma}{}^i_{jk} + h^{ir}\Phi_{;r}t_jt_k \tag{51}$$

(50) can be rewritten as

$$\frac{d^2x_i}{dt^2} + \Gamma^i_{jk}\frac{dx_j}{dt}\frac{dx_k}{dt} = 0 \tag{52}$$

(because $t_j = \partial t/\partial x_j$ for any time-function t). Hence, our equations of motion can be interpreted as geodesic equations for the connection D. This connection is not flat: substituting (51) into (A.19) and using the fact that the connection $\overset{\circ}{D}$ is flat, we find that the curvature tensor of D is given by

$$R^i_{jkl} = 2t_jh^{ir}\Phi_{;r;[k}t_{l]} \tag{53}$$

where the brackets around the last two indices indicate antisymmetrization (see §A.6). We can easily verify that the new connection D still satisfies (38) and (39): D is compatible with dt and h. Note that the possibility of different connections satisfying (39) is due to the singularity of the "metric" h. For nonsingular (semi-Riemannian) metrics there is a unique compatible connection (see (A.25)).

Calculation from (53) shows that the curvature tensor of D satisfies the relations

$$t_{[a}R^i_{j]kl} = 0 \tag{54}$$

and

$$h^{ia}R^j_{kal} = h^{ja}R^i_{lak}. \tag{55}$$

Conversely, it can be shown that (54) and (55) together imply the existence of a scalar field Φ and a flat connection $\overset{\circ}{D}$ such that the components of D satisfy (51) and its curvature tensor satisfies (53).[4] Thus, a formulation of gravitation theory in which the nonflat connection D is basic can yield the more usual version of §III.3 by postulating (54) and (55) as field equations in place of (37a).

Finally, what about Poisson's equation? The Ricci tensor (see §A.6) of the connection D is given by

$$R_{jk} = R^a_{jka} = -h^{ar}\Phi_{;a;r}t_jt_k. \tag{56}$$

---

[4] See Trautman [108].

Comparing (56) with (41a), we see that Poisson's equation is equivalent to

$$R_{jk} = -4\pi k\rho t_j t_k$$

which in free space reduces to $R_{jk} = 0$.

In summary, our second version of Newtonian gravitation theory has four basic geometrical objects: an affine connection D, a scalar field ρ representing mass density, and the familiar co vector field dt and "metric" h. The field equations are:

$$(dt \otimes K)_{1a_2} = 0 \tag{57}$$
$$H_*^2 K(w,v,X,Y) = H_*^2 K(v,w,Y,X) \tag{58}$$
$$\bar{D}(dt) = 0 \tag{59}$$
$$\bar{D}(h) = 0 \tag{60}$$
$$h(dt,w) = 0 \tag{61}$$
$$\text{Ric} = -4\pi k\rho \, dt \otimes dt \tag{62}$$

(for $\otimes$ and $_1a_2$ see §A.3; for H_*^2 see §A.6). The equation of motion is just the geodesic equation

$$D_{T_\sigma} T_\sigma = 0. \tag{63}$$

In terms of components we have

$$t_{[a} R^i_{j]kl} = 0 \tag{57a}$$
$$h^{ia} R^j_{kal} = h^{ja} R^i_{lak} \tag{58a}$$
$$t_{i;j} = 0 \tag{59a}$$
$$h^{ij}_{\;;k} = 0 \tag{60a}$$
$$h^{ij} t_i = 0 \tag{61a}$$
$$R_{jk} = -4\pi k\rho t_j t_k \tag{62a}$$

for the field equations and

$$\frac{d^2 x_i}{dt^2} + \Gamma^i_{jk} \frac{dx_j}{dt} \frac{dx_k}{dt} = 0 \tag{63a}$$

for the equations of motion.

In this formulation of Newtonian gravitation theory there is no well-defined class of privileged inertial coordinate systems, since, in general, there are no coordinate systems in which $\Gamma^i_{jk} = 0$. However, we still have rigid Euclidean systems in which $x_0 = t$ for some time-function t and $(h^{ij}) = \mathrm{diag}(0,1,1,1)$, because we can split the connection D into "inertial" and "gravitational" parts as in (51). And in any coordinate system in which $\overset{\circ}{\Gamma}{}^i_{jk} = 0$ we have

$$\Gamma^i_{jk} = h^{ir}\Phi_{,r}t_jt_k. \tag{64}$$

So (59a) and (60a), together with (61a), imply that $t_{i,j} = 0$ and $h^{ij}{}_{,k} = 0$. Hence, the t_i and h^{ij} are constants, and by a linear transformation we can find a coordinate system in which $x_0 = t$ and $(h^{ij}) = \mathrm{diag}(0,1,1,1)$. It follows that the spatial geometry on any plane of simultaneity remains Euclidean and therefore flat, even though the four-dimensional space-time geometry is non-Euclidean or curved.

The coordinate system we have just constructed is not only rigid Euclidean, it is also Galilean: that is, $\Gamma^i_{00} = \Phi_{,i}$ for some potential Φ; all other Γ^i_{jk} vanish. Nevertheless, the class of rigid Euclidean systems is wider than the class of Galilean systems, because, as we know, any two rigid Euclidean systems are related by a transformation as in (15):

$$y_0 = x_0$$
$$y_i = a_{ij}(t)x_j - b_i(t) \qquad (i,j = 1,2,3).$$

Therefore, if $\langle x_i \rangle$ is Galilean, $\langle y_j \rangle$ is in general not Galilean. For in $\langle y_j \rangle$

$$\Gamma^i_{00} = \Phi_{,i} + a_{ij}\ddot{a}_{kj}y_k + a_{ij}\ddot{b}_j$$
$$\Gamma^i_{j0} = \Gamma^i_{0j} = a_{ik}\dot{a}_{jk} \qquad (i,j = 1,2,3). \tag{65}$$

The system $\langle y_j \rangle$ is Galilean only if $\dot{a}_{jk} = 0$, that is, only if $\langle y_j \rangle$ is not rotating relative to any Galilean system. Thus, any two Galilean systems are related by

$$y_0 = x_0$$
$$y_i = a_{ij}x_j - b_i(t) \qquad (i,j = 1,2,3) \tag{66}$$
$$\dot{a}_{ij} = 0$$

(so the a_{ij} are time-independent). Two Galilean systems can be accelerating relative to one another, but not rotating.

Given a particular Galilean system $\langle x_i \rangle$, we can define a flat connection $\overset{\circ}{D}$ by

$$\overset{\circ}{\Gamma}{}^i_{jk} = \Gamma^i_{jk} - h^{ir}\Phi_{,r}t_jt_k.$$

Relative to $\overset{\circ}{D}$ we can define a class of inertial frames and recover the

102

usual version of Newtonian gravitation theory as in §III.3. However, no fixed class of inertial frames is defined in this way, because the flat connection $\overset{\circ}{D}$ depends on the particular Galilean system chosen. For a second Galilean system $\langle y_j \rangle$ in which $\Gamma^i_{jk} = h^{ir}\Psi_{,r}t_jt_k$, it follows from (65) that

$$\Psi = \Phi + \ddot{b}_k x_k.$$

This gives us a second gravitational potential Ψ and a second flat connection $\overset{\circ}{D}'$. As we know, we can single out a unique potential (and therefore a unique flat connection) only by imposing global conditions. Moreover, even global conditions work only in island universes.

In this new theory, then, we cannot construct inertial frames; but we can construct rigid Euclidean frames and Galilean frames by adapting a coordinate system of the appropriate type to the trajectory $\sigma(t)$ of a given particle. In a rigid Euclidean frame associated with $\sigma(t)$ we have

$$\Gamma^i_{00} = a^i$$
$$\Gamma^i_{0j} = \Gamma^i_{j0} = \Omega^i_j \qquad (i,j = 1,2,3)$$

and all other Γ^i_{jk} vanish along $\sigma(t)$. If the acceleration a^i and the rotation Ω^i_j are both zero, $\sigma(t)$ is a geodesic and $\Gamma^i_{jk} = 0$ all along $\sigma(t)$. If $\langle x_i \rangle$ is such a *local inertial* frame, then along $\sigma(t)$ the law of motion takes the form

$$\frac{d^2 x_i}{dt^2} = 0$$

for freely falling particles and the form

$$m\frac{d^2 x_i}{dt^2} = F^i$$

for particles acted on by a nongravitational force F^i.

This weakening of the notion of inertial frame is a consequence of the fact that the connection D is no longer an absolute object. It is not part of the fixed background space-time in our theory but depends on the distribution of matter ρ. It follows that the symmetry group of our present theory is wider than the Galilean group of §III.2 and §III.3. The symmetry group does not have to leave D invariant; its generators are constrained only by the requirements $L_X dt = 0$ and $L_X h = 0$. Thus, while elements of the Galilean group (as represented in an arbitrary inertial coordinate system) take the form

$$x_0^* = x_0 + c_0$$
$$x_i^* = a_{ij}x_j + c_i \qquad (i = 1,2,3)$$

where the a_{ik} $(i,k = 1,2,3)$ form an orthogonal matrix, the elements of the symmetry group of our present theory (as represented in an arbitrary rigid Euclidean system) take the form

$$x_0^* = x_0 + c_0$$
$$x_i^* = a_{ij}(t)x_j + c_i(t) \qquad (i = 1,2,3) \tag{67}$$

where the $a_{ij}(t)$ and $c_i(t)$ are functions of t subject to the condition that the $a_{ik}(t)$ $(i,k = 1,2,3)$ form an orthogonal matrix for each value of t.

5. Classical Electrodynamics

The Maxwell-Lorentz equations for the electromagnetic field are

$$\vec{\nabla} \cdot \vec{E} = 4\pi\sigma$$
$$\vec{\nabla} \times \vec{B} - \frac{\partial \vec{E}}{\partial t} = 4\pi\vec{j} \tag{68}$$

and

$$\vec{\nabla} \cdot \vec{B} = 0$$
$$\vec{\nabla} \times \vec{E} + \frac{\partial \vec{B}}{\partial t} = 0 \tag{69}$$

where $\vec{E}$ is the electric field intensity, $\vec{B}$ the magnetic induction, σ the charge density, and $\vec{j}$ the current density. ($\vec{\nabla}$ is the vector operator $(\partial/\partial x, \partial/\partial y, \partial/\partial z)$ in R^3, $\cdot$ is the dot product on R^3, and $\times$ is the *cross product* on R^3: $(a_1,a_2,a_3) \times (b_1,b_2,b_3) = (a_2b_3 - a_3b_2, a_1b_3 - a_3b_1, a_1b_2 - a_2b_1)$.) The Lorentz force on a particle with charge q in an external electromagnetic field is given by

$$\vec{F} = q(\vec{E} + \vec{v} \times \vec{B})$$

with $\vec{v}$ the three-velocity of the particle. The Newtonian equation of motion is therefore

$$m\frac{d\vec{v}}{dt} = q(\vec{E} + \vec{v} \times \vec{B}). \tag{70}$$

It is well-known that (68) and (69) imply that electromagnetic wave packets propagate through empty space with constant unit three-velocity.[5] Hence, (68) and (69) can hold only in a limited class of inertial frames, all members of which must be at rest relative to each other. For it is a trivial consequence of the Galilean transformations (33) that if a trajectory has (three-) velocity $\vec{v}_I$ in frame I, and if frame II moves relative to frame I with constant velocity $\vec{v}$, then the velocity of the trajectory in frame II is given by

$$\vec{v}_{II} = \vec{v}_I - \vec{v} \qquad (71)$$

(classical velocity addition law). Therefore, if (68)–(70) hold in frame I, they hold in frame II only if $\vec{v} = 0$.

The equations of classical electrodynamics, unlike the laws of motion of classical particle mechanics or gravitation theory, are valid only in a proper subclass of inertial frames. Since all members of this proper subclass must be at rest relative to each other by (71), it is natural to identify this class with the rest systems of §III.1. And the state of motion relative to which electromagnetic wave packets have constant unit (three-) velocity can be identified with the state of absolute rest. So a four-dimensional version of classical electrodynamics should be formulated in the context of the Newtonian space-time with absolute space of §III.1. It should have the consequence that (68)–(70) hold in rest systems.

The essential new element we need is a symmetric, nonsingular tensor field g^* of type (2,0) defined by

$$g^* = V \otimes V - h \qquad (72)$$

where h is the Newtonian spatial "metric" and V is the vector field of §III.1. The tensor field g^* has components

$$g^{ij} = v^i v^j - h^{ij}. \qquad (72a)$$

So in rest systems, where $(h^{ij}) = \mathrm{diag}(0,1,1,1)$ and $(v^i) = (1,0,0,0)$, g^* takes the form

$$(g^{ij}) = \begin{pmatrix} 1 & 0 & 0 & 0 \\ 0 & -1 & 0 & 0 \\ 0 & 0 & -1 & 0 \\ 0 & 0 & 0 & -1 \end{pmatrix} \qquad (73)$$

---

[5] Compare, e.g., Anderson [3], 271. Note that now and hereafter I am setting $c = 1$.

that is, $(g^{ij}) = \text{diag}(1, -1, -1, -1)$. As we shall see, $g*$ is just the inverse of the Minkowski metric of special relativity.

The only other new objects we need are a contra vector field J and an antisymmetric tensor field F of type (0,2) such that F maps $T_p \times T_p$ into R for every p and $F(X,Y) = -F(Y,X)$. F represents the electromagnetic field, and J represents the charge-current density. Our field equations are the following:

$$K = 0 \tag{74}$$
$$\bar{D}(dt) = 0 \tag{75}$$
$$\bar{D}(h) = 0 \tag{76}$$
$$h(dt,w) = 0 \tag{77}$$
$$\bar{D}(V) = 0 \tag{78}$$
$$dt(V) = 1 \tag{79}$$
$$\text{div}[G_*^1 G_*^2 F] = -4\pi J \tag{80}$$
$$\bar{D}(F)_{1a_3} = 0 \tag{81}$$

A particle of mass m and charge q satisfies the equation of motion

$$mD_{T_\sigma}T_\sigma = qtr^{21}(G_*^1 F \otimes T_\sigma) - dt[qtr^{21}(G_*^1 F \otimes T_\sigma)]V \tag{82}$$

(for tr^{ij} see §A.3). Note that the force appearing on the right-hand side of (82) is just the spatial "projection" of the vector $qtr^{21}(G_*^1 F \otimes T_\sigma)$. In terms of components our field equations are:

$$R^i_{jkl} = 0 \tag{74a}$$
$$t_{i;j} = 0 \tag{75a}$$
$$h^{ij}_{;k} = 0 \tag{76a}$$
$$h^{ij}t_i = 0 \tag{77a}$$
$$v^i_{;j} = 0 \tag{78a}$$
$$t_i v^i = 1 \tag{79a}$$
$$F^{ij}_{;j} = -4\pi J^i \tag{80a}$$
$$F_{[ij;k]} = 0 \tag{81a}$$

and our equation of motion is

$$m\left(\frac{d^2 x_i}{dt^2} + \Gamma^i_{jk}\frac{dx_j}{dt}\frac{dx_k}{dt}\right) = q\left(F^i_j \frac{dx_j}{dt} - t_k F^k_j \frac{dx_j}{dt} v^i\right) \tag{82a}$$

where $F^{ij} = g^{ia}g^{jb}F_{ab}$ and $F^i_j = g^{ia}F_{aj}$.

In rest systems we know that $\Gamma^i_{jk} = 0$, $(t_i) = (1,0,0,0)$, $(v^i) = (1,0,0,0)$, $(h^{ij}) = \text{diag}(0,1,1,1)$, and $(g^{ij}) = \text{diag}(1,-1,-1,-1)$. Equation (80a) therefore becomes

$$\frac{\partial F^{i0}}{\partial x_0} + \frac{\partial F^{i1}}{\partial x_1} + \frac{\partial F^{i2}}{\partial x_2} + \frac{\partial F^{i3}}{\partial x_3} = 4\pi J^i. \tag{83}$$

Substituting $i = 0$ in (83), we obtain the first equation of (68), provided that $F^{0j} = -E_{x_j}$ $(j = 1,2,3)$ and $J^0 = \sigma$. Similarly, substituting $i = 1,2,3$ in (83), we obtain the second equation of (68), provided that $F^{12} = -B_{x_3}$, $F^{13} = B_{x_2}$, $F^{23} = -B_{x_1}$, and $J^i = j_{x_i}$ $(i = 1,2,3)$. Therefore, given that $F^{ij} = g^{ia}g^{jb}F_{ab}$, (83) is equivalent to the first pair of the Maxwell-Lorentz equations, provided that

$$(F_{ij}) = \begin{pmatrix} 0 & E_{x_1} & E_{x_2} & E_{x_3} \\ -E_{x_1} & 0 & -B_{x_3} & B_{x_2} \\ -E_{x_2} & B_{x_3} & 0 & -B_{x_1} \\ -E_{x_3} & -B_{x_2} & B_{x_1} & 0 \end{pmatrix} \tag{84}$$

holds in rest systems and that $(J^i) = (\sigma, \vec{j})$. Further, because of the antisymmetry of F_{ij}, in rest systems (81a) reduces to

$$\frac{\partial F_{ij}}{\partial x_k} + \frac{\partial F_{jk}}{\partial x_i} + \frac{\partial F_{ki}}{\partial x_j} = 0. \tag{85}$$

Given (84), (85) is equivalent to the pair of equations (69). Hence, the Maxwell-Lorentz equations are satisfied by the electromagnetic field in rest systems.

To recover the equations of motion, note that in rest systems (82a) becomes

$$m\frac{d^2 x_i}{dt^2} = q\left(F^i_j \frac{dx_j}{dt} - t_k F^k_j \frac{dx_j}{dt} v^i\right). \tag{86}$$

For $i = 0$ (86) reduces to $0 = 0$; for $i = 1,2,3$ it becomes

$$m\frac{d^2 x_i}{dt^2} = qF^i_j \frac{dx_j}{dt}. \tag{87}$$

Given (84) and $F^i_j = g^{ia}F_{aj}$, (87) is equivalent to

$$m\frac{d\vec{v}}{dt} = q(\vec{E} + \vec{v} \times \vec{B}). \tag{88}$$

107

Equation (88) is the desired form for the law of motion in rest systems.

The theory we have been discussing implies that (68)–(70) hold only in rest systems. This is a direct consequence of the fact that its symmetry group is just the group $O^3 \times T$ of §III.1. The experimental evidence, on the other hand, suggests that (68)–(70) in fact hold in *all* inertial frames. We must conclude that the Galilean transformations linking two inertial frames are incorrect. Since the Galilean transformations are just the symmetries of h and dt, h and dt are suspect as well. Hence, the correct theory of electrodynamics must employ a space-time structure completely different from the space-time of Newtonian physics. As we shall see, the new space-time structure we need is just that of special relativity: we solve the above problem by eliminating h, dt, and V—retaining only the tensor field g^* (now taken as primitive, of course) and the connection D. The symmetry group of g^* will turn out to be the Lorentz group, and the Lorentz group, unlike the Galilean group, will be seen to preserve (68)–(70).[6]

6. Absolute Space and Absolute Time

In what senses do our various Newtonian theories involve absolute space and absolute time? How do these notions function in the different theories? Is their use legitimate or illegitimate, scientifically respectable or "metaphysical"? It will take the rest of this book even to begin to answer these questions, especially the last one. In particular, we must compare the use of these notions in Newtonian physics with their use (and the use of analogous notions) in relativistic physics. Nevertheless, it is a good idea to pause for a moment to see what illumination our diverse formulations of Newtonian kinematics and gravitation theory can shed on these questions.

To begin, let us look at kinematics. In the formulation of §III.1 everything is absolute in every sense of "absolute" (§II.3). There is a frame-independent (absolute in sense (ii)) relation of sameness-of-temporal-position, or simultaneity, and a frame-independent

---

[6] As we shall see later, this picture is slightly misleading. Although (70) indeed holds only in rest systems, (68) and (69) are actually valid in a wider class of frames: the *special relativistic* inertial frames. When we formulate relativistic electrodynamics, we shall have to modify (70) to obtain a law of motion that is also valid in all relativistic inertial systems. See §IV.5.

relation of sameness-of-spatial-position. A unique, well-defined temporal duration and a unique, well-defined spatial distance exist between any two points. There is a frame-independent notion of absolute rest: a trajectory is absolutely at rest just in case any two points on it have the same spatial position. Consequently, there are both absolute (sense (ii)) velocity and absolute (sense (ii)) acceleration. Furthermore, since the objects giving rise to these relations are all absolute in Anderson's sense ((A) of §II.2), they are all absolute in sense (iii): they are fixed independently of the events and processes that occur within space-time. In the formulation of §III.2, on the other hand, although time remains absolute in every sense, space does not. There is a frame-independent spatial metric, but it is defined only for points on the same $t =$ constant hypersurface. No unique, enduring three-dimensional space exists but only a succession of instantaneous three-spaces, because there is no frame-independent relation of sameness-of-spatial-position. There is no unique way to join together the succession of instantaneous three-spaces; different inertial frames join them together in different ways. Consequently, no notion of absolute rest or absolute velocity is possible, although we still have absolute (sense (ii)) acceleration. Only relative to a particular inertial frame is there a well-defined relation of sameness-of-spatial-position and thus a well-defined notion of velocity.

When we consider gravitation theory, things become more interesting. First, note that, although we have formulated both versions of gravitation theory in the Galilean space-time of §III.2, it is perfectly possible to start with the space-time of §III.1 instead. Simply add the vector field v^i and the appropriate field equations, (5a) and (6a), to either theory. If this is done, we automatically add absolute spatial position and absolute velocity to either theory. The gravitation theory of §III.3 adds nothing new to the discussion. The space-time framework remains as in §III.1 or §III.2, depending on which formulation we start with. In the formulation of §III.4, on the other hand, the affine connection is a dynamical object: it depends on the distribution of mass. Hence, while acceleration remains well-defined independently of reference frame and is therefore absolute in sense (ii), it is no longer absolute in sense (iii). What about our other objects? The objects t_i and h^{ij} remain absolute, due to the existence of privileged rigid Euclidean coordinate systems. So time and the instantaneous Euclidean three-spaces remain absolute in both sense (ii) and sense (iii). The status of v^i, however,

109

changes dramatically. In a Galilean system with gravitational potential Φ we have

$$\frac{\partial v^i}{\partial t} = -\Phi_{,i} \qquad (i = 1,2,3).$$

So the points of our enduring three-space—the curves that fit v^i—fall freely in the gravitational field and therefore depend on the distribution of mass. Although the enduring three-space, spatial position, and velocity are all absolute in sense (ii), they are no longer absolute in sense (iii).

A similar situation obtains in the case of classical electrodynamics. In the formulation of §III.5 space and time have the same status as in the space-time of §III.1. Spatial position, temporal position, velocity, and so forth are all absolute in both senses. However, we can also combine the electrodynamics of §III.5 with the gravitation theory of §III.4. Accordingly, the absolute (sense (ii)) enduring three-space (the "aether") again ceases to be absolute in sense (iii). Furthermore, the connection D is neither absolute nor flat. So the Maxwell-Lorentz equations imply the bending of light in a gravitational field, just as in general relativity. Note that if we conjoin the Maxwell-Lorentz equations with the flat connection theory of §III.3, this result is not forthcoming.

How do absolute (in whatever sense) space and time function in our various theories? From Newton's point of view they were essential for stating the laws of motion. Newton did not start, as we do, with a four-dimensional manifold M and a four-dimensional affine structure D. These entities were not basic or fundamental; rather, they were defined in terms of absolute space and absolute time. That is, Newton started by postulating a three-dimensional Euclidean space E^3 and a one-dimensional time R. M was then definable as $E^3 \times R$, and the affine structure of M was definable in terms of the straight lines in $E^3 \times R$. In more Newtonian language, a trajectory σ in $E^3 \times R$ is an affine geodesic just in case

either (i) the projection of σ onto E^3 is a single point (σ is at rest),

or (ii) the projection of σ onto E^3 is a straight line in E^3; and if two segments of σ project onto equal intervals in R, they also project onto equal lengths in E^3 (σ moves uniformly in a straight line).

110

(In our notation, we can define a connection Γ^i_{jk} for $E^3 \times R$ by choosing a Cartesian coordinate system $\langle x_1, x_2, x_3 \rangle$ on E^3 and a coordinate $x_0 = r \in R$. We then set $\Gamma^i_{jk} = 0$ in the coordinate system $\langle x_0, x_1, x_2, x_3 \rangle$ on $E^3 \times R$.)

This affine structure on M ($E^3 \times R$) plays an essential role in the laws of motion. The most vivid way to see this is to consider an arbitrary rigid Euclidean frame $\langle y_j \rangle$. As we know, in such a frame the law of motion becomes

$$ m \frac{d^2 y_i}{dt^2} = \quad F^i \quad - \quad a^i \quad - 2 a_{ik} \dot{a}_{jk} \frac{dy_j}{dt} - \quad a_{ij} \ddot{a}_{kj} y_k $$

| impressed | inertial | Coriolis | centrifugal |
| force | force | force | force |

where $\Gamma^i_{00} = a^i + a_{ij} \ddot{a}_{kj} y_k$ and $\Gamma^i_{0j} = \Gamma^i_{j0} = a_{ik} \dot{a}_{jk}$ $(i,j,k = 1,2,3)$. According to Newton, a^i is the acceleration of our frame relative to absolute space (E^3), and $\dot{a}_{ij}(t)$ is its rotation relative to absolute space.[7] However, as we have also seen above, a^i and $\dot{a}_{ij}(t)$ can equally well be regarded as the acceleration and rotation *relative to an arbitrary inertial frame* (since $\Gamma^i_{jk} = 0$ in *all* inertial frames). So we do not actually need to invoke absolute space (E^3).

The reason for this is twofold. First, only acceleration and rotation appear explicitly in the laws of motion; the notions of absolute velocity and absolute rest simply do not occur. Second, acceleration and rotation are themselves *four-dimensional* notions; they make

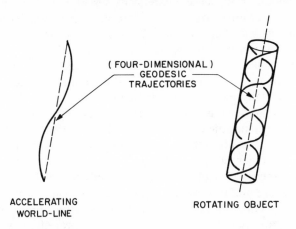

(FOUR-DIMENSIONAL)
—— GEODESIC ——
TRAJECTORIES

ACCELERATING
WORLD-LINE

ROTATING OBJECT

---

[7] This is the essential content of Newton's "rotating bucket" argument for absolute space. See Newton [75], 10–11.

sense independently of all *three-dimensional* spatial structure. Thus, an object is accelerated just in case its world-line deviates from a four-dimensional geodesic trajectory; an object is rotating just in case its world-lines are "twisted" relative to the four-dimensional geodesic trajectories. (Note that we do not need even the succession of *instantaneous* three-spaces to define acceleration and rotation— much less Newton's enduring absolute space E^3). In stating the laws of motion, then, the only objects we need to invoke are the absolute time dt and the affine connection D, now taken as primitive. This, of course, is precisely what happens in the Galilean space-time of §III.2. We retain dt, the Euclidean hypersurfaces $t = $ constant, and the flat affine structure, but we drop the absolute space V. The laws of motion and the explanation of inertial effects (like the rotating bucket) are undisturbed. If absolute space or absolute velocity appeared explicitly in the laws of motion, this move would not be possible. If, for example, the law of motion for free particles were

$$T_\sigma = V$$

(that is, if free particles were required to be at rest relative to absolute space), we clearly could not make do with just dt and D. Such a law would distinguish frames absolutely at rest from all other inertial frames.

Absolute space is not necessary for a Newtonian theory of motion; it is in an important sense theoretically superfluous. Does it follow, then, that Newton's postulation of absolute space was illegitimate, unscientific, and "metaphysical"? Again, we cannot give a fully adequate answer here, but we can make a few preliminary observations. First, if absolute space is unscientific and "bad," this is not because of its absoluteness *per se*. Rather, this is because it is theoretically dispensable: there is an alternative theory—the theory of §III.2—that drops absolute space but does everything we wanted Newton's original theory to do. Note that the temporal co vector field dt is just as absolute in every sense as the rigging V, but it is clearly not theoretically superfluous in this sense. Second, we can see the dispensability of absolute space only by adopting the four-dimensional point of view and using mathematical machinery—the affine connection—that was unknown in Newton's time. From the three-dimensional point of view Newton's procedure was most natural. His theory, absolute space and all, was the best available explanation of the phenomena. Faced with a choice between the theory of §III.1 and the theory of §III.2, *we* can judge that the theory of §III.2 is preferable. However, Newton and his contemporaries simply had no such choice.

But, one wants to object, do we really have to wait for the development of an *alternative* theory (the theory of §III.2) to see that there is something wrong with the absolute space V? Can we see that V is "bad" even from the standpoint of Newton's *original* theory? Interestingly enough, the answer is yes. For it is a consequence of Newton's original theory that inertial motions—motions with constant velocity relative to absolute space—obey a *relativity principle*: all such motions are "equivalent" or "indistinguishable."[8]

What precisely does this *indistinguishability* involve? Suppose we are given a model $\langle M,D,dt,h,V,T_{\hat{\sigma}} \rangle$ for Newtonian kinematics and two inertial frames I and II. The only theoretical difference between I and II concerns the components of V: in I $(v^i) = (1,-\bar{v})$, in II $(v^i) = (1,-\bar{u})$, where $\bar{u}$ and $\bar{v}$ are the absolute velocities of our two frames. I and II agree on the components of D, dt, and h. Can we distinguish between I and II by means of the laws of Newtonian kinematics? Change V to a new vector field V' whose components in I are equal to the components of V in II; that is, $V' = fV$ where f is a Galilean transformation taking II onto I. It is easy to see that $\langle M,D,dt,h,V',T_{\hat{\sigma}} \rangle$ is also a model for Newtonian kinematics. Hence, from the point of view of Newtonian kinematics itself we cannot tell whether we are in a frame with absolute velocity $\bar{v}$ (relative to V) or in a frame with absolute velocity $\bar{u}$ (relative to V'). By contrast, we *can* distinguish absolutely accelerating (or rotating) frames from inertial frames. To see this, let I again be an inertial frame, and let II be a rigid Euclidean frame with absolute acceleration a^i. So in I $\Gamma^i_{jk} = 0$, while in II $\Gamma^i_{00} = a^i$. Now change D to a new connection D' whose components in I are equal to the components of D in II; that is, $D' = fD$, where f is a transformation of the form (15). $\langle M,D',dt,h,T_{\hat{\sigma}} \rangle$ is *not* a model for Newtonian kinematics (§III.2), for any trajectory σ in $\hat{\sigma}$ satisfies

$$\frac{d^2(x_i \circ \sigma)}{du^2} = 0$$

---

[8] Newton's own version of this *Galilean principle of relativity* is *Corollary V* to the Laws of Motion. See Newton [75], 20–21.

in I by hypothesis, and the Newtonian law of motion in I relative to D' is

$$\frac{d^2(x_i \circ \sigma)}{du^2} + a^i = 0.$$

Hence, $T_{\hat{a}}$ does not satisfy the Newtonian law of motion relative to D'.

The equivalence or indistinguishability of different inertial frames in this sense is a direct consequence of the fact that the laws of Newtonian kinematics do not allow us to determine V by fixing the values of the other quantities of the theory. V can vary arbitrarily while the rest of the theoretical quantities remain fixed. Moreover, this arbitrariness—or theoretical indeterminateness—of V is not, of course, a *general* feature of theoretical entities. Consider the absolute time dt, for example. It follows from the equations of Newtonian theory (in either formulation) that any time-function t is an affine parameter on geodesic trajectories. Any two such time-functions, t and t', are therefore related *linearly*: $t' = at + b$ (see (A.15)). Thus, if $\langle M,D,dt,h,(V),T_{\hat{a}} \rangle$ and $\langle M,D,dt',h,(V),T_{\hat{a}} \rangle$ are two models for Newtonian kinematics, then $dt' = adt$. The equations of Newtonian kinematics determine dt up to a multiplicative constant (relative to the other quantities, of course[9]), which simply represents a choice of *units*. This point underscores the fact that the problem with V is not absoluteness *per se*.

Newtonian absolute space therefore suffers from very special difficulties, difficulties that can be clearly seen from the standpoint of Newtonian theory itself. Newtonian theory itself (in the formulation of §III.1) gives us a good reason to look for an alternative theory (the formulation of §III.2) that dispenses with the theoretically indeterminate field V. In practice, however, the elimination of such theoretically indeterminate entities often requires considerable ingenuity. What typically happens is that an indeterminate, and therefore suspect, entity plays a role in the definition of some determinate, and therefore "good," entity. We can then construct our alternative theory by taking the defined (determinate) entity as

[9] Thus, determinateness in the present sense is a different type of property from absoluteness. An object is absolute *simpliciter*: it is determined (up to d-equivalence) by the field equations independently of all other objects. But an object is *determinate* relative to some other objects: fixing these other objects determines (up to a multiplicative constant, say) the object in question.

primitive and simply dropping the old, defining (indeterminate) entity. In the present case, dt and V allow us to define the determinate[10] affine connection D. In our new theory we take D to be primitive and drop V.

Finally, although it is true that the absolute space V is indeterminate and theoretically superfluous in the context of Newtonian kinematics, it does not follow that it is theoretically useless *simpliciter*, that it is an intrinsically unscientific and "metaphysical" entity. For it may be perfectly useful and respectable *in conjuction with other theories*. In fact, as we have seen in §III.5, V is very useful indeed in the context of classical electrodynamics. V appears explicitly in the law of motion (82), and the equations of classical electrodynamics allow us to distinguish a frame at rest relative to V from an arbitrary inertial frame. Thus, if $\langle M,D,dt,h,V,F,J,T_{\hat{a}} \rangle$ is a model for the theory of §III.5, then $\langle M,D,dt,h,V',F,J,T_{\hat{a}} \rangle$ is in general *not* a model. Far from having no "empirical content," V is rich in empirical predictions in §III.5; for example, it predicts that the Michelson-Morley experiment will have a *positive* result. The problem, of course, is that these predictions are false. Optical and electrodynamic phenomena do not in fact distinguish one inertial frame from another—contrary to §III.5. In the correct theory, special relativity, we must eliminate not only absolute space but absolute time as well.

7. The Laws of Motion

We have encountered four laws of motion in this chapter: equation (7), which says that free particles follow timelike geodesics of a flat connection; equation (42), which says that particles acted on only by gravitational forces follow nongeodesic paths determined by the gravitational potential; equation (63), which says that particles acted on only by gravity follow geodesics of a nonflat connection; and equation (82), which says that particles affected only by electromagnetic forces deviate from geodesic paths in accordance with the external electromagnetic field acting upon them. Newton's Second Law (35) is not so much a law of motion as it is a program for possible laws of motion, a program that must be filled out by specific theories of interaction (gravitational, electromagnetic, and so on) that give

---

[10] This follows because two connections are equal if they have the same geodesics and the same torsion tensors (Hicks [59], 65). In the present case, all connections considered are *symmetric*: all have vanishing torsion tensors. See (A.16).

various particular forms for the vector field F^i. From the present point of view such laws of motion have two basic functions. They say something about the geometrical structure (that is, the curvature) of space-time, and they say how this geometrical structure constrains the trajectories of a given class of particles.

In more conventional formulations of Newton's laws of motion, by both physicists and philosophers, the notion of a reference frame—more specifically, an inertial frame—plays a crucial role. The content of Newton's First Law, for example, is taken to be the claim that free particles satisfy equation (10) in inertial reference frames. This approach is more congenial to the operationally minded, because they feel that a reference frame is more concrete and observable than the four-dimensional affine structure of our formulations. (Recall that in our construction a reference frame is a pair consisting of a given physical body or particle and a local coordinate system. If we do not require the trajectory σ defining the frame to be *occupied*, the reference frame approach is equivalent to our approach.) For the same reason, formulations in terms of reference frames fit naturally into relationalist interpretations of motion. The hope is that only relative motion—motion with respect to a given type of reference frame—will figure in the laws of motion. But this approach is faced with several serious difficulties.

First, the approach works smoothly only in the case of flat space-time, for inertial coordinate systems exist only in flat space-times. Only if we are guaranteed reference frames in which $\Gamma^i_{jk} = 0$ can we formulate laws like (7) without explicitly referring to affine structure. In the gravitation theory of §III.4, for example, there are no inertial frames. Of course we still have local inertial frames in which $\Gamma^i_{jk} = 0$ all along the trajectory with which the frame is associated, but this allows us to say only that free particles satisfy (10) on the trajectory itself. Stating the law of motion on any finite neighborhood of the trajectory requires the introduction of nonvanishing Γ^i_{jk}'s. This problem poses no serious difficulties in the present instance, because we know that there are flat space-time versions of Newtonian gravitation theory. The point must be kept in mind in the context of general relativity, however.

A more serious problem involves characterizing the inertial frames. In the absence of affine structure they cannot, of course, be characterized by the vanishing of the Γ^i_{jk}'s. Typically, inertial frames are defined in terms of equation (10) itself; that is, they are defined

116

to be frames in which (10) is satisfied by free particles.[11] This definition obviously renders the First Law circular as it stands, so the usual practice is to construe it as an existence claim, an assertion that there exist reference frames in which (10) is valid for free particles.[12]

If the First Law is taken to be the claim that inertial *coordinate systems* (or unoccupied inertial frames) exist, then it is equivalent to our formulation, since the existence of such coordinate systems is equivalent to the flatness of the affine connection. However, if it is taken to be the claim that occupied inertial frames exist, then we are in deep trouble indeed. For the First Law then implies the existence of free particles, and, in the context of Newtonian physics itself, this is false. According to Newtonian gravitation theory, a nonvanishing gravitational force exists between any two particles in the universe. This problem is raised by N. R. Hanson, who takes it to cast doubt on the meaningfulness of the First Law:

> Meaningfully to claim of that mass that it moves uniformly along a rectilinear path it is necessary to fix physical coordinates by assuming other punctiform masses to be securely anchored. . . . But no particle within [a universe where other massive particles are present] can be force free—a direct inference from another cornerstone of mechanics, the law of universal gravitation . . . [thus] either the law [of inertia] conflicts with our conception of physical meaning or it conflicts with other laws of mechanics. Either way it is difficult to comprehend. ([54], 14)

Actually, Newtonian gravitation theory is not strictly inconsistent with the existence of occupied inertial frames; for all we need are particles with vanishing *net* force, and the gravitational forces may be precisely balanced by other (gravitational or nongravitational) forces. Nevertheless, it is highly unlikely that such precise balancing of forces occurs, at least over any extended period of time. Hence, although the account of Newtonian physics under consideration does not fall into outright inconsistency, it is highly improbable.

It may be objected that, although there are indeed no exactly inertial frames, there are *approximately* inertial frames: for example,

---

[11] For examples, see Bergmann [4], 8; Bunge [7], chap. 3 (esp. Def. 1 on p. 136); and Goldstein [49], 185.

[12] See, e.g., Bergmann [4], 8; Bunge [7], 135.

a frame fixed at the center of the sun.[13] Therefore, frames exist in which (10) is valid to a high degree of approximation. And why should this not suffice? After all, Newtonian physics is only approximately true in any event. This reply is inadequate. Newtonian physics *is* only approximately true, but not because of the existence of *gravity*. From the point of view of a better theory—for example, from the point of view of special relativity—we can and do assert that the universe provides only an approximate model for Newtonian physics. On the account under consideration, however, Newtonian physics itself implies that the universe is very unlikely to provide an exact model for Newtonian physics. The issue is not the truth of our theory but its coherence: Newtonian gravitation theory should not imply the probable falsity of Newtonian kinematics.

The intuitive, and I think correct, response to these difficulties with the notion of inertial frame is that Newtonian kinematics does not require the existence of occupied inertial frames. Newtonian physics is (would be) true even if there are (were) no inertial frames. The First Law deals with the existence of inertial frames only counterfactually: if there were inertial frames (for example, if there were no gravitational forces), free particles would satisfy (10) in them. This intuitive construal cannot, of course, be adopted by the operationalist approach we have been discussing, which regards the above counterfactual as *trivially* true. On the other hand, if we interpret Newtonian physics according to the geometrical, space-time point of view, this intuitive construal of the Law of Inertia is precisely what we get. First, the content of the law is coordinate-independent and therefore independent of reference frame as well: it states that free particles follow geodesics of a flat affine connection. Which curves in space-time are affine geodesics can be specified independently of coordinate systems or reference frames (namely, by equation (7)). Second, the notions of inertial coordinate system and inertial reference frame can be defined independently of the First Law. The assertion that (10) holds in inertial frames (or that it would hold if there were any) is therefore not trivial. In fact, of course, it is equivalent to the geodesic law (7). Hence, the space-time approach provides a better foundation for Newtonian kinematics than does the more traditional operationalist approach. If we view kinematics as a theory about the structure of space-time and the possible tra-

[13] See Goldstein [49], 135.

118

jectories of particles within that structure, rather than as a theory about occupied inertial reference frames, we can give a more satisfactory account of the laws of motion.

Finally, let us remember that these problems about the nature and existence of inertial frames did not arise for Newton. For Newton had a privileged global inertial system—absolute space—relative to which his laws of motion were stated. Moreover, he could easily show that his laws of motion would also hold in any frame moving uniformly relative to absolute space, *if* any such frame *existed*.[14] Thus, as noted above, Newton himself needed neither a primitive notion of inertial frame nor a primitive notion of affine structure (space-time geodesic). However, if we want to dispense with absolute space and also avoid the serious difficulties with the operationalist account of inertial frames, it seems we have no choice but to appeal to the affine structure of space-time. As we shall see in the following chapters, this understanding of the laws of motion makes good sense in relativity theory as well.

The First Law has been thought to be subject to a second kind of circularity problem, this time centering on the notion of a free particle. The First Law states that free particles follow affine geodesics. But what is a free particle? It is sometimes claimed that our only grip on the notion of a free particle, our only criterion for when a particle is subject to no forces, is that it obeys the First Law. The First Law again begins to look like a mere definition, this time a definition of "free particle." An example of this line of argument occurs in Ellis [31]. Ellis claims that our only access to forces is through force-effects and that force-effects are nothing but deviations from some "natural" state of motion (in this case inertial motion). He concludes that since what counts as "natural" motion is conventional, the existence of forces is also conventional; so, consequently, is Newton's First (and Second) Law.

The problem with this argument is that it concentrates exclusively on force-effects and ignores their sources. We have access to forces in two ways: through the effects they produce and through the interactions giving rise to them. Forces must come from somewhere; they are tied to specific types of interaction by theories relating them to the physical properties (mass, charge, and so on) of the objects participating in the interaction. (For example, in the context of Newtonian gravitation theory (§III.3) the spacelike vector field on

---

[14] Newton [75], 20–21.

the right-hand side of equation (34) is tied to the masses of bodies by equations (41) and (42); in the context of classical electrodynamics it is tied to the charges of bodies by equations (80) and (82).) If our best theories of interaction tell us that there are no forces acting on a given particle, we have good reason to think that it is free. Thus, both gravitation theory and electrodynamics assert that a particle is free if it is "sufficiently far" from all other bodies.

It might be objected that our only reason for postulating various kinds of interaction is just observed deviation from the First Law. This might be true. Nevertheless, although deviations from the First Law may motivate us to look for interactions, they cannot guarantee that we will find them. If we repeatedly fail to construct a successful theory of interaction to explain an observed deviation, we have reason to think that no interaction exists and that the First Law is false. A parallel situation occurs in the debate over Mach's Principle. Mach claimed that a stationary earth surrounded by a rotating star shell would experience the same pseudo forces as a rotating earth surrounded by a stationary star shell. However, no one has been able to construct a theory of interaction that would result in such forces. (In particular, as we shall see below, general relativity does not predict such forces.) It is customarily concluded, and I think reasonably, that no such forces exist.

There is no general reason, then, to suppose that laws of motion in the Newtonian style are circular or conventional. Still, specific theories of interaction may be subject to this kind of difficulty if they invoke *theoretically indeterminate* forces. A good example here is the gravitation theory of §III.3. In this particular theory we cannot, given the laws of the theory itself, determine whether the gravitational potential is Φ or $\Psi = \Phi + \ddot{b}_j x_j$. As a result, we do not know whether the gravitational force acting on a particle of mass m is $F_\Phi^i = -mh^{ir}\Phi_{,r}$ or $F_\Psi^i = -mh^{ir}(\Phi_{,r} + \ddot{b}_j x_{j,r})$. So consider a freely falling trajectory σ satisfying $\ddot{b}_j = -\Phi_{,j}$ in $\langle x_i \rangle$. Two interpretations of σ are possible. According to $\langle x_i \rangle$, σ is acted on by a gravitational force $F_\Phi^i = -mh^{ir}\Phi_{,r}$ and is therefore accelerating by an amount $-\Phi_{,j}$. According to $\langle y_j \rangle$, σ is acted on by a gravitational force $F_\Psi^i = 0$ and is therefore free and nonaccelerating. We cannot determine whether σ is really free without making an arbitrary or conventional choice of one Galilean system as our true inertial system. Hence, in this particular theory, the notion of free particle *is* conventional (although the notion of *freely falling* particle is not).

120

This example shows us what is right about Ellis's argument. Theories of interaction and the notion of free—or inertial, or geodesic, or "naturally moving"—particle are intimately connected. If a theory of interaction invokes indeterminate field variables, then the notion of free particle will be correspondingly indeterminate (arbitrary, conventional). But this kind of indeterminateness is not a general feature of theories of interaction and laws of motion. In the electrodynamics of §III.5, for example, the electromagnetic field, the affine connection, the class of inertial frames, and the notion of free particle are all perfectly determinate.

8. Gravitation and Acceleration

As the above considerations suggest, there is a close analogy between the move from the gravitation theory of §III.3 to the gravitation theory of §III.4 and our earlier move from the kinematics of §III.1 to the kinematics of §III.2. Both cases start with a theory that invokes indeterminate entities. In the kinematics of §III.1 the absolute space V is indeterminate: if $\langle M,\overset{\circ}{D},dt,h,V \rangle$ is a model, then so is $\langle M,\overset{\circ}{D},dt,h,V' \rangle$, where V' is related to V by an arbitrary Galilean transformation. In the gravitation theory of §III.3 the gravitational potential Φ (together with the flat connection $\overset{\circ}{D}$) is indeterminate: if $\langle M,\overset{\circ}{D},dt,h,\Phi,\rho \rangle$ is a model, then so is $\langle M,\overset{\circ}{D}{}',dt,h,\Psi,\rho \rangle$, where $\Psi = \Phi + \overset{\cdot\cdot}{b}_j x_j$ and the components of $\overset{\circ}{D}{}'$ vanish in $\langle y_j \rangle$. Moreover, in both cases we move to a new theory that eliminates the indeterminate entities in question by taking formerly definable objects as primitive. In the kinematics of §III.1 the flat connection $\overset{\circ}{D}$ is definable in terms of dt, h, and V; in §III.2 we take $\overset{\circ}{D}$ as primitive and drop V. In the gravitation theory of §III.3 the nonflat connection D is definable in terms of $\overset{\circ}{D}$ and Φ (51); in §III.4 we take D as primitive and drop $\overset{\circ}{D}$ and Φ.[15] Finally, both cases involve a recognition that

---

[15] These questions of definability have an interesting connection with the issues of ontological commitment and theoretical equivalence. One may be tempted to argue as follows: The gravitation theories of §III.3 and §III.4 are obviously equivalent—mere trivial rewritings of one another. For there exist geometrical objects satisfying the theory of §III.3 just in case there exist geometrical objects satisfying the theory of §III.4. (If objects satisfying the former theory exist, then a nonflat connection D satisfying the latter theory can be defined by (51). Conversely, if a nonflat connection satisfying the second theory exists, then (54) and (55) imply the existence of a flat connection $\overset{\circ}{D}$ and a potential Φ satisfying the first theory.)

theoretically distinct reference frames are actually equivalent or indistinguishable. Thus, we can motivate the move from §III.1 to §III.2 by observing that rest frames are indistinguishable from arbitrary inertial frames. We can motivate the move from §III.3 to §III.4 by observing that inertial frames are indistinguishable from arbitrary Galilean frames.

However, there are also disanalogies between the two cases. First of all, the point about equivalent reference frames can mislead one to think that the gravitation theory of §III.4 does for absolute acceleration what the kinematics of §III.2 does for absolute velocity. But, as we have already seen, this is simply not true. By dropping the absolute space V, we eliminate the notion of absolute velocity completely. We retain only the notions of absolute acceleration and rotation induced by the affine connection $\overset{\circ}{D}$. On the other hand, when we drop the gravitational potential Φ in §III.4, we do not simply drop the flat connection $\overset{\circ}{D}$ as well; rather, we change to a new (nonflat) connection D. We eliminate the notions of absolute acceleration and rotation relative to $\overset{\circ}{D}$, but we replace them with new notions of absolute acceleration and rotation relative to D. Hence, the move from §III.3 to §III.4 does not involve a relativization of acceleration parallel to the relativization of velocity effected by the move from §III.1 to §III.2.

Therefore, the two theories are really committed to precisely the same objects and are completely equivalent.

This argument assumes that a space-time theory is committed to all geometrical objects whose existence is implied by its field equations. But this is entirely too generous; far too many geometrical objects exist in this purely mathematical sense. To see this, observe that the above argument applies equally well to our two versions of Newtonian kinematics. The kinematics of §III.2 of course implies the mathematical existence of a vector field V satisfying the theory of §III.1 (the tangent vector field $\partial/\partial x_0$ to the timelike coordinate curves of any inertial coordinate system will do), but the theory of §III.2 is certainly not committed to absolute space. Hence, a more reasonable criterion of ontological commitment requires only that a theory is committed to objects that are *explicitly definable* from its primitive or basic objects. The objects not only exist mathematically; they are distinguished by the formulas of the theory. On this criterion, the formulation of §III.3 remains committed to the nonflat connection of §III.4, since it is definable by (51); but the formulation of §III.4 is no longer committed to the flat connection and gravitational potential of §III.3. In fact, as we have seen, the formulation of §III.4 does not even imply the existence of a unique gravitational potential and flat connection. On a more reasonable criterion of ontological commitment the two theories are not equivalent, for they are not committed to the same objects. This is as it should be. (Here I am indebted to Clark Glymour.)

122

Second, although it is true that the two theoretical transitions are methodologically similar—in both cases we drop unnecessary, indeterminate entities—the transition from the theory of §III.3 to the theory of §III.4 is more complicated. For the advantage that the latter theory enjoys over the former, unlike the advantage of Galilean space-time over absolute space, is dependent on cosmological boundary conditions.

We observed that in an infinite universe with a finite distribution of matter (island universe) we could single out a unique potential Φ in the context of §III.4 by requiring that Φ vanish at infinity. But in addition, given a finite distribution of masses $m_1, m_2, \ldots, m_n$, we can explicitly define Φ by the formula

$$\Phi(p) = -k \sum_{i=1}^{n} \frac{m_i}{r_{ip}} \tag{89}$$

where r_{ip} is the distance between m_i and p in the hypersurface $t = $ constant containing p. If we replace (41) of §III.3 with (89), we obtain a theory that obviously implies the vanishing of $\Phi(p)$ at infinity. (What counts as a boundary condition in one formulation is a consequence of the field equations in another.) Actually, we can do even better. We can replace the theory of §III.3 with a complete action-at-a-distance theory. We put for the gravitational force acting on m_i

$$F = -m_i \operatorname{grad}(\Phi_i) \tag{90}$$

where

$$\Phi_i = -k \sum_{j \neq i} \frac{m_j}{r_{ji}} \tag{91}$$

(r_{ji} is the distance between m_j and m_i on a given spacelike hypersurface). This action-at-a-distance theory avoids the infinite "self-field" that crops up in (89) and in all field-theoretic descriptions of interaction.[16]

On the other hand, in the case of an infinite distribution of matter we cannot single out a unique potential. We cannot, in general, define the potential by (89) or use the action-at-a-distance theory (90), (91). It seems to me, therefore, that we have the following situation. If we have reason to think that the actual distribution of matter is infinite, which appears in fact to be the case, then the theory of §III.4

---

[16] See Anderson [3], 118, 256–262.

is preferable, because it uses fewer primitive objects. If we have reason to think that the actual distribution of matter is finite, then the §III.4 theory loses this advantage. We can use (89) or, equivalently, the boundary conditions at infinity to define a unique potential. Moreover, in this special case, the action-at-a-distance theory (90), (91) is better than either of our two field theories. Finally, if we are agnostic about the actual distribution of matter, then the §III.4 theory is again preferable, because of its greater flexibility.

IV
Special Relativity

1. Kinematics

The space-time of special relativity—hereafter referred to as *Minkowski* space-time—is, like the space-time of Newtonian kinematics, a flat, four-dimensional affine manifold. Minkowski space-time is often required to satisfy global conditions (namely, to be globally Euclidean or homeomorphic to R^4), but, as in my treatment of Newtonian kinematics, I prefer to use only local field equations. Since Minkowski space-time is a flat manifold, our first local condition is the existence of an affine connection D with the field equation $K = 0$. Our equation of motion is the geodesic law for this connection:

$$D_{T_\sigma} T_\sigma = 0$$

and there are always coordinate systems $\langle x_i \rangle$ in which

$$\frac{d^2 x_i}{du^2} = 0.$$

Once again, this is *almost* the desired Law of Inertia. To reconstruct the actual law we have to say more about space and time individually. How does Minkowski space-time "split up" into three-dimensional space and one-dimensional time?

The chief difference between special relativity and Newtonian physics lies here: Minkowski space-time does not possess an absolute time or planes of absolute simultaneity. Instead, the planes of simultaneity are relativized to a choice of inertial trajectory: each inertial trajectory (and therefore each inertial frame) has its own planes of simultaneity. To see how this change comes about, recall the situation in classical electrodynamics (§III.5). Consider the "light cone" at a point p in Newtonian space-time, that is, the set of all trajectories representing the possible paths of light rays emitted at p

125

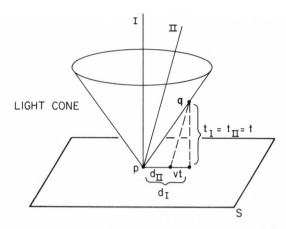

Light cone in Newtonian space-time. The velocity of a light ray *cannot* be the same in frame I and frame II. Since $d_{II} = d_I - vt$, $v_{II} = v_I - v$.

(for example, let the event p be the lighting of a match). Consider two inertial frames adapted to the geodesics I and II respectively. I and II agree on the relation of occurring-at-the-same-time—the plane of simultaneity S—but disagree on the relation of occurring-at-the-same-place. As a result, I and II agree on the time it takes a light ray to travel from p to q but disagree on the distance traversed. Consequently, a light ray will not have the same velocity in the two frames (III.71).

However, all experimental evidence suggests that light rays in fact *do* have the same velocity in all inertial frames. Relativizing the planes of simultaneity—allowing the relation of occurring-at-the-same-time to vary "in tandem" with the relation of occurring-at-the-same-place—enables us to accommodate this fact. Thus, in Minkowski space-time, as the trajectory II tilts "away" from the trajectory I the plane of simultaneity S_{II} tilts "up" from the plane of simultaneity S_I. I and II disagree about both spatial and temporal relations, but these relations co-vary in such a way that the velocity of light remains the same in the two frames.

In Minkowski space-time, then, a light cone exists at every point p in M, but there is no unique plane of simultaneity through p. Rather, for each inertial trajectory σ through p (lying within the light cone at p) there is a distinct plane of simultaneity S_σ. Each plane of simultaneity has its own Euclidean spatial metric, and each inertial *frame* (an inertial trajectory σ together with a corresponding system of planes of simultaneity S_σ) has its own temporal metric

126

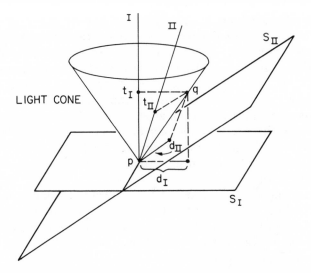

Light cone in Minkowski space-time. The velocity of light *can* be the same in frame I and frame II. $d_{II} \neq d_I$ and $t_{II} \neq t_I$. For a light ray, $v_I = v_{II}$.

(the temporal duration or "distance" between the planes of simultaneity S_σ). Within any single inertial frame, things look precisely the same as in Newtonian kinematics: there is an enduring Euclidean three-space, a global time t, and the inertial law of motion $d^2x_i/dt^2 = 0$. But the different inertial frames are related to one another in a non-Newtonian fashion.

How do we describe this geometrical structure by means of geometrical objects and local field equations? We already have a clue from §III.5 above: we introduce a symmetric, nonsingular tensor field g of type $(0,2)^1$ and signature $\langle 1,3 \rangle$. Since g is symmetric and non-singular (unlike the Newtonian spatial metric h), it is a semi-Riemannian metric tensor. Moreover, we require g to be compatible with D, so Minkowski space-time, unlike Newtonian space-time, is a semi-Riemannian manifold. Since g is indefinite, it induces a division of the tangent space at each point p as follows: $X \in T_p$ is *timelike* just in case $g_p(X,X) > 0$, *spacelike* just in case $g_p(X,X) < 0$, and *null* or *lightlike* just in case $g_p(X,X) = 0$. Furthermore, g gives us a length for any vector in T_p: if X is timelike $|X| = \sqrt{g_p(X,X)}$; if X is spacelike $|X| = \sqrt{-g_p(X,X)}$; if X is null $|X| = 0$. X and Y in

---

[1] The tensor field g is the *inverse* (§A.7) of the tensor g^* of §III.5, which is of type (2,0).

T_p are *orthogonal* just in case $g_p(X,Y) = 0$. Now, it is easy to see that neither the set of timelike vectors nor the set of spacelike vectors forms a subspace of T_p. The set of timelike vectors lies within a "cone" (the light cone at p) that is separated from the set of spacelike vectors by the null vectors. Since there is no unique three-dimensional subspace of spacelike vectors, there are no planes of absolute simultaneity. Instead, given any timelike vector X in T_p, the set of all vectors *orthogonal to* X—all Y in T_p such that $g_p(X,Y) = 0$—forms a three-dimensional spacelike subspace. Relativization enters because distinct timelike vectors induce distinct spacelike subspaces.

A curve in Minkowski space-time is called timelike if its tangent vector field is everywhere timelike, spacelike if its tangent vector field is everywhere spacelike, null if its tangent vector field is everywhere null. We can define the length of a timelike or spacelike curve between p and q by

$$|\sigma|_p^q = \int_a^b |T_{\sigma(u)}|\, du$$

where $\sigma(a) = p$, $\sigma(b) = q$. We can define a temporal distance between any pair of points with *timelike separation* (connected by a timelike curve), and a spatial distance between any pair of points with *spacelike separation*—(connected by a spacelike curve). Thus, if p and q have timelike (spacelike) separation, we define the temporal (spatial) distance between p and q as $|\sigma|_p^q$, where σ is a timelike (spacelike) *geodesic*.

If σ is a timelike curve, we interpret $|\sigma|_p^q$ as the time elapsed from p to q along σ, that is, as the time an ideal clock with trajectory σ would measure from p to q. Time in special relativity is essentially *local*, that is, defined for a *particular* timelike trajectory. In contrast to Newtonian kinematics, there is no coordinate-independent *global* time. We can use the "length" (temporal duration) of timelike curves as a new parameter τ, where

$$\tau(u) = |\sigma|_{\sigma(a)}^{\sigma(u)} = \int_a^u |T_\sigma(u)|\, du$$

with $a \in \mathrm{dom}(\sigma)$ (see (A.29)). Similarly, we can use the spatial length $s(u)$, defined analogously, as a parameter on spacelike curves. Since Minkowski space-time is semi-Riemannian, τ is an affine parameter along timelike geodesics and s is an affine parameter along spacelike geodesics (A.30).

The kinematics of special relativity is simple almost to the point of triviality. We have two geometrical objects: an affine connection

D and a symmetric tensor field g of type (0,2) and signature $\langle 1,3 \rangle$ (this signature characterizes a *Lorentzian* metric). We have two field equations:

$$K = 0 \tag{1}$$

$$\bar{D}(g) = 0. \tag{2}$$

These equations say that Minkowski space-time is a flat, semi-Riemannian manifold. Our equations of motion require that all particles follow timelike curves and that free particles follow timelike geodesics: the trajectories of free particles satisfy

$$D_{T_\sigma} T_\sigma = 0. \tag{3}$$

In terms of components, the field equations take the form:

$$R^i_{jkl} = 0 \tag{1a}$$

$$g_{ij;k} = 0 \tag{2a}$$

while the equations of motion are

$$\frac{d^2 x_i}{d\tau^2} + \Gamma^i_{jk} \frac{dx_j}{d\tau} \frac{dx_k}{d\tau} = 0. \tag{3a}$$

This theory, simple as it is, has all the familiar consequences we expect from special relativistic kinematics.

Since our manifold is flat (1a), we know that around every point p in M there is a coordinate system $\langle y_j \rangle$ in which the components of D vanish: $\Gamma^i_{jk} = 0$. In such a coordinate system covariant differentiation reduces to ordinary differentiation, and (2a) implies that the components of g are constants. Since g has signature $\langle 1,3 \rangle$, we can find a second coordinate system $\langle x_i \rangle$, linearly related to $\langle y_j \rangle$, in which $\Gamma^i_{jk} = 0$ and $(g_{ij}) = \text{diag}(1, -1, -1, -1)$. Such a coordinate system will be called an *inertial* coordinate system. In an inertial coordinate system the curves $x_i = $ constant $(i = 1,2,3)$ have the tangent vector $(dx_0/du, 0, 0, 0)$; so

$$g_{ij} \frac{dx_i}{du} \frac{dx_j}{du} = \left(\frac{dx_0}{du}\right)^2 > 0$$

and such curves are timelike. The curves $x_0 = $ constant, on the other hand, have the tangent vector $(0, dx_1/du, dx_2/du, dx_3/du)$; so

$$g_{ij} \frac{dx_i}{du} \frac{dx_j}{du} = -\left[\left(\frac{dx_1}{du}\right)^2 + \left(\frac{dx_2}{du}\right)^2 + \left(\frac{dx_3}{du}\right)^2\right] < 0$$

129

and such curves are spacelike. Furthermore, the curves $x_0 =$ constant are orthogonal to the curves $x_i =$ constant $(i = 1,2,3)$. Therefore, in an inertial coordinate system x_0 can be properly regarded as a temporal coordinate, and x_1,x_2,x_3 can be properly regarded as spatial coordinates. From now on, I shall usually write t for the x_0 coordinate of an inertial coordinate system. Note that in inertial systems the temporal distance between two points (t,x_1,x_2,x_3) and (t',x_1,x_2,x_3) on a curve $x_i =$ constant $(i = 1,2,3)$ is just $|t' - t|$, while the spatial distance between two points (t,x_1,x_2,x_3) and (t,x'_1,x'_2,x'_3) on a hypersurface $t =$ constant is just

$$\sqrt{(x_1 - x'_1)^2 + (x_2 - x'_2)^2 + (x_3 - x'_3)^2}.$$

Hence, the spatial geometry on a plane of simultaneity is Euclidean as promised.

More generally, the minimal conditions for the y_0 and y_1,y_2,y_3 coordinates of an arbitrary (not necessarily inertial) coordinate system to be temporal and spatial coordinates respectively are that the curves $y_i =$ constant $(i = 1,2,3)$ be timelike and that the curves $y_0 =$ constant be spacelike. The curves $y_i =$ constant will be timelike if

$$g_{00}\left(\frac{dx_0}{du}\right)^2 > 0$$

or $g_{00} > 0$. The curves $y_0 =$ constant will be spacelike if

$$g_{ij}\frac{dx_i}{du}\frac{dx_j}{du} < 0 \qquad (i,j = 1,2,3).$$

So the y_0 and y_1,y_2,y_3 coordinates are suitable as temporal and spatial coordinates respectively (or, as I shall say, the system $\langle y_j \rangle$ is *suitable*) just in case $g_{00} > 0$ and the quadratic form $g_{ij}\alpha^i\alpha^j$ $(i,j = 1,2,3)$ is negative definite. Obviously, then, all inertial systems are suitable.

A timelike geodesic $\sigma(\tau)$ satisfies the equations

$$\frac{d^2(x_i \circ \sigma)}{d\tau^2} = 0 \qquad (4)$$

in an inertial system $\langle x_i \rangle$. The solutions of (4) are $x_i = a_i\tau + b_i$, with a_i and b_i constants. Therefore, $t = x_0 = a_0\tau + b_0$ is an affine parameter on timelike geodesics. Since $\Gamma^i_{jk} = 0$ in inertial systems, timelike geodesics satisfy

$$\frac{d^2 x_i}{dt^2} = 0. \tag{5}$$

Equation (5) is our desired law of motion for free particles.

Clearly, we can use t as a parameter for any timelike curve, and we can write its equations as $x_i(t) = x_i \circ \sigma(t)$ in an inertial system $\langle x_i \rangle$. We thus have two types of time associated with the trajectory of a particle: the coordinate-independent *proper time* τ and the coordinate-dependent *coordinate time* t. Similarly, we have two types of velocity four-vectors: the proper velocity $\mathbf{u}$, with components $u^i = dx_i/d\tau$, and the coordinate velocity $\mathbf{v}$, with components $v^i = dx_i/dt$. Hence, the coordinate velocity $\mathbf{v}$ takes the form $(1,\vec{v})$, with $\vec{v}$ the ordinary three-velocity.[2] To see how these two types of time and velocity are related observe that

$$\tau(t) = \int_a^t \sqrt{g_{ij} \frac{dx_i}{dt} \frac{dx_j}{dt}} \, dt.$$

So

$$\frac{d\tau}{dt} = \sqrt{g_{ij} \frac{dx_i}{dt} \frac{dx_j}{dt}} = \sqrt{1 - \left[\left(\frac{dx_1}{dt}\right)^2 + \left(\frac{dx_2}{dt}\right)^2 + \left(\frac{dx_3}{dt}\right)^2\right]}.$$

Since the quantity in brackets under the rightmost square root sign is just the square of the three-velocity $\vec{v}$, we have

$$\frac{d\tau}{dt} = \sqrt{1 - v^2}. \tag{6}$$

The relationship between the proper velocity $\mathbf{u}$ and the coordinate velocity $\mathbf{v}$ can now be found easily using (6) and the fact that $u^i = dx_i/d\tau = (dx_i/dt)(dt/d\tau)$:

$$u^0 = \frac{dx_0}{d\tau} = \frac{1}{\sqrt{1 - v^2}}$$

$$u^i = \frac{v^i}{\sqrt{1 - v^2}} = \frac{v_i}{\sqrt{1 - v^2}} \qquad (i = 1,2,3) \tag{7}$$

where the v_i are the components of the three-velocity $\vec{v}$.

[2] Both $\mathbf{u}$ and $\mathbf{v}$ are four-dimensional vectors. However, different inertial frames induce different coordinate velocities dx_i/dt, whereas all inertial frames induce the same proper velocity $dx_i/d\tau$. In going from one inertial frame to another, we change the parameter t but not the parameter τ.

Although we cannot use the proper time τ as a parameter on null curves, we can use the coordinate time t of an arbitrary inertial system. In particular, null geodesics must satisfy

$$\left(\frac{dx_0}{dt}\right)^2 - \left[\left(\frac{dx_1}{dt}\right)^2 + \left(\frac{dx_2}{dt}\right)^2 + \left(\frac{dx_3}{dt}\right)^2\right] = 0 \qquad (8)$$

in an arbitrary inertial system. Equation (8) implies that null geodesics have constant unit three-velocity—$v^2 = 1$—in inertial systems. Conversely, any curve with constant unit three-velocity in an arbitrary inertial system satisfies (8) and is a null geodesic.

The natural notion of inertial frame in the context of the present theory is an inertial coordinate system adapted to a moving particle: the trajectory $\sigma(\tau)$ of the particle satisfies the equations $x_0 = \tau$, $x_i = 0$ $(i = 1,2,3)$ in the associated coordinate system. Thus, a particle is at rest at the origin of its associated frame, and the coordinate time $x_0 = t$ of the associated coordinate system agrees with the proper time τ of the particle along its trajectory. There exists an inertial coordinate system adapted to the particle with trajectory $\sigma(\tau)$ just in case $\sigma(\tau)$ is a timelike geodesic. To see this from left to right, suppose there is such an associated coordinate system. Then $\sigma(\tau)$ satisfies the equations $d^2x_i/d\tau^2 = 0$, so $\sigma(\tau)$ is a geodesic. The tangent to $\sigma(\tau)$ has the form $(1,0,0,0)$, so $\sigma(\tau)$ is timelike. For the other direction, suppose $\sigma(\tau)$ is a timelike geodesic. In an arbitrary inertial system $\langle y_j \rangle\, \sigma(\tau)$ satisfies $d^2y_j/d\tau^2 = 0$, so $\sigma(\tau)$ satisfies $y_j = a_j\tau + b_j$. A linear transformation (in fact, a Lorentz transformation) gives us an inertial system $\langle x_i \rangle$ in which $\sigma(\tau)$ satisfies $x_0 = \tau$, $x_i = 0$ $(i = 1,2,3)$. Hence, inertial frames can be associated only with particles performing inertial motion, that is, particles satisfying the law of motion (3).

The problem of constructing noninertial reference frames—frames associated with noninertial motions—is considerably more complicated in special relativity than in the Newtonian case. This is because the connection Γ^i_{jk} and the spatio-temporal metric g_{ij} cannot vary independently as do their Newtonian counterparts. In Minkowski space-time the rigid Euclidean frames coincide with the inertial frames: that is, we cannot associate rigid Euclidean frames with noninertial trajectories. The best we can do is to associate a *semi-Euclidean* frame with a given timelike (not necessarily geodesic) trajectory $\sigma(\tau)$. This is a coordinate system $\langle y_j \rangle$ with the following properties:

132

(i) $\langle y_j \rangle$ is suitable: the curves y_i = constant (i = 1,2,3) are time-like, and the curves y_0 = constant are spacelike.

(ii) $y_0 = \tau$ along $\sigma(\tau)$.

(iii) For each value of τ the curves y_0 = constant are orthogonal to σ at $\sigma(\tau)$.

(iv) (g_{ij}) = diag$(1,-1,-1,-1)$ along σ.

(v) y_1, y_2, y_3 are Cartesian on every y_0 = constant hypersurface: the spatial distance between (y_0, y_1, y_2, y_3) and (y_0, y'_1, y'_2, y'_3) is $\sqrt{(y_1 - y'_1)^2 + (y_2 - y'_2)^2 + (y_3 - y'_3)^2}$.

(vi) $\sigma(t)$ satisfies $y_i = 0$ (i = 1,2,3).

Such a coordinate system can be constructed by choosing for each value of τ along σ an inertial system $\langle x_i \rangle$ whose origin coincides with $\sigma(\tau)$ and whose x_0-axis is tangent to σ at $\sigma(\tau)$. We then assign each point on the spatial hypersurface $x_0 = 0$ the coordinates $y_0 = \tau$, $y_i = x_i$ (i = 1,2,3). Of course, our choice of successive inertial systems must be sufficiently smooth, so that the curves y_i = constant are suitably continuous and differentiable.[3] Note the

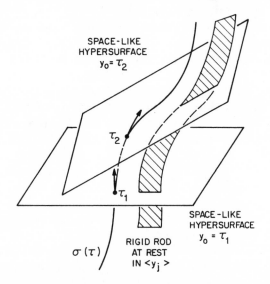

SPACE-LIKE
HYPERSURFACE
$y_0 = \tau_2$

τ_2

τ_1

SPACE-LIKE
HYPERSURFACE
$y_0 = \tau_1$

$\sigma(\tau)$

RIGID ROD
AT REST
IN $\langle y_j \rangle$

Construction of semi-Euclidean frame for noninertial trajectory σ by means of instantaneous tangent inertial systems.

---

[3] This will be the case if the quadruple of vector fields $\langle \partial/\partial y_0, \partial/\partial y_1, \partial/\partial y_2, \partial/\partial y_3 \rangle$ is itself sufficiently continuous and differentiable. See Misner, Thorne, and Wheeler [74], 170–176, 327–332.

important differences between semi-Euclidean frames and the rigid Euclidean frames of §III.1, all of which are direct consequences of the relativization of simultaneity. First, a semi-Euclidean frame can be well-defined only on a finite neighborhood of σ; for at finite distances from σ the hypersurfaces $y_0 = $ constant *intersect* one another. Second, a rod that appears rigid from the point of view of a semi-Euclidean frame (for example, a rod both of whose endpoints are at rest in the frame) will, from the point of view of an inertial frame, experience a Lorentz contraction: its length at τ_2 is measured in a plane of simultaneity different from the one in which its length at τ_1 is measured. Third, spatially separated clocks at rest in a semi-Euclidean frame synchronized at one time τ_1 will not in general remain synchronized at a second time τ_2: the world-line of a clock at rest at one point in our frame will receive a "length" between τ_1 and τ_2 different from the one received by the world-line of a clock at rest at a second point. These last two features make clear the sense in which a semi-Euclidean frame fails to be rigid Euclidean.

The Γ^i_{jk} will not in general vanish in a semi-Euclidean frame. In fact, they again take an especially interesting form along σ itself. In $\langle y_j \rangle$ σ has (proper) four-velocity $(u^i) = (1,0,0,0)$ and four-acceleration $(a^i) = (0,a^1,a^2,a^3)$. Hence, $\mathbf{u} = T_\sigma = \partial/\partial y_0$ and $\Gamma^0_{00} = 0$. If we consider any spatial geodesic $\rho(s)$ that lies within a plane of simultaneity $y_0 = $ constant and intersects $\sigma(\tau)$, $\rho(s)$ must satisfy $d^2 y_i/ds^2 = 0$ (since y_1, y_2, y_3 are Cartesian). Consequently, $\Gamma^i_{jk} = 0$ $(j,k = 1,2,3)$. Therefore, the only nonvanishing components of D are $\Gamma^i_{00}, \Gamma^i_{0j} = \Gamma^i_{j0} (i,j = 1,2,3)$, and $\Gamma^0_{i0} = \Gamma^0_{0i} (i = 1,2,3)$. To see what these represent note that

$$(D_\mathbf{u}\mathbf{u})^i = a^i = \frac{d^2 y_i}{d\tau^2} + \Gamma^i_{jk}\frac{dy_j}{d\tau}\frac{dy_k}{d\tau}.$$

So

$$\Gamma^i_{00} = a^i \qquad (i = 1,2,3). \tag{9}$$

Furthermose, $D_\mathbf{u}g = 0$, so $\Gamma^j_{0i} = -\Gamma^i_{0j}(i,j = 1,2,3)$ and $\Gamma^i_{00} = \Gamma^0_{0i}$. Hence

$$\Gamma^0_{0i} = \Gamma^0_{i0} = a^i \qquad (i = 1,2,3) \tag{10}$$

and

$$\Gamma^i_{j0} = \Gamma^i_{0j} = \Omega^i_j \qquad (i,j = 1,2,3) \tag{11}$$

where Ω^i_j is an antisymmetric rotation matrix. Along σ we have the equation of motion

$$\frac{d^2 y_i}{dy_0^2} + a^i + 2\Omega^i_j \frac{dy_j}{dy_0} - 2a^j \frac{dy_j}{dy_0}\frac{dy_i}{dy_0} = 0. \qquad (12)$$

| inertial | Coriolis | relativistic |
| force | force | correction |

(To derive (12) substitute the values (9)–(11) into the geodesic equation and use (A.15) to change to the [nonaffine] parameter y_0.) If the acceleration a^i and the rotation Ω^i_j both vanish, our semi-Euclidean frame becomes an inertial frame, and the law of motion takes the familiar form (5).

The symmetry group of special relativistic kinematics can now be found in the usual way. It is clear that both objects of our theory—the connection D and the metric g—are absolute objects. So our symmetry group must leave D and g invariant. Invariance of D means that a coordinate representation of our symmetry group in an inertial coordinate system $\langle x_i \rangle$ must be linear, and invariance of g means that the components of the generators of our 1-parameter subgroups must satisfy

$$\frac{\partial a^k}{\partial x_i} g_{kj} + \frac{\partial a^k}{\partial x_j} g_{ik} = 0 \qquad (13)$$

(so these generators are *Killing fields* of g; see (A.49)). Consequently, the a^i must take the form

$$a^i = b_{ij}x_j + c_i. \qquad (14)$$

Since $(g_{ij}) = \text{diag}(1, -1, -1, -1)$, (13) implies that $b_{ij} = -b_{ji}$ $(i, j = 1, 2, 3)$, $b_{i0} = b_{0i}$, and $b_{00} = 0$. Therefore, we have six independent b_{ij} and four independent c_i. The independent b_{ij} form the matrix

$$(b_{ij}) = \begin{pmatrix} 0 & -1 & -1 & -1 \\ -1 & 0 & 1 & 1 \\ -1 & -1 & 0 & 1 \\ -1 & -1 & -1 & 0 \end{pmatrix}$$

(compare with the matrix of the Galilean group in §III.2). Our

symmetry group is thus a 10-parameter Lie group with ten independent generators:

		Lorentz
Translations	*Rotations*	*Transformations*
$(1,0,0,0)$	$(0,x_2,-x_1,0)$	$(-x_1,-x_0,0,0)$
$(0,1,0,0)$	$(0,0,x_3,-x_2)$	$(-x_2,0,-x_0,0)$
$(0,0,1,0)$	$(0,x_3,0,-x_1)$	$(-x_3,0,0,-x_0)$
$(0,0,0,1)$		

(compare with the generators of the Galilean transformations in
§III.2). This symmetry group is just the group $O^3 \times T$ plus the
additional generators of the Lorentz transformations. To see that
these latter really do generate the Lorentz transformations consider
the generator $(-x_1,-x_0,0,0)$, for example. It yields the equations

$$\frac{\partial x_0^*}{\partial s} = -x_1^*, \qquad \frac{\partial x_1^*}{\partial s} = -x_0^*, \qquad \frac{\partial x_2^*}{\partial s} = 0, \qquad \frac{\partial x_3^*}{\partial s} = 0$$

which have the solution

$$\begin{aligned}
x_0^* &= x_0 \cosh(s) - x_1 \sinh(s) \\
x_1^* &= x_1 \cosh(s) - x_0 \sinh(s) \\
x_2^* &= x_2 \\
x_3^* &= x_3
\end{aligned} \qquad (15)$$

(compare with the rotation (A.53)). If we change to a new parameter
$v = \tanh(s)$, (15) takes the more familar form

$$x_0^* = \frac{x_0 - vx_1}{\sqrt{1 - v^2}}$$

$$x_1^* = \frac{x_1 - vx_0}{\sqrt{1 - v^2}} \qquad (16)$$

$$x_2^* = x_2$$

$$x_3^* = x_3.$$

Equation (16) represents a Lorentz transformation along the x_1-axis.
Our full 10-parameter group is called the *Lorentz* group (see Fig. 7).

We can make sense of traditional ways of talking about the
Lorentz group just as we did for the Galilean group. First, we can
reconstrue our group as a group of coordinate transformations.
Every such coordinate transformation preserves the special form of

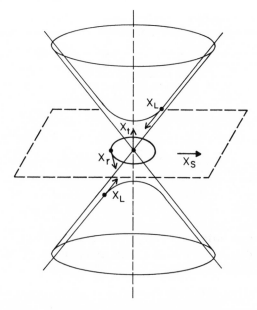

FIGURE 7. The Symmetry Group of Minkowski Space-Time
The inertial trajectories, the light cone, and the Minkowski metric g are invariant. So the transformations in our group are just spatial translations (generated by X_s), temporal translations (generated by X_t), spatial rotations (generated by X_r), and Lorentz transformations (generated by X_L). The generators of the Lorentz transformations are tangent to the hyperbolas $x_0^2 - x_1^2 = \text{constant}^2$ (just as the generators of spatial rotations are tangent to the circles $x_1^2 + x_2^2 = \text{constant}^2$; see Fig. 8 in §A.7).

the components of D and g: that is, it preserves the conditions $\Gamma^i_{jk} = 0$ and $(g_{ij}) = \text{diag}(1, -1, -1, -1)$. These coordinate transformations take inertial coordinate systems onto inertial coordinate systems and preserve the special form (5) of the law of motion. Similarly, we can connect our group with the relative motion of reference frames. Suppose we have two inertial frames, I and II, associated with two timelike geodesics, $\sigma(t)$ and $\rho(t)$, where the parameter t is the coordinate time of frame I. Suppose that the trajectory $\rho(t)$ represents a state of uniform motion with velocity v along the x_1-axis of frame I. Let $\langle x_i \rangle$ be the coordinates of frame I, $\langle y_j \rangle$ the coordinates of frame II, and suppose for simplicity that $\sigma(t)$ and $\rho(t)$ intersect at $(0,0,0,0)$ in both frames and that $x_2 = y_2$, $x_3 = y_3$. Hence, in frame I $\sigma(t)$ has the equations $x_i = 0$ $(i = 1,2,3)$ and $\rho(t)$ has the equations $x_1 = vt$, $x_2 = x_3 = 0$. Let B be an arbitrary

137

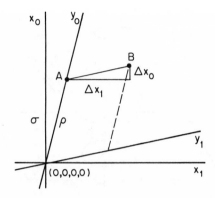

Frame I.

point having coordinates (y_0, y_1, y_2, y_3) in frame II. In frame II point A has coordinates $(y_0, 0, y_2, y_3)$, while in frame I A (by Minkowskian geometry) has coordinates

$$\left(\frac{y_0}{\sqrt{1-v^2}}, \frac{vy_0}{\sqrt{1-v^2}}, x_2, x_3 \right).$$

The x_1-coordinate of B is just the x_1-coordinate of A plus Δx_1, and the x_0-coordinate of B is just the x_0-coordinate of A plus Δx_0. Moreover, $\Delta x_0 / \Delta x_1 = v$, and $\Delta x_1 = y_1 / \sqrt{1-v^2}$. It follows that

$$x_0 = \frac{y_0 + vy_1}{\sqrt{1-v^2}}$$

$$x_1 = \frac{y_1 + vy_0}{\sqrt{1-v^2}} \tag{17}$$

$$x_2 = y_2$$

$$x_3 = y_3.$$

Equation (17) is the relativistic transformation connecting two inertial frames in relative motion with constant velocity v. Thus, the use of the letter "v" for the group parameter in (16) is justified.

2. Digression on Deriving the Lorentz Transformations

The surest and clearest way to derive the Lorentz transformations is to follow the procedure we have used above. We start by postulating the geometrical structure of Minkowski space-time. We then look for the group of transformations of Minkowski space-time onto itself that will preserve that geometrical structure: this turns out to be the Lorentz group (16). Finally, using the geometrical structure of Minkowski space-time, we define the class of inertial frames and show that any two inertial frames are related by a Lorentz transformation (17).

From this point of view we should distinguish three levels of structure preserved by the Lorentz group. The first is *conformal* structure: the structure given by the existence of the light cone at every point. Two Lorentzian metrics, g and g', are said to be conformally equivalent just in case they have the same light cone at every point; that is, just in case $g_p(X,X) = 0$ iff $g'_p(X,X) = 0$. It is not too difficult to show that this condition holds just in case

$$g' = \Omega g \qquad (18)$$

where Ω is a real-valued (sufficiently continuous and differentiable) positive function.[4] Hence, conformal structure is given by an equivalence class of Lorentzian metrics, where the appropriate equivalence relation $g' \simeq g$ is given by (18). The second level is *affine*, or *projective*, structure: the structure given by the class of geodesics of D. Two Lorentzian metrics, g and g', are said to be projectively equivalent just in case they have the same geodesics, that is, just in case $D_{T_\sigma} T_\sigma = 0$ iff $D'_{T_\sigma} T_\sigma = 0$, where D and D' are the unique connections compatible with g and g' respectively. It is not too difficult to show that this condition forces Ω to be constant in (18).[5] That is, g and g' are both projectively and conformally equivalent just in case

$$g' = kg \qquad (19)$$

where k is a positive real number. Hence, conformal plus projective structure is again given by an equivalence class of Lorentzian metrics, where the appropriate equivalence relation $g' \simeq g$ is given by (19). The third level is just the metrical structure itself—the structure given by a Minkowski metric g. Thus, our three levels of structure correspond to three progressively narrower equivalence classes of Lorentzian metrics, and the third class contains just a single member, g.

In order to derive the Lorentz transformations we must make assumptions about both the structure being preserved and the prop-

---

[4] See Malament [69], chap. 3.
[5] Ibid.

erties of the space-time in which that structure is defined. If we are merely given a light cone at each point p, we know only that we are working in the context of *some* Lorentzian manifold $\langle M,g \rangle$. In particular, we do not know whether our manifold is flat or nonflat, since, in general, there will be both flat and nonflat metrics in the equivalence class defined by (18). Accordingly, if we are only given the information that a transformation h preserves the light cone, we can only conclude that

$$hg = \Omega g. \tag{20}$$

We do not know whether g is the flat Minkowski metric, and we cannot conclude that h is an isometry ($hg = g$). Consequently, we cannot conclude that h is a Lorentz transformation. If we are given both the light cone and the class of geodesics, and if we are given a transformation h that preserves both, we can conclude that

$$hg = kg, \tag{21}$$

but we still do not know whether g is flat or whether h is a Lorentz transformation. On the other hand, if we know that our manifold is flat, we can conclude that h is either a Lorentz transformation or a *dilation* of Minkowski space-time (where a dilation multiplies each coordinate x_i in (16) by the same constant factor). If we know that our manifold is flat and that h is an isometry, we can finally conclude that h must be a Lorentz transformation.

The standard practice of textbooks on special relativity is to "derive" the Lorentz transformations without explicit assumptions about four-dimensional geometrical structure.[6] We are merely given two "reference frames" with relative velocity v, and we are to look for those transformations on which light has the same constant velocity c in both frames. Of course, on this basis alone we can *at best* obtain the transformations (20), since we are merely given the light cone and the information that h preserves conformal structure. Therefore, it is customary to invoke various "homogeneity" and "symmetry" assumptions to narrow down our group to the Lorentz group. Let us see how this is done in the simplified two-dimensional case, where we have two reference frames with coordinates t,x and t',x' respectively.[7]

---

[6] A notable exception is Taylor and Wheeler [107].

[7] In the four-dimensional case additional "symmetry" assumptions are needed. See, e.g., Resnick, [90], 59.

In the first place, we need to assume that the transformation taking t,x to t',x' is *linear*: we have

$$t' = a_1 x + a_2 t + a_3$$
$$x' = b_1 x + b_2 t + b_3 \tag{22}$$

where the a_i and b_i are constants depending on the velocity v of t',x' relative to t,x. This assumption is usually justified by appealing to the "homogeneity" of space and time.[8] The notion of homogeneity being used here is seldom made precise, but it is hard to see how homogeneity in any natural sense can be strong enough to yield (22). It would seem that any space-time of *constant curvature* is homogeneous (that is, it has the same geometrical properties at every point), but to obtain linearity we need something stronger. In effect, we need to assume that space-time is *flat* and that our transformation preserves the flat affine structure. Thus, if we assume that t,x and t',x' are *inertial* systems in which $\Gamma^i_{jk} = 0$, it follows from the transformation law for Γ^i_{jk} that they are linearly related.

Next, we suppose that the origins of our two systems coincide; so $t = x = 0$ implies $t' = x' = 0$, and $a_3 = b_3 = 0$. Further, since t',x' has constant velocity v relative to t,x, $x' = 0$ must transform to $x = vt$. It follows that $b_2 = -vb_1$ and (22) becomes

$$t' = a_1 x + a_2 t$$
$$x' = b_1(x - vt). \tag{23}$$

The requirement that light rays through the origin have the same constant velocity c ($=1$ as usual) in our two systems means that

$$x^2 = t^2 \quad \text{iff} \quad x'^2 = t'^2. \tag{24}$$

Using (23) to substitute for x' and t' in (24), we obtain the conditions

$$b_1^2 v + a_1 a_2 = 0$$
$$a_2^2 - b_1^2 v^2 = b_1^2 - a_1^2. \tag{25}$$

Solving (25) for a_2 and a_1, we get

$$a_2 = b_1$$
$$a_1 = -vb_1$$

[8] See Resnick [90], 56–57; Einstein [26], 44.

where b_1 is an arbitrary function of v. If we set $k(v) = b_1\sqrt{1 - v^2}$, (23) becomes

$$t' = k(v) \frac{t - vx}{\sqrt{1 - v^2}}$$

$$x' = k(v) \frac{x - vt}{\sqrt{1 - v^2}}.$$

(26)

In terms of Minkowski space-time, the transformations h_v induced by (26) satisfy

$$h_v g = k(v)g. \tag{27}$$

Hence, the transformations (26) are just those preserving both conformal and affine structure as in (21). They are not yet the isometries.

Narrowing down our group further requires additional assumptions. We can invoke the group character of the transformations (27) to conclude that

$$h_v h_{-v} = h_{v-v} = h_0 = \text{the identity.}$$

Therefore,[9]

$$k(v)k(-v) = 1.$$

To derive the result that $k(v) = 1$—that (26) is a Lorentz transformation—we need the additional "symmetry" assumption:[10]

$$k(v) = k(-v).$$

This assumption is certainly plausible, since without it the traditional relativistic effects—time dilation, length contraction, and so forth—would depend on the direction of velocity as well as its magnitude. Nevertheless, it is important to see that the Lorentz group cannot be derived from linearity and the invariance of the velocity of light alone.[11] We need, in effect, the invariance of the full metrical structure of Minkowski space-time.

3. Dynamics

We know that the law of motion for Newtonian dynamics is

$$m(D_{T_\sigma(t)} T_\sigma(t)) = F \tag{28}$$

[9] Compare Einstein [26], 47.

[10] Ibid.

[11] This mistaken assertion is commonly encountered in works on relativity. See, e.g., Sklar [100], 258–269.

where F is a spacelike vector field, t is any time-function $(dt(X) = Xt$ for all $X)$, and m is the mass of the particle in question. The corresponding law of motion for relativistic dynamics is

$$m_0(D_{T_\sigma(\tau)} T_\sigma(\tau)) = F \tag{29}$$

where F is a spacelike vector field orthogonal to T_σ, τ is proper time, and m_0 is the rest mass of the particle (the mass that would be determined by an ideal measuring device co-moving with the particle). The chief difference between (28) and (29) is that we have changed from the parameter t to the parameter τ: proper time has replaced Newtonian absolute time. To appreciate this change we must be more explicit than we have been about the parameters occurring in our laws of motion.

Suppose we try to replace the parameter of the curves specified in (28) by an *arbitrary* parameter u: that is, we rewrite (28) as

$$m(D_{T_\sigma(u)} T_\sigma(u)) = F. \tag{30}$$

Since F is spacelike in the Newtonian sense $(dt(F) = 0)$, the vector field $(D_{T_\sigma(u)} T_\sigma(u))$ should be spacelike as well. So we have

$$dt(D_{T_\sigma(u)} T_\sigma(u)) = 0.$$

In addition dt is compatible with D: $\bar{D}(dt) = 0$. This implies that

$$(D_{T_\sigma(u)} dt)T_\sigma(u) = 0 = T_\sigma(u)(dt(T_\sigma(u))) - dt(D_{T_\sigma(u)} T_\sigma(u))$$

(see (A.20)). Therefore, $(D_{T_\sigma(u)} T_\sigma(u))$ is spacelike just in case

$$T_\sigma(u)(dt(T_\sigma(u))) = 0.$$

But $dt(T_\sigma(u)) = T_\sigma(u)[(t)] = d(t \circ \sigma)/du$, where t is any time-function. So u is a suitable parameter in (30) just in case

$$\frac{d^2 t}{du^2} = 0$$

and $t = au + b$. Hence, u is a suitable parameter in the Newtonian law of motion just in case u is itself a time-function, and this explains the restriction to time-functions in (28).

What about the relativistic law (29)? In Newtonian dynamics we require that the force vector F be spacelike; that is, F must lie in the unique plane of simultaneity through p for each point p on σ. In relativistic space-time there is of course no unique plane of simultaneity through p. However, for each p on σ there is a unique plane of simultaneity *orthogonal* to σ at p; moreover, this is the

143

plane of simultaneity of the inertial frame that is *tangent* to σ at p. Hence, in relativistic dynamics we require $g(F, T_\sigma(u)) = 0$, and therefore

$$g(D_{T_\sigma(u)} T_\sigma(u), T_\sigma(u)) = 0.$$

What constraints does this condition put on the parameter u? Since g is compatible with D, we have $\bar{D}(g) = 0$, and therefore

$$(D_{T_\sigma(u)} g)(T_\sigma(u), T_\sigma(u)) = 0 = T_\sigma(u)(g(T_\sigma(u), T_\sigma(u)))$$
$$- 2g(D_{T_\sigma(u)} T_\sigma(u), T_\sigma(u))$$

(see (A.20)). So $(D_{T_\sigma(u)} T_\sigma(u))$ is orthogonal to $T_\sigma(u)$ just in case

$$T_\sigma(u)(g(T_\sigma(u), T_\sigma(u))) = 0$$

or

$$\frac{d(g(T_\sigma(u), T_\sigma(u)))}{du} = 0.$$

But

$$\frac{d\tau}{du} = \sqrt{g(T_\sigma(u), T_\sigma(u))}.$$

Hence, u is a suitable parameter for the curves specified in (29) just in case

$$\frac{d^2\tau}{du^2} = 0$$

and $\tau = au + b$: u is a suitable parameter in the relativistic law of motion just in case u is linearly related to the proper time τ.

This restriction on suitable parameters in (29) has important implications. In the Newtonian case the x_0-coordinate of an arbitrary inertial system is itself a suitable time function. So Newton's Second Law (28) becomes

$$m \frac{d^2 x_i}{dx_0^2} = F^i$$

in any inertial coordinate system. This is not true in the relativistic case. The coordinate time $t = x_0$ of an arbitrary inertial coordinate system is linearly related to τ just in case the curve $\sigma(\tau)$ in question has *constant* velocity (see (6)). So t is a suitable parameter in (29) only if $\sigma(\tau)$ is a geodesic (and therefore $F = 0$). In other words,

144

although we can write the law of motion for relativistic *kinematics* indifferently as either $D_{T_{\sigma}(t)}T_{\sigma}(t) = 0$ or $D_{T_{\sigma}(\tau)}T_{\sigma}(\tau) = 0$, t and τ are not interchangeable in relativistic *dynamics*. This fact is responsible for the characteristic differences between relativistic dynamics and Newtonian dynamics.

Thus, for example, in inertial coordinate systems (29) becomes

$$m_0 \frac{d^2 x_i}{d\tau^2} = F^i$$

(*not* $m_0(d^2 x_i/dt^2) = F^i$) or

$$\frac{d}{d\tau}(m_0 u^i) = F^i$$

where u^i is the proper velocity $dx_i/d\tau$. The four-vector $p^i = m_0 u^i$ (*not* the vector $m_0(dx_i/dt)$) is therefore the relativistic counterpart of classical momentum. Using equation (7) we find that p^i can be expressed in the form

$$p^0 = \frac{m_0}{\sqrt{1 - v^2}}$$

$$p^i = \frac{m_0 v_i}{\sqrt{1 - v^2}} \qquad (i = 1,2,3)$$

(31)

where v_i is the three-velocity of the particle in our given inertial coordinate system. The quantity

$$p = \frac{m_0 v}{\sqrt{1 - v^2}}$$

is the magnitude of the relativistic three-momentum.

Moreover, if we define the relativistic mass as $m = p/v$ we find that

$$m = \frac{m_0}{\sqrt{1 - v^2}}.$$

(32)

Equation (32) is the traditional formula for the dependence of mass on velocity. If we define the relativistic energy of our particle by the equation

$$\frac{dE}{dt} = v \frac{dp}{dt}$$

we find that

$$E = \frac{m_0}{\sqrt{1 - v^2}} + E_0 \tag{33}$$

with E_0 a constant of integration. The usual practice is to set $E_0 = 0$ in (33) and to define the relativistic energy by

$$E = \frac{m_0}{\sqrt{1 - v^2}} = m. \tag{34}$$

This is the traditional relation between relativistic energy and relativistic mass (with $c = 1$). Comparing (34) and (31) we see that the four-vector p^i has components $(E, \vec{p})$, where $\vec{p}$ is the relativistic three-momentum. This is why p^i is called the *energy-momentum* vector.

4. Electrodynamics

In §III.5 we saw how to formulate Maxwell-Lorentz electro-dynamics in the context of Newtonian space-time. The key object needed was the tensor g^{ij} defined in terms of the spatial metric h^{ij} and the absolute space vector field v^i. In Minkowski space-time we no longer have h^{ij} and v^i. However, we do have the Minkowski metric g_{ij} and its *inverse* g^{ij}, which satisfies $g^{ik}g_{kj} = \delta_{ij}$ (see §A.6). Hence, in inertial coordinate systems we have

$$(g^{ij}) = \begin{pmatrix} 1 & 0 & 0 & 0 \\ 0 & -1 & 0 & 0 \\ 0 & 0 & -1 & 0 \\ 0 & 0 & 0 & -1 \end{pmatrix} \tag{35}$$

that is, $(g^{ij}) = \text{diag}(1, -1, -1, -1)$ (compare (35) with (III.73)). Just as in §III.5, we also introduce an antisymmetric tensor field F of type (0,2), which represents the electromagnetic field, and a vector field J, which represents the charge-current density. Our field equations are

$$K = 0 \tag{36}$$

$$\bar{D}(g) = 0 \tag{37}$$

$$\text{div}[G^1_* G^2_* F] = -4\pi J \tag{38}$$

$$\bar{D}(F)_{1a_3} = 0. \tag{39}$$

146

A particle of rest mass m_0 and charge q satisfies the equation of motion

$$m_0 D_{T_\sigma} T_\sigma = q tr^{21}(G^1_* F \otimes T_\sigma). \qquad (40)$$

The vector field on the right-hand side of (40) is spacelike, and, indeed, orthogonal to T_σ. So there is no need to apply a projection as in the Newtonian case (III.82). In terms of components, our field equations are

$$R^i_{jkl} = 0 \qquad (36a)$$

$$g_{ij;k} = 0 \qquad (37a)$$

$$F^{ij}_{\ ;j} = -4\pi J^i \qquad (38a)$$

$$F_{[ij;k]} = 0 \qquad (39a)$$

while our equation of motion is

$$m_0 \frac{d^2 x_i}{d\tau^2} + \Gamma^i_{jk} \frac{dx_j}{d\tau} \frac{dx_k}{d\tau} = q F^i_j \frac{dx_j}{d\tau} \qquad (40a)$$

where $F^{ij} = g^{ia} g^{jb} F_{ab}$ and $F^i_j = g^{ia} F_{aj}$.

Now $(g^{ij}) = \text{diag}(1, -1, -1, -1)$ in all inertial systems. Therefore, we can follow the procedure of §III.5 to show that in all inertial systems (38a) reduces to the first pair of the Maxwell-Lorentz equations

$$\vec{\nabla} \cdot \vec{E} = 4\pi\sigma$$
$$\vec{\nabla} \times \vec{B} - \frac{\partial \vec{E}}{\partial t} = 4\pi \vec{j} \qquad (41)$$

provided that

$$(F_{ij}) = \begin{pmatrix} 0 & E_{x_1} & E_{x_2} & E_{x_3} \\ -E_{x_1} & 0 & -B_{x_3} & B_{x_2} \\ -E_{x_2} & B_{x_3} & 0 & -B_{x_1} \\ -E_{x_3} & -B_{x_2} & B_{x_1} & 0 \end{pmatrix} \qquad (42)$$

holds in inertial systems and that $(J^i) = (\sigma, \vec{j})$. Similarly, (39a) reduces to the second pair of the Maxwell-Lorentz equations

$$\vec{\nabla} \cdot \vec{B} = 0$$
$$\vec{\nabla} \times \vec{E} - \frac{\partial \vec{B}}{\partial t} = 0 \qquad (43)$$

147

in all inertial systems. Hence, unlike in classical electrodynamics, the Maxwell-Lorentz equations are valid in all inertial frames.[12]

To recover the equations of motion, note that in inertial systems (40a) takes the form

$$m_0 \frac{d^2 x_i}{d\tau^2} = qF^i_j \frac{dx_j}{d\tau}$$

or

$$\frac{dp^i}{d\tau} = qF^i_j \frac{dx_j}{d\tau} \tag{44}$$

where p^i is the energy-momentum four-vector. Using the chain rule and multiplying both sides of (44) by $d\tau/dt$, we obtain

$$\frac{dp^i}{dt} = qF^i_j \frac{dx_j}{dt}. \tag{45}$$

Substituting $i = 0$ in (45) and using (42) and the fact that $F^i_j = g^{ia}F_{aj}$, we get

$$\frac{dE}{dt} = q\vec{E} \cdot \vec{v} \tag{46}$$

where $E = p^0 = m_0/\sqrt{1 - v^2}$ is the relativistic energy, and $\vec{v}$ is the three-velocity of the particle in question. Equation (46) is the traditional relation between energy and work. Substituting $i = 1,2,3$ in (45), we obtain

$$\frac{d\vec{p}}{dt} = q(\vec{E} + \vec{v} \times \vec{B}) \tag{47}$$

with $\vec{p}$ the relativistic three-momentum. Equation (47) is the desired relativistic law of motion.

As we observed in §III.5, the Maxwell-Lorentz equations imply that electromagnetic wave packets propagate through empty space with constant unit three-velocity. Moreover, as we have just seen, the Maxwell-Lorentz equations hold in every inertial frame. Therefore, electromagnetic wave packets have constant unit velocity in every inertial frame and are consequently propagated along null geodesics in Minkowski space-time. Hence, if we identify light signals with electromagnetic wave packets, it is a *consequence* of

---

[12] Compare Chapter III, note 6.

relativistic electrodynamics that light signals follow null geodesics and have constant unit velocity in every inertial frame.

Finally, what is the symmetry group of relativistic electrodynamics? As before, both the electromagnetic field tensor F and the charge-current density J are dynamical objects, since there can be everywhere-vanishing and not-everywhere-vanishing solutions to our field equations for both F and J. The absolute objects of our theory are just D and g; the fixed background space-time structure of our theory is just Minkowski space-time. It follows that the symmetry group of relativistic electrodynamics is just the Lorentz group.

5. The Special Principle of Relativity

What is the special principle of relativity? How does it function as a methodological guide in the transition from Newtonian mechanics to relativity theory? What does it have to do with the relativity of motion and the "equivalence" of reference frames? The most popular way of formulating the principle goes like this:

(R) The laws of nature are the same (or take the same form) in all inertial reference frames.[13]

According to tradition, (R) functions as a methodological guide in the following way. We observe that in classical physics two inertial frames are related by a Galilean transformation and that (R) is satisfied by the laws of *mechanics* (namely, kinematics and gravitation theory). However, (R) is not satisfied by the laws of *electrodynamics*: when we subject Maxwell's equations to a Galilean transformation, we obtain equations of a quite different form. Therefore, assuming the correctness of Maxwell's equations, (R) forces us to change the transformations connecting two inertial frames. We find, in fact, that (R) plus Maxwell's equations leads us to the Lorentz transformations.[14] Moreover, (R) is thought to express the relativity of all inertial motions in a straightforward way. In classical physics all inertial motions are not "equivalent," because the laws of electrodynamics do not take the same form in all inertial frames: Maxwell's equations hold only in rest systems. When we

[13] See, e.g., Einstein [26], 41; Einstein [29], 25; Resnick [90], 35; Bergmann [4], 15.
[14] See, e.g., d'Abro [17], 135–136.

change to the Lorentz transformations, on the other hand, we restore the "equivalence" of all inertial frames. The underlying assumption of this traditional line of thought is that two reference frames are "equivalent" (or "indistinguishable") if the laws of nature take the same form in both.

The trouble with (R) is that the notion of "the form" of a physical law—and, therefore, the notion of a physical law being "the same" in two reference frames—is ambiguous. In general, the laws of a given space-time theory are expressible in a variety of different forms. Relativistic electrodynamics, for example, can be expressed in coordinate-free form, as in (36)–(40), or in generally covariant form in terms of components, as in (36a)–(40a), or in the familiar three-dimensional form with respect to an arbitrary inertial co-ordinate system, as in (41), (43), and (47). If we express relativistic electrodynamics intrinsically, as in (36a)–(40a), then (R) is *trivially* true; in fact, (36a)–(40a) hold in all coordinate systems whatsoever. Further, classical electrodynamics can also be expressed in generally covariant form, as in (III.74a)–(III.81a). Accordingly, (R) is trivially satisfied by classical electrodynamics as well.

What has gone wrong? The requirement that laws of nature take the same form in a given class of reference frames is much too weak to express a relativity principle. All known space-time theories can be written in generally covariant form; so all known space-time theories trivially satisfy any relativity principle on this construal. It is simply not true that "the covariance of equations is the mathematical property which corresponds to the existence of a relativity principle for the physical laws expressed by those equations."[15] As we have seen in §II.2, the connection between covariance requirements, on the one hand, and relativity principles, on the other, is a much more subtle matter. For the same reason, requirements like (R) cannot straightforwardly imply the "equivalence" or "indistinguishability" of different states of motion. That the laws (III.74a)–(III.81a) of classical electrodynamics take the same form in all reference frames does not guarantee that all reference frames are physically "equivalent." As we have seen, it does not even guarantee that all *inertial* frames are so "equivalent."

Is it possible to formulate a better notion of physical *equivalence* or *indistinguishability*? According to another traditional line of thought, two reference frames are equivalent just in case no "me-

[15] Bergmann [4], 15.

150

chanical experiment" can distinguish between them.[16] Thus, for example, in the context of classical mechanics we can distinguish accelerated reference frames from inertial frames by observing the trajectories of free particles. In an accelerated reference frame free particle trajectories will appear as curved paths; in inertial frames they will appear straight. Similarly, in classical electrodynamics we can distinguish rest frames from inertial frames moving relative to the "aether" by observing the trajectories of light rays. In rest frames light rays will have velocity c; in an arbitrary inertial frame they will have velocity $c - u$, where u is the velocity of the inertial frame relative to the "aether." On the other hand, in neither classical mechanics nor relativistic electrodynamics can we distinguish between different inertial frames in this way. In all inertial frames free particles travel straight paths, and light rays have velocity c.

This notion of equivalence can be made more precise.[17] Suppose we are operating in the context of a given space-time theory T. We are given two reference frames, F and F', each of which possesses a definite state of motion according to T. The state of motion of a frame is given by the components of the affine connection and, if present, the absolute space vector field V in that frame. The absolute acceleration is given by Γ^i_{00}, the absolute rotation by Γ^i_{0j}, and the absolute velocity, if present, by $-v^i$. Now, F and F' will be related by a coordinate transformation $\langle x_i \rangle \to \langle y_j \rangle$, which gives rise to a manifold transformation h such that $x_i(hp) = y_i(p)$.[18] We know further that h takes any geometrical object Φ onto a second geometrical object $h\Phi$ whose components in F (at hp) are equal to the components of Φ in F' (at p). In particular, h maps the connection D onto hD and, if present, the absolute space vector field V onto hV in such a way that hD (hV) attributes the same state of motion to F that D (V) attributes to F'. When we ask whether a mechanical experiment can distinguish F from F' in the context of T, we are asking whether a mechanical experiment can distinguish $D(V)$ from hD (hV), given the laws of T.

---

[16] See, e.g., d'Abro [17], 104–105.

[17] Here I follow Earman [22], although I am not sure that the present account of equivalence is quite the same as his. See also Jones [62].

[18] We assume here that F and F' overlap sufficiently. We can suppose without significant loss of generality that the world-lines defining the origins of F and F' intersect at a point $p \in M$ and that p has coordinates $\langle 0,0,0,0 \rangle$ in both frames. If $F = \langle \sigma, \phi \rangle$ and $F' = \langle \rho, \psi \rangle$, then the map $\phi^{-1} \circ \psi$ is a well-behaved manifold transformation h on a neighborhood of $p = \sigma \cap \rho$. See Chapter 2, note 5.

What is given to us by a mechanical experiment? Suppose we are given the trajectories of the particles whose world-lines T attempts to specify in its laws of motion (including, perhaps, the trajectories of light rays) and the matter fields or source variables (mass density, charge density, and so on) giving rise to the interactions described by T. These entities are relatively observable, and they are precisely the entities that the traditional relationalist is willing to admit. So they are suitable (and uncontroversial) outcomes of a mechanical experiment. When we try to distinguish between F and F' by a mechanical experiment, then, we fix particle trajectories and matter fields (for example, we fix $T_{\hat{a}}$ and the mass density ρ in the context of Newtonian gravitation theory), and we ask whether both $D\ (V)$ and $hD\ (hV)$ are consistent with the laws of T. In the context of Newtonian gravitation theory, for example, we ask if there are both models of T containing $\langle D,(V),\rho,T_{\hat{a}}\rangle$ and models of T containing $\langle hD,(hV),\rho,T_{\hat{a}}\rangle$. Thus, if F and F' are equivalent or indistinguishable in this sense, we cannot ascertain (in the context of T itself) which frame we are occupying by measuring matter fields and observing trajectories.

Let us look at some examples of how this notion of equivalence works. We have already applied it to Newtonian kinematics in §III.6. If $\langle M,D,dt,h,(V),T_{\hat{a}}\rangle$ is a model for Newtonian kinematics and f is a Galilean transformation, then $\langle M,fD,dt,h,(fV),T_{\hat{a}}\rangle$ is also a model (of course $fD = D$). Similarly, if $\langle M,D,dt,h,(V),\Phi,\rho,T_{\hat{a}}\rangle$ is a model for Newtonian gravitation theory (§III.3) and f is a Galilean transformation, then $\langle M,fD,dt,h,(fV),\Phi,\rho,T_{\hat{a}}\rangle$ is also a model; so all inertial frames are equivalent in Newtonian gravitation theory as well. By contrast, if f is a transformation to an arbitrary rigid Euclidean frame (of the form III.15), then $\langle M,fD,dt,h,(fV),T_{\hat{a}}\rangle$ is not a model for Newtonian kinematics, and $\langle M,fD,dt,h,(fV),\Phi,\rho,T_{\hat{a}}\rangle$ is not a model for Newtonian gravitation theory. Inertial frames are not in general equivalent to accelerating and rotating frames. Note, however, that in the context of Newtonian gravitation theory (§III.3) merely *accelerating* frames are equivalent to inertial frames: if $\langle M,D,dt,h,\Phi,\rho,T_{\hat{a}}\rangle$ is a model for gravitation theory, then so is $\langle M,fD,dt,h,\Psi,\rho,T_{\hat{a}}\rangle$, where f is of the form (III.66) and Ψ satisfies (III.47).

What about electrodynamics? In classical electrodynamics all inertial frames are not equivalent: rest frames are distinguishable from arbitrary inertial frames. For, if $\langle M,D,dt,h,V,F,J,T_{\hat{a}}\rangle$ is a model for the theory of §III.5 and f is a Galilean transformation, then $\langle M,\mathrm{f}D,dt,h,\mathrm{f}V,F,J,T_{\hat{a}}\rangle$ is not a model. In a rest frame, $T_{\hat{a}}$ satisfies

$$m \frac{d^2(x_i \circ \sigma)}{dt^2} - qF_k^i \frac{d(x_k \circ \sigma)}{dt} = 0$$

while, relative to $\mathrm{f}V$, the law of motion becomes

$$m \frac{d^2(x_i \circ \sigma)}{dt^2} - qF_k^i \frac{d(x_k \circ \sigma)}{dt} = qF_k^0 \frac{d(x_k \circ \sigma)}{dt} u^i$$

where $-u^i = \mathrm{f}v^i$. By contrast, in relativistic electrodynamics all inertial frames are equivalent once again. If $\langle M,D,g,F,J,T_{\hat{a}}\rangle$ is a model for the theory of §IV.4 and f is a Lorentz transformation, then $\langle M,\mathrm{f}D,g,F,J,T_{\hat{a}}\rangle$ is also a model (in fact, $\langle M,D,g,F,J,T_{\hat{a}}\rangle = \langle M,\mathrm{f}D,g,F,J,T_{\hat{a}}\rangle$). In particular, the law of motion for relativistic electrodynamics (40a), unlike the law of motion for classical electrodynamics (III.82a), does not enable us to distinguish one inertial frame from another.

The experimental equivalence or indistinguishability of two reference frames does not at all follow from the laws of nature taking the same form in both, for in all our examples the laws of nature take the same *generally covariant* form in every reference frame. We should therefore replace (R) with

(R1) All inertial reference frames are physically equivalent or indistinguishable,

where the notion of physical equivalence is understood as above, of course. However, (R1) by itself does not yet capture the full methodological force of the special principle of relativity. To see this, observe that (R1) is satisfied by *both* versions of Newtonian kinematics: the theory of §III.1 with absolute space and the theory of §III.2 without absolute space. To capture our methodological preference for the theory of §III.2 we should supplement (R1) with

(R2) If two frames are indistinguishable according to T, they should be theoretically identical according to T.

Thus, the Newtonian kinematics of §III.1 satisfies (R1) but not (R2): inertial frames are indistinguishable, yet they differ concerning the components of V. In the theory of §III.2 we eliminate this undetectable difference: all inertial frames have precisely the same state of absolute motion (namely, zero absolute acceleration). An equivalent way to put the point is this: in the theory of §III.1 the indistinguishable models $\langle M,D,dt,h,V,T_{\hat{a}}\rangle$ and $\langle M,\mathrm{f}D,dt,h,\mathrm{f}V,T_{\hat{a}}\rangle$ are still

153

different; in the theory of §III.2 the indistinguishable models $\langle M,D,dt,h,T_{\hat{a}}\rangle$ and $\langle M,fD,dt,h,T_{\hat{a}}\rangle$ are identical (Galilean transformations are symmetries of D but not V). (R2) requires us to prefer theories of the latter type.[19]

We can now see how Anderson's notion of the *symmetry group* of a theory connects with our interpretation of the special principle of relativity as the conjunction of (R1) and (R2). The symmetry group of T is the largest group of manifold transformations that will preserve the absolute objects of T. Equivalent reference frames in the above sense induce a second group associated with T—a group we might call the *indistinguishability group* of T—namely, the group of all transformations connecting two equivalent reference frames and therefore leading to indistinguishable models of T. That is, such a frame transformation h will be in the indistinguishability group of T only if there are models of T with fixed matter fields and particle trajectories containing both D (V) and hD (hV). If T satisfies (R1), and therefore has inertial reference frames, D will be an absolute object of T. Similarly, V, if present, will also be an absolute object of T, since V is compatible with D. But if T satisfies (R2) as well, then elements of the indistinguishability group of T must preserve D and, if present, V. So if T satisfies (R1) and (R2), the indistinguishability group of T = the symmetry group of T in Anderson's sense.[20]

We call a theory *well-behaved* if it satisfies (R2). Thus, the Newtonian kinematics of §III.1 is not well-behaved, since indistinguishable inertial frames have different absolute velocities. Similarly, the Newtonian gravitation theory of §III.3 is not well-behaved, since indistinguishable Galilean frames have different absolute accelerations. In such theories the indistinguishability group is larger than the symmetry group: the indistinguishability group of §III.1 is the Galilean group, while its symmetry group is just $O^3 \times T$; the indistinguishability group of §III.3 is the group of

[19] If we apply (R2) *alone* to gravitation theory, we are moved to drop the flat connection $\hat{D}$, and to replace the theory of §III.3 with the nonflat space-time theory §III.4. As we shall see in Chapter V, this is precisely the methodological force of the "principle of equivalence."

[20] What about the other absolute objects of T? We are assuming here that the equivalent reference frames in question have the same spatio-temporal geometry: for example, in the Newtonian case we are assuming that the two frames are both rigid Euclidean. So elements of the indistinguishability group must preserve the metrical structure of T as well: for example, in the Newtonian case, elements of the indistinguishability group necessarily preserve h and dt.

all "transformations to accelerated frames" (III.66), while its symmetry group is just the Galilean group. In well-behaved theories, on the other hand, the indistinguishability group is contained in the symmetry group. Hence, when we use (R2) as a methodological guide in moving from a non-well-behaved theory to a well-behaved one, we have to increase the symmetry group and therefore decrease the number of absolute objects. Moreover, we know that the absolute objects of the new theory should admit all elements of the indistinguishability group of the old theory as symmetries. For example, the Newtonian kinematics of §III.2 admits all Galilean transformations as symmetries; the Newtonian gravitation theory of §III.4 admits all "transformations to accelerated frames" (III.66) as symmetries.

Finally, how do covariance and "sameness of form" come into the picture? Suppose T describes a flat space-time in which inertial coordinate systems exist. Let $\mathcal{D}(\Phi^\alpha, \Theta^\beta)$ be a system of differential equations for the components of the *nonabsolute* objects Φ, Θ induced by the laws of T in some particular inertial system. Thus, $\mathcal{D}(\Phi^\alpha, \Theta^\beta)$ is what we called the *standard formulation* of T in §II.2. Suppose T satisfies the relativity principle $\ulcorner(R1) \,\&\, (R2)\urcorner$. Then we know that T is well-behaved and that the indistinguishability group of T = the symmetry group of T = the group of all transformations from one inertial frame to another. But, as we saw in §II.2, the symmetry group of T = the covariance group of the standard formulation of T. So the covariance group of $\mathcal{D}(\Phi^\alpha, \Theta^\beta)$ = the indistinguishability group of T, and $\mathcal{D}(\Phi^\alpha, \Theta^\beta)$ holds in all and only the inertial frames of T. The laws of nature referred to in (R), then, are the *extrinsic* differential equations $\mathcal{D}(\Phi^\alpha, \Theta^\beta)$, not the *intrinsic* laws of T. In Newtonian gravitation theory, for example, the laws in question are equations (III.43) and (III.45) for Φ, ρ, and $T_{\hat{a}}$; in relativistic electrodynamics the laws in question are equations (41), (43), and (44) for F, J, and $T_{\hat{a}}$. Hence, in the case of *flat* space-time theories in which inertial coordinate systems exist, relativity principles, the equivalence or indistinguishability of reference frames, covariance of laws, and "sameness of form" go together in the way they are supposed to. As we shall see when we discuss general relativity, the situation is otherwise in the case of *nonflat* space-time theories in which no inertial coordinate systems exist.

Let us now look in detail at how (R1) and (R2) function in the transition from classical physics to special relativity. We know what happens when we apply (R1) and (R2) to classical kinematics or gravitation theory: we drop the absolute space vector field V, and we enlarge our symmetry group from $O^3 \times T$ to the full Galilean group. To motivate the transition to special relativity and the Lorentz group, we have to apply (R1) and (R2) to classical electrodynamics. In the electrodynamics of §III.5 the Maxwell-Lorentz equations (III.68)–(III.70) hold only in rest systems, where the rest systems are characterized by $\Gamma^i_{jk} = 0$, $(t_i) = (1,0,0,0)$, $(v^i) = (1,0,0,0)$, $(h^{ij}) = \text{diag}(0,1,1,1)$, and (therefore) $(g^{ij}) = \text{diag}(1,-1,-1,-1)$. However, there is an important difference between the field equations (III.68) and (III.69) and the law of motion (III.70). Whereas (III.70) holds only in rest systems, (III.68) and (III.69) actually hold in a much wider class of coordinate systems. To see this, note that the only conditions we actually use in deriving (III.68) and (III.69) from the generally covariant (III.79a) and (III.80a) are the conditions $\Gamma^i_{jk} = 0$ and $(g^{ij}) = \text{diag}(1,-1,-1,-1)$. So (III.68) and (III.69) hold in all coordinate systems in which these last two conditions hold. By contrast, when we derive (III.70) from the generally covariant (III.82a), we need the additional requirements $(t_i) = (1,0,0,0)$ and $(v^i) = (1,0,0,0)$.

Which are the coordinate systems satisfying $\Gamma^i_{jk} = 0$ and $(g^{ij}) = \text{diag}(1,-1,-1,-1)$? These are precisely the special relativistic inertial systems. Suppose we subject a rest system $\langle x_i \rangle$ to an arbitrary Lorentz transformation (17). The component of dt change to $(\bar{t}_i) = (1/\sqrt{1 - v^2}, v/\sqrt{1 - v^2}, 0, 0)$, the components of V change to $(\bar{v}^i) = (1/\sqrt{1 - v^2}, -v/\sqrt{1 - v^2}, 0, 0)$, and the components of h change to

$$
(\bar{h}^{ij}) = \begin{pmatrix} \dfrac{v}{1 - v^2} & \dfrac{-v}{1 - v^2} & 0 & 0 \\[2mm] \dfrac{-v}{1 - v^2} & \dfrac{1}{1 - v^2} & 0 & 0 \\[2mm] 0 & 0 & 1 & 0 \\[2mm] 0 & 0 & 0 & 1 \end{pmatrix}
$$

Nevertheless, the components of g^* remain unchanged: $(\bar{g}^{ij}) = \text{diag}(1,-1,-1,-1)$. Moreover, we know that the only transformations preserving this form for $(\bar{g}^{ij})$ are the Lorentz transformations, since g^* is just the inverse of the Minkowski metric g. Hence, the coordinate systems in which Maxwell's equations (III.68) and (III.69) are valid are all and only the special relativistic inertial systems, and the covariance group of (III.68) and (III.69) is precisely the Lorentz group.

In other words, we have *three* classes of privileged coordinate systems in classical physics, all of which satisfy the condition $\Gamma^i_{jk} = 0$; and so all are inertial in this minimal sense. First, there is the class of Newtonian inertial systems in which $(t_i) = (1,0,0,0)$ and $(h^{ij}) = \text{diag}(0,1,1,1)$: these are just the systems in which traditional Newtonian gravitation theory (III.43), (III.45) is valid. Second, there is the class of special relativistic inertial systems in which $(g^{ij}) = \text{diag}(1,-1,-1,-1)$: these are just the systems in which Maxwell's equations are valid. Finally, there is the class of rest systems in which $(v^i) = (1,0,0,0)$: these belong to the *intersection* of the first two classes and comprise just the systems in which the Lorentzian

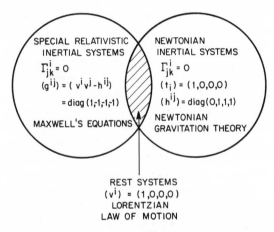

REST SYSTEMS
$(v^i) = (1,0,0,0)$
LORENTZIAN
LAW OF MOTION

law of motion (III.70) is valid. If we start with a rest system and apply arbitrary Galilean transformations, we generate the Newtonian inertial systems. If we apply arbitrary Lorentz transformations, we generate the special relativistic inertial systems. The group of transformations that keeps us within the class of rest systems is just $O^3 \times T = $ (the Galilean group) $\cap$ (the Lorentz group).

How do we move from classical physics to special relativity? We first notice that (R1) holds: in particular, we observe that electrodynamic phenomena do not enable us to distinguish one inertial frame from another. But classical electrodynamics does *not* satisfy (R1). The indistinguishability group of classical electrodynamics is just $O^3 \times T$, which contains no "velocity transformations": all frames generated by $O^3 \times T$ are at rest relative to one another. We conclude that classical electrodynamics is incorrect. However, we also observe that there is an important subpart of classical

157

electrodynamics, the Maxwellian field equations (III.80) and (III.81), that picks out a wider class of privileged coordinate systems than does the whole theory. These coordinate systems are related by elements of the Lorentz group, which does contain "velocity transformations." So we decide that we were wrong about how the inertial frames are to be characterized. The inertial frames in the real world, which we construct via "ideal" clocks and measuring rods, for example, are not the Newtonian inertial frames but the special relativistic inertial frames.

Thus far we have not changed the *laws* of classical electrodynamics one bit. We are using a different characterization of the inertial frames, but these new frames are just as much a part of classical electrodynamics as they are of relativistic electrodynamics (the standard formulation (41), (42) of Maxwell's equations is "already" Lorentz covariant). So (R1) does not force us to change any of the laws (III.74)–(III.81). What must be changed is the law of motion (III.82). Two factors stand in the way of Lorentz-indistinguishability and Lorentz-covariance. First, the right-hand side of (III.82) contains dt and V, and we know that these objects are not preserved by a Lorentz transformation. Second, we know that the $t = x_0$ coordinate of an arbitrary special relativistic inertial frame is not a suitable parameter (§IV.3). Therefore, we use proper time τ as our parameter (note that τ is also definable in classical physics, since we have the metric g^*) and drop dt and V from the right-hand side of (III.82). The result, of course, is just the relativistic law of motion (40). The theory we get by conjoining (40) to (III.74)–(III.81) does in fact satisfy (R1): its indistinguishability group is just the Lorentz group.

This last theory is a rather uneasy mixture of classical and relativistic ideas.[21] The metrical structure of Minkowski space-time is superimposed on that of Newtonian space-time, but only the former is put to any real use. In particular, we use the relativistic proper time τ, not the Newtonian absolute time dt, in stating the laws of motion. The problematic character of this theory is precisely expressed by its failure to satisfy (R2): its indistinguishability group is the Lorentz group, but its symmetry group remains $O^3 \times T$. However, it is now easy to move to a well-behaved theory in which the symmetry group coincides with the indistinguishability group. We simply drop the Newtonian objects dt, h, and V and the cor-

---

[21] It bears a strong resemblance, in fact, to the early "aether" theories of Lorentz and Fitzgerald. See, e.g., Lorentz [67].

responding field equations (III.75)–(III.79). We retain only the Minkowski metric g, with its corresponding equation (37) (which is a consequence of (III.76) and (III.78)). The resulting theory is just special relativistic electrodynamics (36)–(40). So (R1) and (R2), working in tandem, do in fact lead from classical electrodynamics to special relativity.

6. *The Velocity of Light*

Traditionally, three principles concerning the velocity of light have been thought to play foundational roles for special relativity:

(P1) The constancy of the velocity of light. Light is propagated with a constant velocity c independent of the velocity of its source.

(P2) The invariance of the velocity of light. Light has the same constant velocity c in all inertial reference frames.

(P3) The limiting character of the velocity of light. No "causal" signal can propagate with velocity greater than that of light.

Some authors use (P1) conjoined with the special principle of relativity as a basis for our theory.[22] Some use (P2) conjoined with the special principle of relativity.[23] Others take (P2) alone to be fundamental.[24] In "causal" theories like Reichenbach's [85] (P3) is fundamental.

What are the logical relations among (P1)–(P3)? Let us look first at (P1) and (P2). It is clear that (P2) implies (P1): if the velocity of light is independent of the inertial frame in which it is evaluated, it is certainly independent of the velocity of its source (consider an inertial frame whose velocity coincides with the velocity of the source). On the other hand, (P1) does not imply (P2): light can have a source-independent velocity that is equal to c in some inertial frames without having the same velocity c in all inertial frames. This, in fact, is precisely the situation in classical electrodynamics. The light cone is quite independent of the source of the light emitted: light rays follow null geodesics of g regardless of how they are

---

[22] Einstein [26], 41; Anderson [3], 136.

[23] Resnick [90], 35–36.

[24] Bunge [7], 185.

produced.[25] Nevertheless, the magnitude of the velocity of light varies as we move from one inertial frame to another. Yet the conjunction of (P1) with the special principle of relativity does imply (P2). This is how Einstein proceeds in his 1905 paper: he uses the principle of relativity together with (P1) to obtain (P2)[26] and then uses (P2) together with various "homogeneity" and "symmetry" assumptions to derive the Lorentz transformations.

We can think of the relation between (P1) and (P2) as follows. (P1) is necessary if the light cone is to be a single, well-defined geometrical structure preserved by the fundamental symmetries of our theory. If (P1) is false, there is no such thing as *the* light cone at any point $p \in M$; rather, each trajectory has its own light cone.

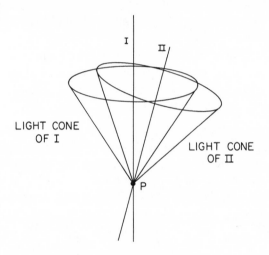

Space-time in which (P1) fails.

Whether (P2) also holds, however, depends on what other geometrical structures are preserved by our fundamental symmetries. In particular, if we require that the planes of simultaneity are also preserved, we get a classical space-time in which (P2) fails. If we do not require that the planes of simultaneity are preserved, we get a relativistic space-time in which (P2) holds. Thus, (P1) concerns the light cone alone; (P2) concerns the relationship between the light cone and the planes of simultaneity.

[25] Hence, if we want to deny (P1), we must modify Maxwell's equations, as in the so-called "emission theory" of Ritz. Cf. Anderson [3], 45.

[26] Einstein [26], 45.

What about (P3)? In our formulation this principle finds expression in the requirement that all physical trajectories are time-like curves: the tangent vector to any such trajectory at any point p must lie inside the light cone of T_p. If this requirement is to pick out a well-defined class of "causal" signals or "causal" curves, then (P1) must hold as well. Otherwise we have put no restriction at all on "causal" signals, since any curve will lie within its own light cone. So (P3) is pointless without (P1). Nevertheless, (P1) does not imply (P3): (P1) is satisfied in classical electrodynamics, but (P3) does not hold. On the other hand, we can easily add (P3) to classical electrodynamics: we can require that physical trajectories ("causal" signals) always lie inside the light cone in the space-time of §III.5. In this space-time the velocity of light will be a *limiting* velocity but not an *invariant* velocity. In rest frames the limiting velocity will be c; in frames moving with absolute velocity $v < c$ it will be $c - v$. This shows that (P3) does not imply (P2).

The other direction is more interesting. (P2) does not strictly entail (P3): we can drop (P3) from special relativity and allow spacelike curves to represent possible physical trajectories ("causal" signals). But various peculiarities result. First, it is easy to see that any two events with spacelike separation will be indeterminate in their temporal order, in the sense that they will receive different

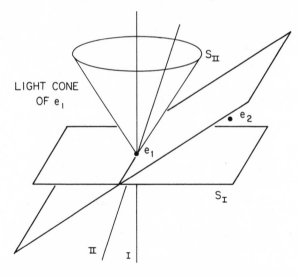

Indeterminacy of temporal order. e_2 lies outside the light cone of e_1: according to S_I, e_1 occurs earlier than e_2; according to S_{II}, e_2 occurs earlier than e_1.

161

temporal orderings in different inertial frames. This cannot happen, of course, for events with timelike separation. So if we think that "causal" processes cannot be indeterminate in their temporal order ("the cause must always precede the effect"), we have motivation for restricting causal processes to timelike curves.[27]

Second, and more dramatically, if "causal" signals with superluminal velocities are allowed, then we can send signals "backwards in time" in Minkowski space-time. That is, the following is possible: an inertial "observer" I sends out a superluminal but finite signal at τ_1; the signal is received by inertial "observer" II, who immediately sends back a second superluminal but finite signal; this second signal is received by I at τ_2; $\tau_2 < \tau_1$. This possiblity leads to the usual paradoxes of time travel.[28] Hence, the characteristic structure

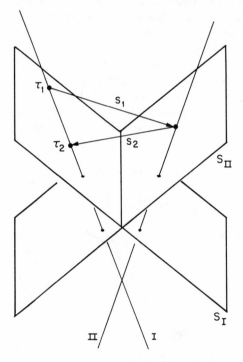

Sending signals backward in time. s_1 has finite superluminal velocity in I; s_2 has finite superluminal velocity in II.

[27] See Anderson [3], 190–191.
[28] See Earman [21]; Horwich [60].

of the planes of simultaneity required by (P2) makes (P3) a natural principle in Minkowski space-time. No such motivation for (P3) exists in the space-time of classical electrodynamics.

How do (P1)–(P3) function as foundational principles? It is clear that neither (P1) nor (P3) (nor their conjunction) can serve as a foundation for special relativity on its own. (P1) is just as true in classical electrodynamics, and (P3) is quite consistent with the latter theory (although there is no particular point in adding (P3) to the theory of §III.5). Only (P2) is inconsistent with classical electrodynamics, since only (P2) forces us to relativize the planes of simultaneity. It is not surprising, then, that (P2) is the basis for most traditional formulations of our theory. One uses (P1), (P3), or their conjunction to conclude that the light cone is an invariant structure (of course, we know this already from classical electrodynamics). One then applies (P2) to conclude that the planes of simultaneity do *not* constitute an invariant structure. Accordingly, one looks for the transformations preserving only the light cone, and one tries to demonstrate that these are just the Lorentz transformations.

As we saw in §IV.2, however, this procedure cannot work without additional assumptions about the geometrical properties of space-time. In particular, we have to assume explicitly that space-time is *flat* if we are even to come close to the Lorentz transformations. Traditional derivations smuggle in this assumption under the rubrics of "homogeneity" and "symmetry." In this respect, therefore, (P2) is quite inferior to the special principle of relativity. As we saw in §IV.5, this principle can serve as a foundation for special relativity all by itself: applying (R1) to electrodynamics does lead to the Lorentz group, since that group is precisely the covariance group of the standard formulation of Maxwell's equations. (R2) then eliminates the unwanted classical structures from electrodynamics, and we indeed end up with special relativity (including (P2) as a *consequence*). This situation is just what we should expect. For (P2) results from applying the special principle of relativity to a consequence of Maxwell's equations (P1); applying the principle to Maxwell's equations themselves naturally leads to stronger results.

Nevertheless, there is something right about the traditional reliance on (P2). For it can be shown that if our space-time is flat, then the transformations preserving the light cone are just the Lorentz transformations and the dilations. Given the assumption of flatness, we do not need the additional assumption that our transformations preserve affine or projective structure as in §IV.2. In

163

Minkowski space-time the transformations preserving conformal structure automatically preserve affine structure as well; in fact, affine structure turns out to be *explicitly definable* from conformal structure in Minkowski space-time.[29] This fact about Minkowski space-time allows us to prove an interesting *representation theorem*. We start with a set of points M and a two-place relation λ on M: intuitively, $p\lambda q$ just in case p and q are connectible by a light ray (alternatively, we can start with a two-place relation $p\tau q$ of timelike separation or a two-place relation $p\sigma q$ of spacelike separation). We then construct a set of axioms for λ characterizing the properties it has in Minkowski space-time, and we show that for any model $\langle M,\lambda \rangle$ of our axioms there is a one-one map h from M onto R^4 such that

$$p\lambda q \quad \text{iff} \quad (\mathrm{h}(p)_0 - \mathrm{h}(q)_0)^2 - (\mathrm{h}(p)_1 - \mathrm{h}(q)_1)^2 - (\mathrm{h}(p)_2 - \mathrm{h}(q)_2)^2$$
$$- (\mathrm{h}(p)_3 - \mathrm{h}(q)_3)^2 = 0$$

(where $\mathrm{h}(p)_i$ is the ith coordinate of $\mathrm{h}(p)$). Further, h is unique up to a Lorentz transformation or a dilation (which now can be viewed as a choice of scale).

The above representation theorem captures the sense in which the light cone structure of Minkowski space-time determines *all* its geometrical properties. Projective structure is definable from conformal structure, and so are the planes of simultaneity (relativized to an inertial trajectory, of course), the spatial and temporal congruence relations (also relativized to an inertial trajectory), and so on.[30] Contrast this situation with the one in classical electrodynamics, where the geometrical structure of Newtonian space-time is certainly not determined by the conformal structure of electrodynamics. In particular, neither the planes of simultaneity nor the spatial and temporal congruence relations are definable from λ. To see this, note that there are automorphisms of λ—namely, Lorentz transformations—that do not preserve any of these Newtonian relations. Hence, if we want to prove an analogous representation theorem for the space-time of classical electrodynamics (where the representation function h should be unique up to an element of $O^3 \times T$ or a dilation), we must add a simultaneity relation and

---

[29] The fact about light cone preserving transformations in Minkowski space-time is proved in Zeeman [120]. It is implicit in Robb [94], who actually constructed the definition of affine structure. See Malament [69] for a very clear exposition of both results.

[30] These results and the representation theorem are also due to Robb [94]. Again, see Malament [69] for a clear exposition.

164

spatial and temporal congruence relations to the structure $\langle M,\lambda \rangle$ (along with appropriate axioms for these new primitives).

These considerations explain why special relativity is especially well-suited to "causal" theories of spatio-temporal structure. For we can use the relation $p\tau q$ to construct all the geometrical structure of Minkowski space-time; and, if (P3) holds, $p\tau q$ is just the relation of "causal" connectibility: $p\tau q$ just in case a "causal" signal can be sent from p to q. It is not that (P3) by itself can somehow lead to the rest of our theory; rather, (P3) is necessary if the reduction to the relation $p\tau q$ is to count as a "causal" reduction. Still, it is not clear why *this* is so important. The relation $p\lambda q$—connectibility by a light signal— is at least as "observational" as the relation $p\tau q$—connectibility by a "causal" signal. But λ can serve as a basis for reduction whether or not (P3) holds. So if our motivations for a "causal" theory are operationalist (what else could they be?), the relation $p\lambda q$ is just as good as the relation $p\tau q$. So who needs (P3)?

7. Nonstandard ($\varepsilon \neq \frac{1}{2}$) Simultaneity Relations

As we have seen, the central difference between Newtonian physics and special relativity concerns the status of time and simultaneity. In Newtonian space-time there is a unique (up to a linear transformation) global time t and a unique simultaneity relation S generated by the planes of absolute simultaneity: the t = constant hypersurfaces. Special relativity has no unique global time. Instead, for each trajectory σ there is a well-defined local time for σ—namely, proper time τ along σ—and this local time allows one to make temporal comparisons only between events on σ itself. Nevertheless, if σ is an inertial trajectory, we can extend τ to a unique (up to a linear transformation) global-time-relative-to-σ: the coordinate time t_F of some inertial frame F adapted to σ. This global-time-relative-to-σ induces the unique simultaneity-relation-for-σ S_F, where S_F is generated by the t_F = constant hypersurfaces. (We can extend the local time τ of a noninertial trajectory $\sigma(\tau)$ as well, but our extension will be well-defined only on a finite neighborhood of σ; see §IV.1.)

What we have just described is the well-known and uncontroversial *relativity of simultaneity* in Minkowski space-time. Inertial frames or trajectories in motion relative to one another do not agree on simultaneity, so simultaneity has to be relativized to a choice of frame or trajectory. However, a second claim or thesis about simultaneity has insinuated itself into the literature in a very confusing way. This second thesis says nothing about relative motion

165

and the comparison of different inertial frames; rather, it concerns the status of simultaneity within a *single* inertial frame. The claim is that, even given a particular inertial trajectory σ, there is still no unique objective simultaneity-relation-for-σ. Even the choice of simultaneity-relation-for-σ (and not just the choice of σ itself) is arbitrary or conventional. Thus, according to this view, only the local time τ has objective status in special relativity. Let us call this thesis, which has been introduced into the literature by Reichenbach and Grünbaum, the *conventionality of simultaneity*.

How does a problem about the simultaneity relation within a single frame F arise? Consider two spatial positions A and B in F, and suppose that there is a clock at each position. A light signal is sent from A to B and reflected back to A. The light signal leaves A

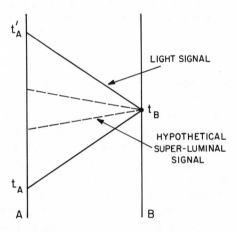

Determining simultaneity within a single frame.

at t_A, as determined by the clock at A, and returns at t'_A. Our problem is to synchronize the clock at A with the clock at B, to say when, according to A-time, the light signal arrives at B. In other words, we must determine which event between t_A and t'_A at A is simultaneous (in F) with t_B.

That a choice exists here is encouraged by the operationalist flavor of Einstein's 1905 paper,[31] where he says that we "establish *by definition*" that

$$t_B - t_A = t'_A - t_B$$

[31] Einstein [26], 40.

166

or

$$t_B = t_A + \tfrac{1}{2}(t'_A - t_A).$$ (48)

If (48) is a mere "definition," then it is natural to suppose that there must be equally good, alternative "definitions." This is precisely what Reichenbach and Grünbaum contend. They claim that

$$t_B = t_A + \varepsilon(t'_A - t_A)$$ (49)

where ε is any real number such that $0 < \varepsilon < 1$, is just as good a "definition" of synchrony or simultaneity (relative to F) as (48). Only computational simplicity can favor the choice $\varepsilon = \tfrac{1}{2}$ over any other admissible value of ε; there are no facts that make (48) any truer or more objective than (49). Of course, (48) is equivalent to the assumption that the velocity of light is the same on the trip from A to B as on the return trip. Reichenbach and Grünbaum argue, however, that any claim about the *one-way* velocity of light—as distinct from its *round-trip* velocity—is arbitrary or conventional as well. On their account, (P2) should be interpreted as a claim about the round-trip velocity of light in any inertial frame: it says nothing at all about one-way velocity.

Clearly, the conventionality of simultaneity within a single inertial frame is quite distinct from the relativity of simultaneity between different inertial frames. Nevertheless, Reichenbach systematically confuses the issue by using "the relativity of simultaneity" to refer to the conventionality of simultaneity. Similarly, he uses "absolute simultaneity" to mean *nonconventional simultaneity*, even though in its standard usage "absolute simultaneity" means *invariant* (as between different inertial frames) *simultaneity*. A good example of the kind of problem created by this systematic confusion is the following misleading line of thought.[32] Suppose there were no limit to the velocity of causal signals. Then it would be possible to establish the simultaneity of events by means of infinitely fast (or, at least, arbitrarily fast) signals. Therefore simultaneity would be absolute. Of course, this last sentence does not follow under the usual interpretation of "absolute," because an infinite velocity is not necessarily invariant: infinite velocities are invariant in Galilean space-time but not in Minkowski space-time. Moreover, as we saw in §IV.6, what distinguishes Minkowski space-time is a finite invariant velocity, not a finite limiting velocity.

[32] See Reichenbach [86], 129.

Reichenbach's systematic confusion explains why he puts so much emphasis on (P3). If we could send signals with velocities arbitrarily greater than the velocity of light, no choice of ε in (49) would remain. We could asymptotically approach a unique event at A simultaneous with t_B (see diagram). Hence, if (P3) fails, no problem about the conventionality of simultaneity can even arise. This confusion also explains why Reichenbach makes the startling claim that "the relativity of simultaneity has nothing to do with the relativity of motion."[33] Reichenbach's point is that if we are allowed to choose different values of ε for different inertial trajectories, we can make all inertial trajectories *agree* on simultaneity: simply set $\varepsilon = \frac{1}{2}$ for a given inertial trajectory σ, and if ρ has velocity v relative to σ, set $\varepsilon = \frac{1}{2}(v + 1)$ for ρ. But, of course, it is the conventionality of simultaneity, not the relativity of simultaneity (as ordinarily understood), that has nothing to do with the relativity of motion.

Let us try to clarify this issue in a preliminary way by getting a sense of what nonstandard ($\varepsilon \neq \frac{1}{2}$) simultaneity relations look like. (In later chapters we shall tackle the philosophical problems about conventionality and objectivity.) We know what the standard ($\varepsilon = \frac{1}{2}$) simultaneity relation looks like. Given an inertial trajectory σ and a

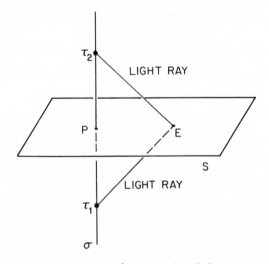

Standard ($\varepsilon = \frac{1}{2}$) simultaneity relation.

[33] Ibid., 146.

point p on σ, the set of points E that are ($\varepsilon = \frac{1}{2}$)-simultaneous with p lie on a spacelike plane S *orthogonal* to σ at p. This spacelike plane satisfies the condition $t = \tau(p) = \tau_1 + \frac{1}{2}(\tau_2 - \tau_1)$, where t is the coordinate time of an inertial coordinate system adapted to σ.

What happens if $\varepsilon \neq \frac{1}{2}$? Our event E lies on a unique inertial trajectory ρ at rest relative to σ, that is, the trajectory whose spatial coordinates in any inertial frame adapted to σ are equal to the spatial coordinates of E. Suppose the distance between σ and ρ in any such inertial frame is r. Since the round-trip velocity of light $= 1$, $(\tau_2 - \tau_1) = 2r$. Now suppose we ascribe the time

$$t_E = \tau_1 + \varepsilon(\tau_2 - \tau_1) \tag{50}$$

to E. According to (50), the event q on σ simultaneous with E is not in general identical to p. Whereas p has proper time $\tau(p) = \tau_1 + r$, q has proper time $\tau(q) = \tau(p) - \delta r$, where $\delta = 2(\frac{1}{2} - \varepsilon)$. Hence, the

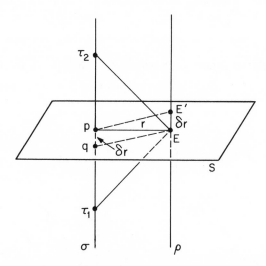

event E' on ρ ε-simultaneous with p is not in general identical to E. Whereas E has coordinate time $t(E) = \tau(p) = \tau_1 + r$ in any inertial system adapted to σ, E' has coordinate time $t(E') = t(E) + \delta r$. (Note that the condition $0 < \varepsilon < 1$ is just the condition that E' have *spacelike* separation from p.)

So far we have only specified which event E' on the particular trajectory ρ is ε-simultaneous with p. To characterize a full non-standard simultaneity relation we must specify an event ε-simultaneous with p on every trajectory ρ' at rest relative to σ.

In general, then, ε (and therefore δ) will depend on both the distance r between ρ' and σ and the direction θ of ρ'. To make this dependence explicit, let us introduce polar coordinates on the planes S; that is, we transform our inertial coordinate system $\langle t,x_1,x_2\rangle$ adapted to σ to a coordinate system $\langle t,r,\theta\rangle$, where

$$x_1 = r\cos(\theta)$$
$$x_2 = r\sin(\theta)$$

(we suppress one spatial dimension as usual). Our ε-simultaneity relation results from transforming any point E with coordinates $\langle t,r,\theta\rangle$ to a point E' with coordinates $\langle t + \delta(r,\theta)r,r,\theta\rangle$. In other words, we transform any standard plane of simultaneity S to a nonstandard spacelike hypersurface h(S), where h is a manifold transformation given by

$$t^* = t + \delta(r,\theta)r$$
$$r^* = r \qquad\qquad (51)$$
$$\theta^* = \theta.$$

Equivalently, we introduce a *coordinate* transformation

$$\bar{t} = t - \delta(r,\theta)r$$
$$\bar{r} = r \qquad\qquad (52)$$
$$\bar{\theta} = \theta.$$

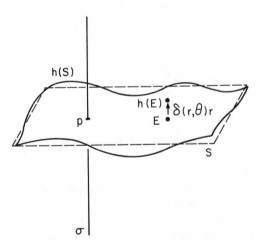

Nonstandard simultaneity surface h(S). The point $E(\langle t,r,\theta\rangle)$ on S is transformed to the point $E(\langle t + \delta(r,\theta)r,r,\theta\rangle)$ on h(S).

170

Our nonstandard simultaneity surfaces are just the hypersurfaces $\bar{t}$ = constant (while the standard simultaneity surfaces satisfy t = constant).

What does h(S) look like? Every point on S is moved a "distance" $\delta(r,\theta)r$ "parallel" to σ. Since $\delta(r,\theta)$ varies with r and θ, h(S) will in general be nonflat. Therefore, the spatial geometry on h(S) will in general be non-Euclidean. A sufficient condition for the geometry on h(S) to be Euclidean is that δ (and therefore ε) should be independent of r: $\partial\delta/\partial r = 0$. A simple example of this type of Euclidean nonstandard simultaneity surface is obtained by tilting a standard plane of simultaneity at an arbitrary (nonorthogonal) angle from σ. Thus, we can tilt S along the x_1-axis by

$$\begin{aligned}
\bar{t} &= t - \delta_{x_1}x_1 \\
\bar{x}_1 &= x_1 \\
\bar{x}_2 &= x_2
\end{aligned}$$ (53)

(where δ_{x_1} is the (constant) value of δ along the x_1-axis) to obtain an h(S) that looks like this:

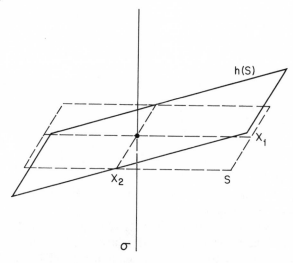

Planar simultaneity surface. $\delta = \delta_{x_1}\cos(\theta)$.

A $\langle \bar{t}, r, \theta \rangle$ system (52) for h(S) obviously satisfies

$$\begin{aligned}
\bar{t} &= \bar{t} + \delta_{x_1}\cos(\theta)r \\
x_1 &= r\cos(\theta) \\
x_2 &= r\sin(\theta).
\end{aligned}$$ (54)

So $\delta(r,\theta) = \delta_{x_1} \cos(\theta)$. Such a tilted h(S) is itself a standard simultaneity surface for some *other* inertial trajectory, though not the standard simultaneity surface *for σ*.

If δ (and therefore ε) is independent of θ as well—if $\partial\delta/\partial\theta = 0$— then h(S) is a cone around σ or S itself. A conical system $\langle \bar{t}, r, \theta \rangle$

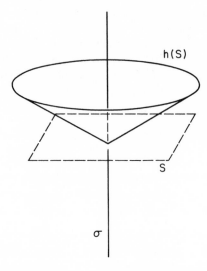

Conical simultaneity surface. δ is constant (independent of both r and θ).

is related to an inertial system $\langle t, x_1, x_2 \rangle$ by

$$
\begin{aligned}
t &= \bar{t} + \delta_{x_1} r \\
x_1 &= r \cos(\theta) \\
x_2 &= r \sin(\theta)
\end{aligned}
\tag{55}
$$

where δ_{x_1} is again the (constant) value of δ along the x_1-axis. So $\delta(r,\theta) = \delta_{x_1} =$ constant.

Now, special relativity receives its standard formulation in inertial coordinate systems $\langle t, x_1, x_2 \rangle$ whose $t =$ constant hypersurfaces define the standard ($\varepsilon = \frac{1}{2}$) simultaneity relations. We can translate special relativity into $\varepsilon \neq \frac{1}{2}$ language by performing the coordinate transformation

$$
\begin{aligned}
t &= \bar{t} - \delta(r,\theta) r \\
x_1 &= r \cos(\theta) \\
x_2 &= r \sin(\theta)
\end{aligned}
\tag{56}
$$

172

and by rewriting everything in terms of the noninertial coordinates $\langle \bar{t}, r, \theta \rangle$. In general, because (56) is nonlinear, this procedure results in horribly complicated formulas that cannot be surveyed. However, certain special cases of (56) are simple enough to be worth discussing in more detail.

Consider, for example, the tilting transformation (53). The new coordinates $\langle \bar{t}, \bar{x}_1, \bar{x}_2 \rangle$ are linearly related to the inertial system $\langle t, x_1, x_2 \rangle$. So $\bar{\Gamma}^i_{jk} = 0$ in $\langle \bar{t}, \bar{x}_1, \bar{x}_2 \rangle$ and $\bar{g}_{ij} = \text{constant}$. In fact, it follows from the transformation law for g_{ij} (A.7) that

$$(\bar{g}_{ij}) = \begin{pmatrix} 1 & \delta_{x_1} & 0 \\ \delta_{x_1} & \delta^2_{x_1} - 1 & 0 \\ 0 & 0 & -1 \end{pmatrix} \tag{57}$$

(where one spatial dimension is suppressed). Let us characterize the ε_{x_1}-systems by $\bar{\Gamma}^i_{jk} = 0$ and (57), where $\delta_{x_1} = 2(\frac{1}{2} - \varepsilon_{x_1})$. When $\varepsilon_{x_1} = \frac{1}{2}$, then, (57) reduces to $\text{diag}(1, -1, -1)$ and our ε_{x_1}-system becomes an inertial system. An ε_{x_1}-system is *suitable* just in case the quadratic form $\bar{g}_{ij}\alpha^i\alpha^j$ ($i, j = 1, 2$) is negative definite. It follows from (57), therefore, that an ε_{x_1}-system is suitable just in case

$$\bar{g}_{11} = \delta^2_{x_1} - 1 < 0. \tag{58}$$

Substituting $2(\frac{1}{2} - \varepsilon_{x_1})$ for δ_{x_1} in (58), the condition for suitability becomes

$$\varepsilon^2_{x_1} < \varepsilon_{x_1}. \tag{59}$$

And (59) holds just in case $0 < \varepsilon_{x_1} < 1$.

Next, consider a trajectory with constant velocity $v = dx_1/dt$ along the x_1-axis of our inertial system. Its "velocity" $\bar{v} = d\bar{x}_1/d\bar{t}$ along the $\bar{x}_1$-axis of our ε_{x_1}-system satisfies

$$v = \frac{\bar{v}}{1 + \delta_{x_1}\bar{v}}. \tag{60}$$

Moreover, whereas the relation between the proper time τ along this trajectory and the coordinate time t is given by

$$\frac{d\tau}{dt} = \sqrt{1 - v^2}$$

the analogous relation between τ and the coordinate "time" $\bar{t}$ is

$$\frac{d\tau}{d\bar{t}} = \sqrt{1 + 2\delta_{x_1}\bar{v} + (\delta^2_{x_1} - 1)\bar{v}^2} \tag{61}$$

or

$$\frac{d\tau}{d\bar{t}} = \sqrt{[1 - \bar{v}(2\varepsilon_{x_1} - 1)]^2 - \bar{v}^2}. \tag{62}$$

Consequently, whereas $\sqrt{1 - v^2}$ is the basic *dilation term* in standard formulations of special relativity, $\sqrt{[1 - \bar{v}(2\varepsilon_{x_1} - 1)]^2 - \bar{v}^2}$ is the basic dilation term in $\varepsilon_{x_1} \neq \frac{1}{2}$ formulations.

Thus, for example, we know that two inertial frames are related by a Lorentz transformation:

$$t^* = \frac{t - vx_1}{\sqrt{1 - v^2}}$$

$$x_1^* = \frac{x_1 - vt}{\sqrt{1 - v^2}} \tag{63}$$

$$x_2^* = x_2$$

where $\langle t^*, x_1^*, x_2^* \rangle$ moves with velocity v along the x_1-axis of $\langle t, x_1, x_2 \rangle$. What is the analogous relationship between two ε_{x_1}-systems? Let there be given two ε_{x_1}-frames, I and II, with coordinates $\langle \bar{t}, \bar{x}_1, \bar{x}_2 \rangle$ and $\langle \bar{t}^*, \bar{x}_1^*, \bar{x}_2^* \rangle$ respectively. Let the respective values of ε_{x_1} in the two frames be ε_1 and ε_2, and let frame II move with constant "velocity" $\bar{v}$ along the $\bar{x}_1$-axis of frame I. We can use the following procedure to find the transformation connecting the two frames: first, use (53) to transform I into an inertial system $\langle t, x_1, x_2 \rangle$; second, use (60) and (63) to transform $\langle t, x_1, x_2 \rangle$ into a second inertial system $\langle t^*, x_1^*, x_2^* \rangle$ moving with velocity v relative to the first; finally, use (53) to obtain the ε_2-system $\langle \bar{t}^*, \bar{x}_1^*, \bar{x}_2^* \rangle$.

This procedure results in some tedious algebra and

$$\bar{t}^* = \frac{\bar{t}[2\bar{v}(1 - \varepsilon_1 - \varepsilon_2) + 1] - \bar{x}_1[2(\varepsilon_1 - \varepsilon_2) + 4\bar{v}\varepsilon_1(1 - \varepsilon_1)]}{\sqrt{[1 - \bar{v}(2\varepsilon_1 - 1)]^2 - \bar{v}^2}}$$

$$\bar{x}_1^* = \frac{\bar{x}_1 - \bar{v}\bar{t}}{\sqrt{[1 - \bar{v}(2\varepsilon_1 - 1)]^2 - \bar{v}^2}} \tag{64}$$

$$\bar{x}_2^* = \bar{x}_2.$$

Note that when $\varepsilon_1 = \varepsilon_2 = \frac{1}{2}$, $\bar{v} = v$ and (64) reduces to a Lorentz transformation.

174

The transformations (64) were first derived by John Winnie, although from a completely different point of view.[34] (At the end of his paper Winnie briefly alludes to the possibility of obtaining his transformations in something like the above manner.)[35] Winnie's central claim is that special relativity as formulated using the standard synchronization rule (48) is "kinematically equivalent" to a formulation using the nonstandard rule (49) with $\varepsilon \neq \frac{1}{2}$, thus vindicating the Reichenbach-Grünbaum thesis of the conventionality of simultaneity. However, we have to be very careful with this equivalence claim, for there is one sense in which it is obviously true but completely trivial, and there is a second sense in which it is not at all obvious and completely unsupported by Winnie's arguments.

The sense in which the equivalence claim is obviously true is that Minkowski space-time can be described equally well from the point of view of ε-coordinate systems as from the point of view of inertial coordinate systems. Formulations of special relativity in ε-systems say the same thing about Minkowski space-time as formulations in inertial systems. Indeed, they are nothing but different (extrinsic) coordinate representations of the same (intrinsic) theory. Hence, the two formulations cannot disagree about the behavior of light— light rays follow null geodesics independent of coordinate system— or about the behavior of clocks—ideal clocks measure proper time along their trajectories independent of coordinate system—and so on. But note that in this sense of equivalence there is no need to restrict ourselves to ε-coordinate systems (which, for example, are at least *suitable*). Minkowski space-time can be described equally well from the point of view of any coordinate system; our theory can be represented in arbitrary coordinate systems. Therefore, the equivalence of ε-systems and inertial systems in this sense reveals no deep facts about Minkowski space-time and special relativity; rather, it is simply a trivial consequence of general covariance.

If Winnie's equivalence claim is to have any real content, then, it should say something about the intrinsic space-time structures described by our theory, not just about the different coordinate systems in which these structures are represented. Consider, for example, the transition from Newtonian theory to special relativity

---

[34] Compare (64) with Winnie's *ε-Lorentz transformations* (remembering that we have set $c = 1$); Winnie [118], 234.

[35] Ibid., 236–237.

in its usual $\varepsilon = \frac{1}{2}$ version. According to Newtonian theory, space-time has the structure $\langle M,S \rangle$, where S is the unique simultaneity relation determined by the $t = $ constant hypersurfaces of absolute time. According to special relativity, no such unique S exists; rather, space-time has the structure $\langle M,S_\sigma^{(1/2)} \rangle$, where for each inertial trajectory σ, $S_\sigma^{(1/2)}$ is the standard simultaneity relation for σ. A similar transition occurs when we move from the Newtonian theory with absolute space to Galilean space-time. According to the former theory, space-time has the structure $\langle M,V \rangle$, where V is a distinguished rest frame determining a unique absolute velocity for every object. According to the latter theory, no such unique V exists; rather, space-time has the structure $\langle M,V_\sigma \rangle$, where for each inertial trajectory σ, V_σ is an inertial frame adapted to σ.

Hence, if the thesis of the conventionality of simultaneity is not to be trivial, it should recommend an analogous transition from the space-time $\langle M,S_\sigma^{(1/2)} \rangle$ to the space-time $\langle M,S_\sigma^{(\varepsilon)} \rangle$, where we have relativized our simultaneity relation S to both an inertial trajectory and a choice of ε. The point, of course, is that no such transition follows from the mere observation that inertial systems and ε-systems are equally good coordinate representations of the space-time $\langle M,S_\sigma^{(1/2)} \rangle$. For this reason arguments like Winnie's, which show only that special relativity as formulated in inertial systems can be translated into a formulation in ε-systems (a fact that is obvious in generally covariant formulations of our theory), cannot support a nontrivial version of the thesis. What we need are methodological/experimental arguments, analogous to the considerations prompting the transitions $\langle M,S \rangle \rightarrow \langle M,S_\sigma^{(1/2)} \rangle$ and $\langle M,V \rangle \rightarrow \langle M,V_\sigma \rangle$, for making the transition $\langle M,S_\sigma^{(1/2)} \rangle \rightarrow \langle M,S_\sigma^{(\varepsilon)} \rangle$. I shall investigate the prospects for such arguments and reconsider the conventionality of simultaneity in a more general setting in Chapter VII.

V
General Relativity

General relativity, unlike the theories we have discussed so far, is standardly presented from the space-time point of view. The theory describes a four-dimensional differentiable manifold; its field equations and laws of motion are written in generally covariant tensor form; the field equations and laws of motion together determine the geometrical structure of space-time and how that structure constrains physical events and processes; and so on. Traditionally, of course, these were thought to be distinctive features that separated general relativity from previous theories. From our present point of view, however, these aspects of the general theory simply represent a new way of formulating dynamical theories, one that, as we have seen, applies equally well to such theories as Newtonian physics and special relativity. In any case, I can be brief in giving a formulation of general relativity within our present framework. On the other hand, although the form of the field equations of general relativity is a matter of common agreement, there is considerable controversy over its foundations and philosophical motivations. The bearing of our theory on the "relativity" of motion has been especially obscure. Consequently, I shall devote most of this chapter to the latter questions.

1. The Theory

General relativity is a theory of gravitation formulated in the context of the conceptions of space and time due to special relativity. To arrive at such a theory, we take a cue from the Newtonian gravitation theory of §III.4. That is, we represent the gravitational field by a nonflat connection D that determines the trajectories of freely falling particles (particles subject only to gravitational forces) via a geodesic law of motion. The source of the gravitational field determines the curvature of space-time in a manner analogous to equation (III.62), and the curvature of space-time in turn determines the behavior of freely falling particles.

The gravitation theory of §III.4 embodies Newtonian conceptions of space and time by virtue of its absolute time dt and Euclidean spatial metric h. To preserve relativistic space-time we replace these with the metric g of Chapter IV. Since this metric, unlike the Newtonian spatial and temporal metrics, is nonsingular, our new space-time is a semi-Riemannian manifold. Hence, the metric g and nonflat connection D cannot play the independent roles their counterparts did in the Newtonian case. On the contrary, the condition $\bar{D}(g) = 0$ implies

$$\Gamma^i_{jk} = \{{}^{\ i}_{jk}\} = \tfrac{1}{2}g^{ir}\left(\frac{\partial g_{rk}}{\partial x_j} + \frac{\partial g_{rj}}{\partial x_k} - \frac{\partial g_{jk}}{\partial x_r}\right)$$

(see (A.25)). Of course, this is just as true in *special* relativity. The metric g, like the connection D, is nonflat: in general no coordinate systems exist in which $(g_{ij}) = \mathrm{diag}(1,-1,-1,-1)$. However, we can have $(g_{ij}) = \mathrm{diag}(1,-1,-1,-1)$ at any given *point* $p \in M$ (but not necessarily on any finite *neighborhood* of p) so long as g has signature $\langle 1,3 \rangle$. A semi-Riemannian manifold whose metric g has this signature is called a *Lorentzian* manifold (hence, Minkowski space-time is just a flat Lorentzian manifold). A Lorentzian metric g allows us to divide the tangent space at $p \in M$ into timelike, spacelike, and null vectors; to define timelike and spacelike curves; to define proper time τ along timelike curves and proper distance s along spacelike curves; and so forth.

The shift from a *non*-semi-Riemannian (Newtonian) theory to a semi-Riemannian (relativistic) theory has far-reaching implications. In a Newtonian theory the spatial and temporal metrics remain absolute even though the affine connection is dynamical. In a relativistic theory this cannot happen: if the connection D is dynamical, then so is the spatio-temporal metric g. So our relativistic theory of gravitation has no absolute objects. In the Newtonian case the four-dimensional geometry is non-Euclidean (nonflat), but the three-dimensional geometry on spacelike hypersurfaces is Euclidean (flat). In the relativistic case spatial Euclideanism can fail, for the three-dimensional geometry on spacelike hypersurfaces is necessarily affected by the four-dimensional curvature. Finally, a relativistic setting offers no choice between a curved space-time and a flat space-time formulation of gravitation theory. In contrast to the

178

Newtonian situation, where both the nonflat connection D and the flat connection $\overset{\circ}{D}$ are compatible with h and dt, only one connection is compatible with the Lorentzian metric g.

Our problem is to give a field equation for the gravitational field, an equation relating a geometrical object representing the curvature of space-time to a geometrical object representing the source of the gravitational field. In Newtonian theory we have used the mass density ρ as the source of the gravitational field. However, as we shall see in the next section, we can also regard the *momentum tensor* $P = \rho(\mathbf{u} \otimes \mathbf{u})$, where $\mathbf{u}$ is the velocity field of the mass distribution in question, as our source variable. The relativistic counterpart of this tensor is the *energy*-momentum tensor, since the relativistic counterpart of classical momentum is the energy-momentum four-vector (§IV.3). But, if the mechanical energy of our mass distribution can serve as a source for the gravitational field, why not include other forms of energy (for example, electromagnetic energy) as well? This is exactly what we do: we form the total *stress-energy tensor* $\mathbf{T}$, which includes the energy-momentum of our mass distribution plus any forms of energy density arising from interactions within our distribution. Specific forms for $\mathbf{T}$ must be supplied by specific theories of interaction (for example, electrodynamics), each of which supplies a contribution to the total energy density.[1] The basic point is that, due to the relativistic relation between mass and energy (IV.34), energy density as well as mass density must be included in the source of the gravitational field.

Comparison with the Newtonian theory of §III.4 suggests that the Ricci tensor is the appropriate geometrical object for representing the curvature of space-time: our gravitational field equation should relate the Ricci tensor to the stress-energy tensor. The simplest possible relation is proportionality: we could set $\text{Ric} = \kappa \mathbf{T}$. However, since $\mathbf{T}$ represents the *total* mass and energy density, we want it to satisfy the *conservation law* $\text{div}(\mathbf{T}) = 0$. Hence, it is natural to look for a conservative tensor built up from the Ricci tensor. Fortunately, there is such a tensor: the *Einstein tensor* $\mathcal{G} = G^1_* G^2_* \text{Ric} - \frac{1}{2} R g^*$ satisfying $\text{div}(\mathcal{G}) = 0$ (see (A.26)). R, the *curvature scalar*, is given by $R = g^{ab} R_{ab}$. So we take as our field equation $\mathcal{G} = \kappa \mathbf{T}$. To ensure that we can derive Newtonian theory as

---

[1] See, e.g., Misner, Thorne, and Wheeler [74], chap. 5.

a limiting case we set $\kappa = -8\pi k$, where k is the Newtonian gravitational constant of Chapter III.[2]

Our theory thus has three geometrical objects: an affine connection D, a semi-Riemannian metric tensor g of signature $\langle 1,3 \rangle$, and a symmetric tensor field **T** of type $(2,0)$. We have two field equations:

$$G_*^1 G_*^2 \text{Ric} - \tfrac{1}{2}Rg^* = -8\pi k\mathbf{T} \tag{1}$$

$$\bar{D}(g) = 0. \tag{2}$$

Our equation of motion is just the geodesic law

$$D_{T_\sigma} T_\sigma = 0. \tag{3}$$

In terms of components we have

$$R^{jk} - \tfrac{1}{2}g^{jk}R = -8\pi k T^{jk} \tag{1a}$$

$$g_{ij;k} = 0 \tag{2a}$$

where $R^{jk} = g^{ja}g^{kb}R_{ab}$, and

$$\frac{d^2 x_i}{d\tau^2} + \Gamma^i_{jk}\frac{dx_j}{d\tau}\frac{dx_k}{d\tau} = 0 \tag{3a}$$

where τ is of course proper time.

Equation (1a) has various equivalent forms. First, we can lower all indices to obtain

$$R_{jk} - \tfrac{1}{2}g_{jk}R = -8\pi k T_{jk} \tag{4}$$

where $T_{jk} = g_{ja}g_{jb}T^{ab}$. Second, by multiplying both sides of (4) by g^{jk}, we find that $R = 8\pi kT$, where $T = g^{ab}T_{ab}$. Substituting this into (4), we obtain

$$R_{jk} = -8\pi k(T_{jk} - \tfrac{1}{2}g_{jk}T). \tag{5}$$

In free space (5) reduces to $R_{jk} = 0$, just as in the Newtonian case. Let us also note in passing that we can generalize (1a) slightly to

$$R^{jk} - \tfrac{1}{2}g^{jk}R + \Lambda g^{jk} = -8\pi k T^{jk} \tag{6}$$

where Λ is constant and $|\Lambda|$ is small. This is Einstein's field equation with *cosmological term* Λ. The left-hand side of (6), like the left-hand side of (1a), is conservative; but (6), unlike (1a), does not reduce to

---

[2] See Fock [35], 199; Anderson [3], 361; Bergmann [4], 185; Adler, Bazin, and Schiffer [1], 279. The most sophisticated contemporary treatment of this limiting process is Malament [72]. An interesting consequence of Malament's work is that the limiting process *forces* the geometry on the Newtonian planes of simultaneity to be Euclidean, or flat. Intuitively, as the light cones in a general relativistic space-time "collapse," all *spatial* curvature is "squeezed out."

$R_{jk} = 0$ for free space. For simplicity, we shall almost always be concerned with (1a), with the case $\Lambda = 0$.

We have constructed the equation of motion (3a) independently of the field equations (1a) and (2a). It is common practice, however, to derive the equation of motion from the field equations. The usual derivation proceeds as follows. Consider a finite distribution of mass ρ acted on only by gravitational forces. The stress-energy tensor of this distribution reduces to the energy-momentum tensor P with components

$$P^{jk} = \rho \, \frac{dx_j}{d\tau} \frac{dx_k}{d\tau}. \tag{7}$$

The mass-energy conservation law, which follows from (1a) and the conservative character of the Einstein tensor, in this case takes the form

$$P^{jk}{}_{;k} = 0. \tag{8}$$

It is then shown that (7) and (8) imply equation (3a) in the limit as the volume of our mass distribution becomes indefinitely small.[3] Thus, what is crucial to this derivation is the conservation law (8), and, since the Einstein tensor is itself conservative (this is why we chose it), the conservation law follows from the field equation (1a).

In this theory we have no privileged inertial coordinate systems, owing to the nonvanishing curvature, nor do we have rigid Euclidean systems, owing to the semi-Riemannian character of the metric. However, we can associate a particularly useful kind of coordinate system—a *normal* coordinate system—with any given timelike (not necessarily geodesic) trajectory $\sigma(\tau)$. This is a coordinate system $\langle y_j \rangle$ with the following properties:

(i) $\langle y_j \rangle$ is suitable: the curves $y_i = $ constant ($i = 1,2,3$) are timelike, and the curves $y_0 = $ constant are spacelike.
(ii) $y_0 = \tau$ along $\sigma(\tau)$.
(iii) the curves $y_0 = $ constant are orthogonal to σ at $\sigma(\tau)$.
(iv) $(g_{ij}) = \text{diag}(1, -1, -1, -1)$ along $\sigma(\tau)$.
(v) any purely spatial geodesic through σ on a $y_0 = $ constant hypersurface satisfies $d^2 y_i / ds^2 = 0$.
(vi) $\langle y_j \rangle$ is adapted to σ: σ satisfies $y_i = 0$ ($i = 1,2,3$).

The difference between a normal frame and a semi-Euclidean frame (§IV.1) lies in (v). In a semi-Euclidean frame (v) is true for *any*

---

[3] See e.g., Fock [35], 238–240; Adler, Bazin, and Schiffer [1], 296–302. Geroch and Jang [42] is a sophisticated contemporary treatment.

purely spatial geodesic, not just for spatial geodesics *through* σ. Semi-Euclidean frames have Euclidean spatial geometry on every y_0 = constant hypersurface; normal frames do not necessarily have Euclidean spatial geometry.

A normal coordinate system can be constructed by choosing a quadruple of "orthonormal" vectors $\langle X_j \rangle$ in $T_{\sigma(\tau)}$ for each value of τ. That is, $(g_{\sigma(\tau)}(X_i, X_j)) = \text{diag}(1, -1, -1, -1)$ for each τ. So $X_0 = T_\sigma(\tau)$. From each point σ(τ) we send out a family of spatial geodesics orthogonal to X_0 whose parameters are defined by their proper distance *s* from σ(τ). We assign each point *p* on such a geodesic the coordinates

$$y_0 = \tau$$
$$y_i = -g_{\sigma(\tau)}(n, X_i)s \qquad (i = 1,2,3)$$

where *n* is the unit tangent vector to our geodesic at σ(τ) and *s* is the proper distance along the geodesic from σ(τ) to *p*. Again, the choice of "orthonormal" quadruples ("infinitesimal inertial frames") must be sufficiently smooth.[4] Also, this coordinate system is well-defined only on a finite neighborhood of σ, owing to the intersection of spatial geodesics associated with different values of τ.

The Γ^i_{jk} will not in general vanish in such a frame. We can use exactly the same reasoning as in §IV.1 to find their values along σ itself. That is, we find that

$$\Gamma^0_{0i} = \Gamma^0_{i0} = \Gamma^i_{00} = a^i \qquad (i = 1,2,3) \tag{9}$$

and

$$\Gamma^i_{j0} = \Gamma^i_{0j} = \Omega^i_j \qquad (i,j = 1,2,3) \tag{10}$$

where Ω^i_j is the usual antisymmetric rotation matrix. All other Γ^i_{jk} vanish. Therefore, along σ we have the equation of motion

$$\frac{d^2 y_i}{dy_0^2} + a^i + 2\Omega^i_j \frac{dy_j}{dy_0} - 2a^j \frac{dy_j}{dy_0} \frac{dy_i}{dy_0} = 0. \tag{11}$$

	inertial force	Coriolis force	relativistic correction

The law of motion along σ in a general relativistic normal frame is identical to the law of motion along σ in a special relativistic semi-Euclidean frame (IV.12), but the two differ at finite distances from σ. If the acceleration a^i and the rotation Ω^i_j both vanish, σ is a geodesic, and our normal frame becomes a *local inertial* frame. In a local

---

[4] See Misner, Thorne, and Wheeler [74], 327–330.

inertial frame the law of motion becomes

$$\frac{d^2 y_i}{dy_0^2} = 0 \tag{12}$$

along σ.

Finally, it is clear that the symmetry group of general relativity is just $\mathcal{M}$ itself: the group of all differentiable transformations of our manifold. This is because all objects of our theory are dynamical objects. None of the geometrical structure of general relativistic space-time is fixed independently of its material content. Rather, all geometrical structure depends on the distribution of stress-energy $\mathbf{T}$.

2. *Comparison with Newtonian Theory*

In §III.4 we formulated Newtonian gravitation theory using the spatial metric h and the temporal co vector field dt. Comparison with general relativity is facilitated by the introduction of the singular temporal metric $\mathbf{t} = dt \otimes dt$. This object is clearly a symmetric tensor field of type (0,2). It assigns a nonzero length to every timelike vector: $|X|^2 = \mathbf{t}(X,X) = dt(X) \otimes dt(X)$; allows us to define a temporal length for every timelike curve; and induces a nondegenerate temporal interval between any two points not lying on the same $t = $ constant hypersurface. The components of $\mathbf{t}$ are given by

$$t_{jk} = t_j t_k = \frac{\partial t}{\partial x_j} \frac{\partial t}{\partial x_k}. \tag{13}$$

If we now introduce the momentum tensor $P = \rho(\mathbf{u} \otimes \mathbf{u})$ with components

$$P^{jk} = \rho u^j u^k = \rho \frac{dx_j}{dt} \frac{dx_k}{dt} \tag{14}$$

we can rewrite the Newtonian field equation (III.62a) as follows. Equations (13) and (14) imply that the "lowering" of P^{jk} by t_{jk} is given by

$$P_{jk} = t_{ja} t_{kb} P^{ab} = \rho t_j t_k. \tag{15}$$

So (III.62a) can be written as

$$R_{jk} = -4\pi k P_{jk} \tag{16}$$

and the momentum tensor P can be seen as the source of the gravitational field as promised.

We can make (16) look even more like (5) if we notice that $t_{ab}P^{ab} = \rho$, so that $t_{ja}t_{kb}P^{ab} = t_{jk}t_{ab}P^{ab} = \rho t_{j}t_{k}$. It follows that (16) is equivalent to

$$R_{jk} = -8\pi k(t_{ja}t_{kb}P^{ab} - \tfrac{1}{2}t_{jk}t_{ab}P^{ab}). \tag{17}$$

On the other hand, writing (5) in explicit form, we get

$$R_{jk} = -8\pi k(g_{ja}g_{kb}T^{ab} - \tfrac{1}{2}g_{jk}g_{ab}T^{ab}). \tag{18}$$

Equation (18) differs from (17) only in that the space-time metric g_{jk} has replaced the temporal metric t_{jk}, and the total stress-energy tensor T^{ab} has replaced the momentum tensor P^{ab}. In an important sense, then, general relativity is just Newtonian gravitation theory plus special relativity.

Furthermore, we can introduce the classical stress-energy tensor $T = P + \theta$, where $-\text{div}(\theta) = F$ and F is a spacelike vector field representing the total nongravitational force. The Newtonian field equation (17) does not imply a conservation law for the stress-energy tensor, but we can certainly postulate such a law independently. That is, we can require $\text{div}(T) = 0$, or, in terms of components,[5]

$$T^{ij}_{\;;j} = \rho u^{i}u^{j}_{\;;j} + \theta^{ij}_{\;;j} = 0. \tag{19}$$

This conservation law allows us to derive the Newtonian equations of motion (III.63a) in a manner precisely analogous to the derivation of the general relativistic equations of motion from the general relativistic conservation law.

Looking at the two theories in this way gives us a much clearer view of the differences between the Newtonian and general relativistic theories of gravitation. First, Newtonian theory is formulated in the context of a Newtonian space-time manifold: at each point p

---

[5] The meaning of (19) becomes clearer if we switch to the formulation of gravitation theory of §III.3 and incorporate the gravitational energy into θ. In such a formulation we have a well-defined notion of inertial coordinate system, and in inertial coordinate systems $(u^{i}) = (1,\vec{v})$ and $(-\theta^{ij}_{\;;j}) = (F^{i}) = (0,\vec{F})$, where F is the total (gravitational plus nongravitational) force. Substituting $i = 0$ in (19) yields

$$\frac{\partial\rho}{\partial t} + \text{div}(\rho\vec{v}) = 0.$$

This is the classical mass conservation law. Substituting $i = 1,2,3$ in (19) results in

$$F^{i} = \frac{\partial(\rho v^{i})}{\partial t} + \frac{\partial(\rho v^{i}v^{1})}{\partial x_{1}} + \frac{\partial(\rho v^{i}v^{2})}{\partial x_{2}} + \frac{\partial(\rho v^{i}v^{3})}{\partial x_{3}}$$

This is the law of motion for classical continuum mechanics. See, e.g., Bergmann [4], 122–123.

the tangent space T_p possesses the characteristic Newtonian structure induced by the temporal metric **t** and spatial metric *h*. General relativity is formulated in the context of a relativistic space-time manifold: at each point *p* the tangent space T_p possesses the characteristic Minkowskian structure induced by the Lorentzian spatio-temporal metric *g*. Moreover, since Newtonian space-time is not semi-Riemannian, **t** and *h* remain absolute, and spatial Euclideanism is retained. In general relativistic space-time *g* is not absolute, and spatial Euclideanism fails. Second, in the Newtonian case the mass density ρ or the momentum tensor *P* is the source of the gravitational field. Due to the relativistic relation between mass and energy, we take the total stress-energy tensor **T** as the source of the gravitational field in general relativity. Finally, as a consequence, our general relativistic field equations imply the conservation law div(**T**) = 0 for the total stress-energy, whereas our Newtonian field equations do not imply the corresponding conservation law. However, such a law can be postulated independently in the Newtonian case, and the equations of motion can be derived just as they are in general relativity.

3. Comparison with Special Relativity

We have already seen that general relativity and special relativity have much in common. They agree that space-time has a Lorentzian metric tensor *g*, that the affine connection *D* is semi-Riemannian (IV.2;V.2); and that the trajectories of "free" particles are geodesics of *D* (IV.3;V.3). They even agree about the laws of electrodynamics: the general relativistic theory of electrodynamics is obtained by adding the laws (IV.38)–(IV.40) to (V.1)–(V.3).[6] Where the two theories differ, of course, is over the *curvature* of space-time. Special relativity results from adding the condition of flatness (IV.1) to (IV.2), (IV.3), (IV.38)–(IV.40). General relativity results from adding Einstein's field equation (V.1) to these same laws.

Hence, at each point *p* in a general relativistic space-time the tangent space T_p looks just like Minkowski space-time. There is a distinguished light cone of null vectors *X* such that $g_p(X,X) = 0$, and there are always "orthonormal" bases $\langle X_i \rangle$ such that $(g_p(X_i,X_j)) = \operatorname{diag}(1,-1,-1,-1)$. Moreover, any two such bases are related by a rotation of the spacelike subbasis or a Lorentz

---

[6] Of course, we also have to incorporate the electromagnetic stress-energy into **T**. See Misner, Thorne, and Wheeler [74], 141.

transformation. For example, if $\langle X_i \rangle$ is "orthonormal," then so is $\langle Y_j \rangle$, where

$$Y_0 = \frac{X_0 - vX_1}{\sqrt{1 - v^2}}$$

$$Y_1 = \frac{X_1 - vX_0}{\sqrt{1 - v^2}}$$

$$Y_2 = X_2$$
$$Y_3 = X_3$$

for any scalar v such that $-1 < v < 1$. In Minkowski space-time, however, the tangent spaces at nearby points all look the same. All light cones have the same "width" and are "tilted" at

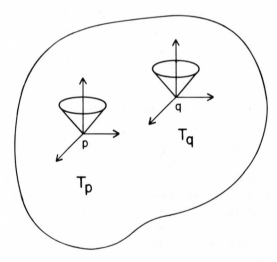

Tangent spaces in Minkowski space-time.

the same "angle." In other words, since Minkowski space-time is flat, it is itself a vector space within which the tangent spaces at different points fit together in a characteristically Euclidean manner.

On the other hand, a general relativistic space-time has *variable* curvature; hence, the tangent spaces at nearby points do not look the same. In particular, the light cone "tilts," "expands," and "contracts" as we move from one tangent space to another, and this feature of general relativistic space-times is responsible for some of their most interesting and dramatic properties. For example, a

186

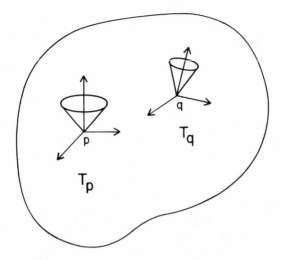

Tangent spaces in general relativistic space-time.

black hole in general relativistic space-time is just a finite region within which the light cones are "tilted" so far that no lightlike (and *a fortiori* no timelike) trajectory can escape:[7]

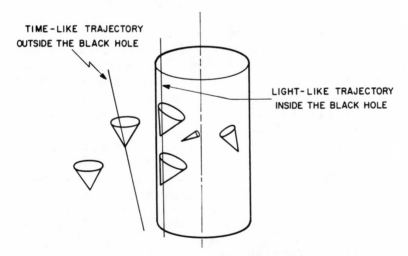

Black hole in general relativistic space-time. No lightlike or timelike curve can escape the cylindrical region.

[7] Here I follow the beautifully intuitive treatment of Geroch [41], 190.

Thus, although general relativity and special relativity agree about the structure of T_p at each point p, they disagree about how the tangent spaces at different points fit together.

More generally, we can distinguish two levels of structure in any affine manifold: *first-order structure*, which determines the properties of the tangent space at each point p; and *second-order structure*, which determines how the tangent spaces at different points are related to one another. First-order structure is given by various vector and tensor fields on M that pick out distinguished vectors, subspaces, and so on in each tangent space T_p. Second-order structure is given by a derivative operator on M that picks out *parallel* vectors in nearby tangent spaces. We can rephrase what we have just said as follows: if M is a general relativistic manifold, then it has the same first-order structure as Minkowski space-time but not (in general) the same second-order structure.

How are first-order structure and second-order structure related? Think of what we can do with first-order structure alone. At each point p we can consider arbitrary tensors in $T^{(r,s)}(p)$—the set of all tensors of type (r,s) defined on T_p (see §A.3). We can perform the operations of addition, scalar multiplication, contraction, tensor product, and so forth on elements of $T^{(r,s)}(p)$. Moreover, if f is an arbitrary scalar field on a neighborhood of p, we can take the directional derivative $X_p f$ for any vector X_p in T_p. What we cannot do in general is take directional derivatives of tensor fields of higher types, for this operation requires the derivative operator (see (A.20)). However, if our manifold happens to be semi-Riemannian, we know that the derivative operator D is definable from the metric g. In such a manifold the derivative operator is determined by the first-order structure: we can construct a general directional derivative for T_p using first-order structure alone.

Let us look more closely at how this is done. On any manifold we have a derivative $D_{X_p}^p$ in T_p for scalar fields f around p: $D_{X_p}^p(f) = X_p f$. Our problem is to extend $D_{X_p}^p$ to tensor fields of arbitrary type: given any tensor field Θ of type (r,s) around p, we want a tensor $D_{X_p}^p(\Theta)$ in $T^{(r,s)}(p)$. If $D_{X_p}^p(\Theta)$ exists, it is completely determined by its components with respect to any basis for T_p. Let us suppose that our manifold is semi-Riemannian, and let $\langle X_i \rangle$ be an "orthonormal" basis for T_p: $(g_p(X_i, X_j)) = \mathrm{diag}(\pm 1, \pm 1, \pm 1, \pm 1)$. Choose a coordinate system $\langle x_i \rangle$ around p such that (i) $\partial/\partial x_i = X_i$ at p and (ii) $g_{ij,k} = 0$ at p. Define the components of $D_{X_p}^p(\Theta)$ with respect to

188

$\langle X_i \rangle$ to be the numbers

$$X_p(\Theta_1), X_p(\Theta_2), \ldots, X_p(\Theta_r)$$

where the Θ_k are the components of Θ in the coordinate system $\langle x_i \rangle$. Of course, this definition makes sense only if the numbers $X_p(\Theta_k)$ are independent of the particular coordinate system $\langle x_i \rangle$. That is, if $\langle y_j \rangle$ is a second coordinate system around p satisfying (i) and (ii), we must have

$$X_p(\bar{\Theta}_k) = X_p(\Theta_k)$$

where the $\bar{\Theta}_k$ are the components of Θ in $\langle y_j \rangle$. It is easy to see that this condition will be met just in case $\langle x_i \rangle$ and $\langle y_j \rangle$ are *linearly related* at p, that is, just in case

$$\left. \frac{\partial^2 y_i}{\partial x_j \, \partial x_k} \right|_p = 0. \tag{20}$$

But if g is a semi-Riemannian metric tensor, (20) follows from (i) and (ii). Hence, on a semi-Riemannian manifold $D^p_{X_p}$ is a well-defined map from tensor fields of type (r,s) around p into $T^{(r,s)}(p)$; and, of course, $D^p_{X_p}(\Theta) = (D_X\Theta)_p$, where D is the unique connection compatible with g and X is any vector field such that $X(p) = X_p$.

It is instructive to see how the above construction breaks down in non-semi-Riemannian manifolds. Consider, for example, a Newtonian space-time $\langle M,D,dt,h \rangle$. We can construct "orthonormal" bases $\langle X_i \rangle$ for T_p such that $(dt_p(X_i)) = (1,0,0,0)$ and $(h_p(w_i,w_j)) = \mathrm{diag}(0,1,1,1)$, where $w_i(X_j) = \delta_{ij}$; and we can, of course, define $D^p_{X_p}(f)$ for scalar fields f around p as $X_p f$. But the conditions (i) $\partial/\partial x_i(p) = X_i$, (ii) $t_{i,j}|_p = 0$, (iii) $h^{ij}{}_{,k}|_p = 0$ do *not* imply that the linear relationship (20) holds between two such coordinate systems $\langle x_i \rangle$ and $\langle y_j \rangle$ (intuitively, the two systems are merely rigid Euclidean at p and so can have arbitrary mutual acceleration and rotation). Hence, the numbers $X_p(\Theta_k)$ depend on the particular coordinate system $\langle x_i \rangle$, and $D^p_{X_p}$ does not extend to a well-defined map from tensor fields of type (r,s) around p into $T^{(r,s)}(p)$. This is why first-order structure does not determine second-order structure in general.

We can now describe the relationship between special relativity and general relativity more exactly. Let T_p be the tangent space at a point p in an arbitrary manifold. Let us say that an equation is first-order if it is built up (i) from the operations of addition, contraction, tensor product, and so forth on $T^{(r,s)}(p)$ and (ii) from

189

the application of the derivative operator $D^p_{X_p}$ to tensor fields around p—*insofar as this operation is well-defined.* Thus, there are two ways in which an equation may fail to be first-order: (a) like the geodesic equation $D_{T_\sigma}T_\sigma = 0$ in a Newtonian manifold, it may involve the application of a derivative operator to tensor fields of type higher than (0,0), where $D^p_{X_p}$ does *not* extend to such fields; (b) like the equations governing curvature in any manifold, it may involve *iterations* of a derivative operator and hence $D^q_{x_q}$ for q in a neighborhood of p (see (A.17)). In a semi-Riemannian manifold $D^p_{X_p}$ is always well-defined, so the only second-order equations are those involving iterations of the derivative operator. But both special relativity and general relativity are semi-Riemannian theories, so equations (IV.3), (IV.38)–(IV.40), which are common to the two theories, are all first-order. For example, the geodesic equation (IV.3; V.3) says that if T_σ is the tangent vector field of any "free" trajectory σ through p, then

$$D^p_{T_\sigma(p)}(T_\sigma) = 0.$$

Equation (IV.39) says that if F is the electromagnetic field around p, then

$$D^p_{X_p}(F)_{1a_3} = 0$$

for all X_p in T_p. On the other hand, the equations governing curvature (IV.1; V.1) essentially involve $D^q_{X_q}$ for q in a neighborhood of p; they, unlike the preceding, are clearly second-order. This is the precise sense in which special relativity and general relativity agree about first-order structure but disagree about second-order structure.

That special relativity and general relativity have the same first-order structure (the division of T_p at every point p by a light cone) but very different second-order structures (flat vs. nonflat) has interesting philosophical consequences. Recall that special relativity is especially well suited for Robb-type "causal" theories of spatiotemporal structure, because all the basic relations of the theory—simultaneity (relative to an inertial frame), spatial congruence (relative to an inertial frame), temporal congruence (relative to an inertial frame)—are definable from the relation $p\tau q$ of "causal" connectibility (§IV.6). This is because in Minkowski space-time the transformations h that will preserve τ (or λ) satisfy

$$hg = kg$$

where k is a positive constant, and therefore preserve all spatio-temporal relations. However, this result, in turn, holds only because Minkowski space-time is *flat*: if $\langle M,g \rangle$ is an *arbitrary* Lorentzian manifold and h preserves τ (or λ), then we can conclude only that

$$hg = \Omega g$$

where Ω is a positive *function* (§IV.2). Hence, in a general relativistic manifold the transformations that will preserve τ (or λ) do not automatically preserve the spatial or temporal congruence relations (relativized to a local inertial frame), so these relations are not in general definable from τ (or λ). General relativity, unlike special relativity, does not admit a Robb-type "causal" reduction.

4. The Principle of Equivalence

A proper understanding of the principle of equivalence is one of the keys to a proper understanding of general relativity, for it plays a basic, though usually obscure, role in almost all expositions of the theory, from Einstein's fundamental 1916 paper [28] to Weinberg's sophisticated contemporary text [115]. We know that in Einstein's own thought the principle of equivalence functioned as an argument for the general principle of relativity: he wanted to extend the special relativity principle that holds for uniform motions in classical physics and in special relativity to cover nonuniform, accelerated motions as well. We also know that this argument is invalid: a dynamical space-time theory of gravitation (which does result from the principle of equivalence) does not automatically relativize acceleration in the way that special relativity (like the Newtonian theory of §III.2) relativizes velocity. Let us begin by looking more closely at why Einstein's argument fails.

In the second section of Einstein's 1916 paper, entitled "The Need for an Extension of the Postulate of Relativity," the principle of equivalence is introduced in the following passage:

there is a well-known physical fact which favors an extension of the theory of relativity. Let K be a[n inertial] system of reference, i.e., a system relatively to which (at least in the four dimensional region under consideration) a mass, sufficiently distant from other masses, is moving with uniform motion in a straight line. Let K' be a second system of reference which is moving relatively to K in *uniformly accelerated* translation. Then, relatively to K', a mass sufficiently distant from other masses would have an accelerated

191

motion such that its acceleration and direction of acceleration are independent of the material composition and physical state of the mass.

Does this permit an observer at rest relatively to K' to infer that he is on a "really" accelerated system of reference? The answer is in the negative; for the above-mentioned relation of freely movable masses to K' may be interpreted equally well in the following way. The system of reference K' is unaccelerated, but the space-time region in question is under the sway of a gravitational field, which generates the accelerated motion of the bodies relatively to K'. ([28], 114)

Here, of course, Einstein is referring to the fact that, in the context of conventional Newtonian gravitation theory (§III.3), an acceleration $\ddot{b}_i$ can be incorporated into the gravitational potential Ψ as in §III.4.[8]

Einstein's point is perfectly correct. In conventional Newtonian gravitation theory (without global boundary conditions) it is impossible to distinguish an inertial reference frame K from an arbitrary Galilean reference frame K'. The theory of §III.3 satisfies the following version of the principle of equivalence.

(E) All Galilean reference frames are physically equivalent or physically indistinguishable,

where the notion of physical equivalence is understood as in §IV.5. But (E) bears a close resemblance to the first conjunct of the special principle of relativity (R1) (§IV.5). Since the Galilean frames are just the arbitrarily accelerated frames, it is natural to suppose that (E) in conjunction with (R2) (§IV.5) does in fact lead to something like a general principle of relativity, that ⌜(E) & (R2)⌝ does for acceleration what ⌜(R1) & (R2)⌝ does for velocity.

What is wrong with this line of thought? The first thing to notice is that ⌜(E) & (R2)⌝ cannot yield a truly *general* relativity principle, for, although the class of Galilean frames contains all arbitrarily accelerating frames, it does not contain arbitrarily *rotating* frames. Including the rotating frames widens the class under consideration to the rigid Euclidean frames, and within this wider class of frames no principle like (E) holds. Rotating reference frames are physically

---

[8] Although Einstein mentions only uniform acceleration in this passage, non-uniform acceleration can obviously be accommodated as well, by using time-dependent gravitational fields.

distinguishable from nonrotating frames: for example, rotation induces Coriolis forces, which, unlike gravitational forces or inertial forces, depend on the velocity of the particle acted upon. Second, even if we forget about rotation, ⌜(E) & (R2)⌝ still does not amount to a generalized relativity principle: it does not even give us a relativization of *acceleration*.

Consider once again how (R1) and (R2) lead us from the kinematics of §III.1 to the kinematics of §III.2. Applying (R1) to the theory of §III.1, we observe that its indistinguishability group is the Galilean group; its symmetry group, however, is the smaller group $O^3 \times T$. This theory is therefore not well-behaved, and (R2) tells us to look for a well-behaved theory with which to replace it. The theory of §III.2 fits the bill perfectly: its symmetry group = its indistinguishability group = the Galilean group. So the frames that are merely indistinguishable in §III.1 are actually theoretically identical in this new theory. Hence, the theory of §III.2 does implement a thoroughgoing relativity principle for uniform motions: all uniform motions are both physically indistinguishable and theoretically identical (that is, they have all their theoretical properties in common).

Now consider the transition from the gravitation theory of §III.3 to the gravitation theory of §III.4. Applying (E) to the theory of §III.3, we observe that its indistinguishability group is the group of all "transformations to accelerated frames" (III.66); its symmetry group, however, is the smaller Galilean group. This theory is therefore not well-behaved, and (R2) tells us to look for a well-behaved theory with which to replace it. Moreover, the theory of §III.4 is well-behaved. But here the analogy between the two cases breaks down. For the symmetry group of our new theory is not the indistinguishability group of our old theory; rather, it is the much larger group of "transformations to arbitrary rigid Euclidean frames" (III.67). What about the indistinguishability group of our new theory? This group is *smaller* than the indistinguishability group of our old theory! To see this, let F be a freely falling (local inertial) frame, and let f map F to an arbitrarily accelerating frame via (III.66). The indistinguishability group of the theory of §III.3 contains f, because if $\langle M, \overset{\circ}{D}, dt, h, \Phi, \rho, T_{\hat{a}} \rangle$ is a model, then so is $\langle M, f\overset{\circ}{D}, dt, h, \Psi, \rho, T_{\hat{a}} \rangle$, where Ψ satisfies (III.47). On the other hand, f is *not* in the indistinguishability group of the theory of §III.4: if $\langle M, D, dt, h, \rho, T_{\hat{a}} \rangle$ is a model, then $\langle M, fD, dt, h, \rho, T_{\hat{a}} \rangle$ is not a model, because if σ is a geodesic of D, then σ is not a geodesic of fD. In the theory of §III.3

193

variations in $\overset{\circ}{D}$ are "counterbalanced" by variations in Φ; in the theory of §III.4, where we have dropped Φ, variations in D "stick." This is why such variations are detectable in §III.4 but not in §III.3.

In the theory of §III.4, therefore, (E) is actually *false*. A privileged subclass of Galilean frames, the class of freely falling or local inertial frames, is distinguishable from the rest.[9] Furthermore, each state of absolute acceleration relative to D is distinguishable from all other states of absolute acceleration relative to D. In other words, we do not obtain a well-behaved theory by widening the symmetry group to coincide with the indistinguishability group induced by (E); rather, we narrow the indistinguishability group (III.66) to include only the transformations from one local inertial frame to another. Instead of widening the class of theoretically identical frames, we narrow the class of theoretically indistinguishable frames. In addition, since our new derivative operator D is not absolute, the symmetry group of our new theory has no particular connection with the equivalence of frames. The indistinguishability group of §III.4 $\neq$ its symmetry group; rather, the former is a proper subgroup of the latter.

The difference between our two theoretical transitions can be explained as follows. In the theory of §III.1 we have an absolute object, the connection $\overset{\circ}{D}$, that is definable from three other absolute objects: dt, h, and V. When we move to §III.2, we drop one of these defining objects, the absolute space V, and take the absolute $\overset{\circ}{D}$ as primitive. Since $\overset{\circ}{D}$ is absolute and the theory of §III.2 is well-behaved, we have the standard "nice" connections between indistinguishability, symmetry, and relativity: $\ulcorner$(R1) & (R2)$\urcorner$ yields a relativity principle. On the other hand, in the theory of §III.3 we have a dynamical object, the nonflat connection D, that is definable from two other objects: the absolute connection $\overset{\circ}{D}$ and the dynamical

---

[9] To appreciate fully the difference between the two theories on this score, the reader should be careful about the technical sense of "indistinguishability." Even in the context of §III.3, one can of course pick out the freely falling frames from the wider class of Galilean frames, that is, as the frames whose origins satisfy the law of motion. What one cannot do is distinguish the *state of motion* (absolute acceleration) of such a frame from that of an arbitrary Galilean frame. In the theory of §III.4, on the other hand, one can do precisely this, for one defines absolute acceleration relative to the nonflat connection D instead of the flat connection $\overset{\circ}{D}$. Thus, trajectories satisfying the law of motion of §III.4 have a definite, and determinable, absolute acceleration, namely, $a^i = 0$. Trajectories satisfying the law of motion of §III.3 have a definite absolute acceleration, too, but we cannot determine whether $a^i = -\Phi_{,i}$ (relative to $\overset{\circ}{D}$) or $a^i = -\Psi_{,i}$ (relative to f$\overset{\circ}{D}$).

potential Φ. When we move to §III.4, we drop both of these defining objects and take the *nonabsolute D* as primitive. So even though the theory of §III.4 is well-behaved, we do not have the standard "nice" connections between indistinguishability, symmetry, and relativity: $\ulcorner$(E) & (R2)$\urcorner$ does not yield a relativity principle.

So far we have compared standard, flat space-time Newtonian gravitation theory (§III.3) with the revised, nonflat Newtonian theory of §III.4 Everything we have said remains true if we compare standard Newtonian theory with general relativity. Indeed, as far as the present issue is concerned, general relativity is structurally identical to the theory of §III.4 (see §V.2). That is, in general relativity, just as in §III.4, we define absolute acceleration relative to the new, nonflat derivative operator D, and each state of absolute acceleration relative to D is distinguishable from all others. In particular, a local inertial frame F is distinguishable from an arbitrarily accelerating frame hF. The indistinguishability group of general relativity is the same as the indistinguishability group of §III.4; both indistinguishability groups are narrower than the indistinguishability group of §III.3. The difference between general relativity and the theory of §III.4, of course, is that in general relativity we have widened the symmetry group even further. Since the connection D is defined on a Lorentzian manifold, we are left with *no* absolute objects. In neither theory do we have the standard (flat space-time) connections between indistinguishability, symmetry, and relativity.

In general relativity, then, (E) is false as well. Arbitrarily accelerating frames are not physically equivalent or indistinguishable; only freely falling or gravitationally "accelerating" (that is, *non*accelerating relative to D) frames are so equivalent. Hence, if we interpret the principle of equivalence as (E), this principle is not actually satisfied by our theory. Rather, it functions as a methodological guide, in conjunction with (R2), in moving from standard Newtonian gravitation theory (the only theory in which (E) *is* true) to our new theory. Are there versions of the principle of equivalence that are satisfied by general relativity? I shall consider three such versions, paying particular attention to the uses to which they are put and the extent to which they distinguish general relativity from Newtonian gravitation theory.

(1) The Equivalence of Gravitational and Inertial Mass.

Typical field theories of interaction describe the interaction between a source and a particle in two steps. The field variables of

195

the theory are related to the source variables in a field equation; then the field variables are related to the possible trajectories of our particle by equations of motion. These equations of motion have the general form

$$m_0\left(\frac{d^2x_i}{du^2} + \Gamma^i_{jk}\frac{dx_j}{du}\frac{dx_k}{du}\right) = [\text{coupling factor}] \times [\text{field variables}]$$

where the coupling factor represents some physical property of the particle in question. For example, in electrodynamics the coupling factor is the charge q of the particle, and the relativistic equation of motion is

$$m_0\left(\frac{d^2x_i}{d\tau^2} + \Gamma^i_{jk}\frac{dx_j}{d\tau}\frac{dx_k}{d\tau}\right) = qF^i_j\frac{dx_j}{d\tau}.$$

If Newtonian gravitation theory is formulated as this type of field theory, as it is in §III.3, we find that our field variable is the gravitational potential Φ, and our equation of motion is

$$m_0\left(\frac{d^2x_i}{dt^2} + \overset{\circ}{\Gamma}{}^i_{jk}\frac{dx_j}{dt}\frac{dx_k}{dt}\right) = -m_0 h^{ir}\Phi_{;r}.$$

Here we have an interesting phenomenon: the coupling factor in this case is just the mass m_0 of the particle, the very same mass that appears on the left-hand side of our equation of motion as well. Thus, all bodies move the same way in a gravitational field regardless of their different masses. An analogous result does not hold for other types of interaction. The motions of charged bodies in an electromagnetic field, for example, depend very definitely on their individual charges.

Actually, mass plays three conceptually distinct roles in classical physics. First, it figures on the left-hand side of all our equations of motion: this is the so-called *inertial* mass m_i. Second, it figures as the coupling factor in our equations of motion for gravitational interaction: let us call this m_c. Third, it serves as the source for the gravitational field in Poisson's equation

$$h^{ij}\Phi_{;i;j} = 4\pi k\rho.$$

Let us call this m_s. It so happens that these three conceptually distinct masses are all equal: $m_i = m_c = m_s$. The principle of the equivalence of inertial and gravitational mass need only apply to the first two, however. That is, $m_i = m_c$ is sufficient for the result that all bodies follow the same trajectories in a gravitational field (of course, the strict *proportionality* of m_i and m_c suffices for this result).

The fact that $m_i = m_c$ enables us to "geometrize away" gravitational forces by incorporating the gravitational field variables (that is, the gravitational potential Φ) into the affine connection. It allows us to define a new connection

$$\Gamma^i_{jk} = \overset{\circ}{\Gamma}{}^i_{jk} + h^{ir}\Phi_{;r}t_jt_k$$

and to rewrite the equations of motion for the gravitational interaction as

$$\frac{d^2x_i}{dt^2} + \Gamma^i_{jk}\frac{dx_j}{dt}\frac{dx_k}{dt} = 0.$$

The equality of inertial and gravitational mass allows us to reconstrue the trajectories of gravitationally affected particles as geodesics of a nonflat connection. It implies the existence of a connection D such that freely falling bodies follow geodesics of D. This is not the case for other types of interaction, where the ratio of m_i to the coupling factor is not the same for all bodies. Thus, the trajectories of charged particles in an electromagnetic field cannot be construed as the geodesics of any *single* connection, for the ratio m_0/q varies from particle to particle.

The principle of the equivalence (strict proportionality) of inertial and gravitational mass must therefore be true if any theory of gravitation like general relativity, in which gravitational interaction is explained by the dependence of a nonflat connection on the distribution of matter, is to be possible. But, of course, general relativity is not the only theory of this type. Classical gravitation theory can also be formulated in this way by taking advantage of the very same equivalence $m_i = m_c$. However, the mere use of a nonflat connection D is essentially trivial if, as in the theory of §III.3, it exists alongside of a flat connection $\overset{\circ}{D}$. If the principle of equivalence is to reflect a "deep" fact, it must imply that the nonflat connection D is *the* connection of space-time. The nonflat connection D, whose existence is guaranteed by the equivalence $m_i = m_c$, must also be unique.

(2) The Uniqueness of the Gravitational Connection.

Our second version of the principle of equivalence asserts that the nonflat connection D determined by the freely falling trajectories is unique. In particular, there is no distinguishable flat connection $\overset{\circ}{D}$, and there is no distinguishable class of inertial frames following the geodesics of $\overset{\circ}{D}$. There are only the local inertial frames following the geodesics of D. Of course, we already know that there is no *distinguishable* flat connection $\overset{\circ}{D}$ in the gravitation theory of §III.3: this is precisely the import of (E). But our second version of the principle of equivalence goes beyond this fact in two ways. First, it says that the flat connection $\overset{\circ}{D}$ is indistinguishable or indeterminate in all theories: not only does gravitational interaction fail to distinguish $\overset{\circ}{D}$, but all other forms of interaction fail to do so as well. So, for example, electrodynamics must also be formulated in terms of the gravitational connection D, and, consequently, light rays must follow geodesics of D, not $\overset{\circ}{D}$.[10] Second, our principle says that if $\overset{\circ}{D}$ is indeterminate or indistinguishable (in all theories), then $\overset{\circ}{D}$ should be dropped: there exists only the nonflat connection D. Hence, both general relativity and the Newtonian theory of §III.4 satisfy this version of the principle of equivalence; the theory of §III.3 does not.

Our second version is therefore rich in consequences. It puts constraints on the laws of all possible new theories—no such laws should enable us to distinguish a flat connection $\overset{\circ}{D}$—and, using the "parsimonious" methodological principle (R2), it tells us that we can in fact dispense with $\overset{\circ}{D}$ in all possible new theories. Moreover, this version of the principle allows us to make sense of Einstein's often repeated claim that general relativity *explains* the equivalence of inertial and gravitational mass in a way that Newtonian theory does not.[11] For the principle tells us that the fact $m_i = m_c$ is necessarily integrated with other important physical facts. The connection D, whose existence is guaranteed by the equivalence of inertial and gravitational mass, necessarily plays a role in the explanation of other phenomena (for example, electrodynamic phenomena). In the theory of §III.3, on the other hand, D is a mere curiosity; in this sense $m_i = m_c$ is purely accidental.[12]

What exactly does this version of the principle of equivalence have to do with *equivalence*? Viewed as a constraint on *laws*, our principle tells us that the laws of any theory containing $\overset{\circ}{D}$ fail to

[10] Hence, light follows curved paths in a gravitational field—one of the main results that the principle of equivalence is supposed to produce. See Einstein [27].

[11] See, e.g., Einstein [27], 100–101.

[12] I shall have more to say about the relationship between theoretical *explanation* and this kind of *integration* or *unification* in Chapters VI and VII.

198

distinguish $\overset{\circ}{D}$. Hence, inertial frames of $\overset{\circ}{D}$ are equivalent (in the sense of §IV.5) to arbitrarily accelerating frames of $\overset{\circ}{D}$. However, viewed as a constraint on *objects*, our principle tells us to drop such an indistinguishable $\overset{\circ}{D}$. After this is done, we no longer have inertial frames, and we can no longer assert their equivalence with any other frames. Instead, we *replace* the inertial frames of $\overset{\circ}{D}$ with the local inertial frames of D, which are all equivalent to one another but not to arbitrarily accelerating frames of D. This is why the principle of equivalence does not eliminate privileged reference frames and privileged states of motion. Rather, the difference between theories that obey the principle of equivalence (general relativity and the theory of §III.4) and theories that do not (special relativity and the theory of §III.3) is only this: in the former theories the privileged reference frames (local inertial frames) follow geodesics of a nonflat connection instead of geodesics of a flat connection. Consequently, the law of motion characterizing such a privileged reference frame takes the simple form (12) only along the trajectory defined by the origin of the frame.

(3) The "Local" Equivalence of Freely Falling Frames and Special Relativistic Inertial Frames.

Freely falling, nonrotating, normal reference frames (that is, local inertial frames) look just like special relativistic inertial frames along the trajectory σ to which they are adapted. In such a frame $(g_{ij}) = \text{diag}(1, -1, -1, -1)$ along σ, $\Gamma^i_{jk} = 0$ along σ, and, therefore, the law of motion takes the simple form (12) along σ. Moreover, the laws of electrodynamics take the form

$$F^{ij}{}_{,j} = -4\pi J^i \tag{21}$$

$$F_{[ij,k]} = 0 \tag{22}$$

and

$$m_0 \frac{d^2 y_i}{d\tau^2} = qF^i_j \frac{dy_j}{d\tau} \tag{23}$$

along σ. This equivalence along σ between general relativistic local inertial frames and special relativistic inertial frames follows from our second version of the principle of equivalence (applied to a special relativistic manifold). Freely falling frames follow geodesics of the unique connection D, the very same connection that figures in the laws of motion and the electrodynamic field equations. The

199

vanishing of the components of this connection along σ yields the special relativistic equations for an inertial frame.

It must be emphasized that this equivalence holds only on a single trajectory σ. At finite distances from σ nonvanishing Γ^i_{jk}'s appear, and the special relativistic laws cease to hold. Thus, although physics texts often claim that freely falling frames are "locally" equivalent to inertial frames, this assertion is strictly false if "local" has its usual mathematical meaning: local $=$ *on some neighborhood*. If there is a nonvanishing gravitational field at $p \in M$, then $K \neq 0$ at p, and there is *no* neighborhood of p in which $\Gamma^i_{jk} = 0$. Freely falling frames are only "infinitesimally" equivalent to inertial frames: only at a single point or on a single trajectory. (Of course, freely falling frames *approximate* inertial frames on a neighborhood of σ: the smaller the neighborhood, the better the approximation.)

For the same reason, gravitational forces cannot be equated with "apparent" or "fictitious" inertial forces. Inertial forces arise from adopting noninertial (or non-local-inertial) accelerating frames of reference. They can therefore be "transformed away" by a suitable coordinate transformation, because inertial effects reflect no change in the affine connection itself but only a change in its *components* brought about by adopting a noninertial (or non-local-inertial) coordinate system. The presence of a gravitational field, on the other hand, is evidenced by the nonvanishing of the curvature tensor, which implies that the connection is nonflat and therefore cannot be "transformed away" by any coordinate transformation (except at a single point). But if there is merely an "inertial" field at p, then $K = 0$, and there is a finite neighborhood of p in which $\Gamma^i_{jk} = 0$: an inertial coordinate system exists around p. In other words, gravitational forces, unlike inertial forces, are intrinsic to the space-time manifold. They do not depend on the choice of a particular reference frame.

Several authors try to use our third version of the principle of equivalence to derive the laws of general relativity from the laws of special relativity.[13] The idea is this. The principle of equivalence guarantees that at every point $p \in M$ there is a coordinate system in which the laws that hold in special relativistic inertial systems are valid at p. Call this coordinate system $SR(p)$. We can find the laws that hold in any given coordinate system at any point p by transforming the special relativistic laws from $SR(p)$ to our given

[13] See Misner, Thorne, and Wheeler [74], chap. 16; and especially Weinberg [115], chap. 3.

coordinate system. For example, we know that the law of motion in $SR(p)$ at p is

$$\left.\frac{d^2 y_i^p}{d\tau^2}\right|_p = 0$$

(the superscript indicates the dependence of y_i on p). Transforming to an arbitrary system $\langle x_i \rangle$ results in

$$\frac{d^2 x_i}{d\tau^2} + \frac{\partial x_i}{\partial y_m^p} \frac{\partial^2 y_m^p}{\partial x_k \partial x_j} \frac{dx_j}{d\tau} \frac{dx_k}{d\tau} = 0.$$

This is just (3a), because

$$\frac{\partial x_i}{\partial y_m^p} \frac{\partial^2 y_m^p}{\partial x_k \partial x_j} = \Gamma^i_{jk}(p) \tag{24}$$

(A.12). Similarly, transforming (21) from $SR(p)$ to $\langle x_i \rangle$ results in

$$F^{ij}_{,j} + \Gamma^i_{mj} F^{mj} + \Gamma^i_{mj} F^{im} = -4\pi J^i.$$

That is,

$$F^{ij}_{;j} = -4\pi J^i$$

(A.22). The connection (24) is guaranteed to be the general relativistic gravitational connection, because the coordinate systems $SR(p)$ in which $\Gamma^i_{jk} = 0$ at p are just those associated with freely falling local inertial frames through p. Thus, (24) enables us to derive the laws that hold locally (on a neighborhood of p) from the laws that hold "infinitesimally" (at p itself).

There is a problem with this procedure, however, because the phrase "the laws of special relativity" is ambiguous. Thus, for example, in special relativity (21) is equivalent to

$$F^{ij}_{,j} = -4\pi J^i + R \tag{25}$$

where R is the curvature scalar, because $R = 0$ in special relativity. But transforming (25) to $\langle x_i \rangle$ results in the incorrect

$$F^{ij}_{;j} = -4\pi J^i + R \tag{26}$$

In general, since the curvature vanishes in special relativity, we can add arbitrary terms involving curvature to our laws. These laws will then give incorrect results when transplanted to general relativity, where the curvature does not vanish.[14] Hence, if we formulate our

[14] This problem is well known of course. For interesting discussions, see Misner, Thorne, and Wheeler [74], 388–391; Weinberg [115], 133.

third version of the principle of equivalence as: at every point p there is a coordinate system in which the laws that hold in special relativistic inertial systems are valid at p—we cannot unambiguously derive the correct general relativistic laws. We need to state our principle more carefully.

What we need is a distinction between "nice" laws like (21), which can be transplanted from special relativity to general relativity, and "bad" laws like (25), which cannot. Such a distinction is already available: it is just the distinction between *first-order* laws and *second-order* laws of §V.3. First-order laws describe only the structure of the tangent space T_p at a single point p: they are built up from tensor operations on $T^{(r,s)}(p)$ and the application of the derivative operator $D^p_{X_p}$ at p. Second-order laws put constraints on how the tangent spaces fit together on a neighborhood of p: they essentially involve $D^q_{X_q}$ for q on a neighborhood of p. In particular, all laws involving curvature at p are second-order. Thus, the laws of special relativity referred to in our third version of the principle of equivalence can only be the *first-order* laws of special relativity. The principle should read: at every point p the first-order laws describing T_p are precisely the first-order laws of special relativity.

Standard formulations of the principle of equivalence characteristically obscure this crucial distinction between first-order laws and second-order laws by blurring the distinction between "infinitesimal" laws, holding at a single point, and local laws, holding on a neighborhood of a point. They lose the distinction between the structure of the tangent space T_p and the configuration of the tangent spaces T_q for q in a neighborhood of p. (This is one place where the physicist's casual attitude toward the "infinitesimal" gets him into real trouble!) What the principle of equivalence says, then, is that special relativity and general relativity have the same "infinitesimal" structure, not that they have the same local structure. This latter claim is of course false, for the local structure of general relativistic space-time is governed by Einstein's field equation (1), whereas the local structure of Minkowski space-time is governed by the condition of flatness (IV.1). Both these equations are second-order. It is therefore a serious mistake to think that one can derive Einstein's field equation itself from the principle of equivalence.[15] The principle of

---

[15] See Weinberg [115], 151–155, for such an attempt. What Weinberg actually produces is a version of the usual heuristic argument to the effect that the field equation is the only equation—with certain additional desirable properties—that reduces to the Newtonian Poisson equation in the nonrelativistic limit.

equivalence cannot serve as a foundation for general relativity on its own.

What makes our third version of the principle of equivalence work is the fact that both special and general relativistic space-times are semi-Riemannian manifolds: the affine connection D is determined by the tangent space structure induced by the metric g. In such a manifold we can extend the derivative operator $D^p_{X_p}$ to tensor fields of arbitrary type, as in §V.3, and so formulate an interesting set of first-order laws. By contrast, this procedure fails in the Newtonian theory of §III.4. To be sure, we still have freely falling, nonrotating, local inertial frames that look just like the inertial frames of §III.2 along the trajectory σ to which they are adapted. In such frames $(t_i) = (1,0,0,0)$ along σ, $(h^{ij}) = \mathrm{diag}(0,1,1,1)$ along σ, $\Gamma^i_{jk} = 0$ along σ; and, consequently, the law of motion takes the simple form

$$\frac{d^2 x_i}{dt^2} = 0$$

along σ. But this fact does not allow us to derive laws of Newtonian gravitation theory like the law of motion (III.63). The reason is that Newtonian space-time is not semi-Riemannian, and so we cannot extend $D^p_{X_p}$ beyond the scalar fields of type (0,0). Hence, none of the interesting laws of Newtonian gravitation theory are first-order. Note, however, that if we add the absolute space vector field V to the theory of §III.4 (for example, by combining classical electrodynamics and classical nonflat gravitation theory), we *can* extend $D^p_{X_p}$ beyond the scalar fields.[16] There is a unique connection D compatible with dt, h, and V. Therefore, if we add the vector field V to the theory of §III.4 we *do* obtain an interesting set of first-order laws. Moreover, we can derive these first-order laws, including the laws of electrodynamics, from the structure of the tangent space at each point p—just as in general relativity.

In sum, there are three distinct versions of the principle of equivalence that are actually satisfied by general relativity, and these three versions can be ordered in terms of increasing strength. The first version says that the trajectories of freely falling particles define an affine connection D. This version is satisfied by all theories of gravitation we have considered: general relativity, the Newtonian theory

---

[16] To see this, observe that an "orthonormal" basis $\langle X_i \rangle$ for T_p now satisfies $(V_p(w_i)) = (1,0,0,0)$, where $w_i(X_j) = \delta_{ij}$; and the condition (iv) $v^i_{,j}|_p = 0$, added to the conditions (i)–(iii) on p. 189, does ensure the linear relationship (20).

of §III.4, and even the Newtonian theory of §III.3. It merely expresses the equality (strict proportionality) of gravitational and inertial mass. The second version says that the connection D determined by freely falling particles is the unique connection of space-time: D governs all physical phenomena, not just gravitation. This version is satisfied by general relativity and the Newtonian theory of §III.4, but not by the theory of §III.3, where D exists alongside of a flat connection $\mathring{D}$. Finally, the third version says that D is uniquely determined by the structure of the tangent space at each point in space-time. This version is satisfied by general relativity, but not by the theory of §III.4. However, the addition of absolute space to §III.4 results in a Newtonian theory in which even the third version of the principle is satisfied.

5. *The General Principle of Relativity*

Einstein's central philosophical motivation in developing the general theory was his desire to implement a fully relational, or relativistic, conception of motion, thereby vindicating Leibniz's relationalist attitude toward space and time over Newton's absolutist attitude. Indeed, this philosophical vision is responsible for both the name of our theory and much of its intellectual notoriety. We now know that general relativity, despite its name, does not in fact fulfill these relationalist ambitions. Motion (and therefore space-time) remains just as absolute in the relevant sense as it does in the Newtonian gravitation theory of §III.4. The space-time structure used to define absolute motion—the derivative operator D—becomes *dynamical*, but it is not thereby eliminated in favor of relations between concrete physical objects and events. Nevertheless, it is well worth our while to get as clear as possible about the reasons that led Einstein to think (and hope) that his theory realized a relationalist conception, for these reasons present a fascinating tangle of physical, mathematical, and philosophical ideas that is perhaps unique in the history of science.

Let us begin by reviewing the situation in which Einstein found himself after the invention of special relativity. Special relativity, like Newtonian kinematics and gravitation theory, but unlike classical electrodynamics, satisfies a relativity principle for inertial motions. All inertial motions are equivalent or indistinguishable, and there is no need to invoke anything more than the perfectly symmetrical concept of relative motion to explain variations within this class. But special relativity, like Newtonian theory, does not satisfy a relativity principle for noninertial motions. Variations in acceleration and rotation produce nonsymmetrical, differential effects that are explained in terms of motion relative to a class of privileged reference frames—frames that are quite unlikely to be "occupied" or "anchored" by concrete physical objects. Hence, special relativity still involves absolute motion in precisely the sense that is most objectionable to the philosophical (Leibnizean) relationalist.

At the beginning of his 1916 paper Einstein formulates this philosophical objection to his own theory of 1905 with beautiful clarity:

In classical mechanics, and no less in the special theory of relativity, there is an inherent epistemological defect which was, perhaps for the first time, clearly pointed out by Ernst Mach. We will elucidate it by the following example:—Two fluid bodies of the same size and nature hover freely in space at so great a distance from each other and from all other masses that only those gravitational forces need be taken into account which arise from the interaction of different parts of the same body. Let the distance between the two bodies be invariable, and in neither of the bodies let there by any relative movements of the parts with respect to one another. But let either mass, as judged by an observer at rest relatively to the other mass, rotate with constant angular velocity about the line joining the masses. This is a verifiable relative motion of the two bodies. Now let us imagine that each of the bodies has been surveyed by means of measuring instruments at rest relatively to itself, and let the surface of S_1 prove to be a sphere, and that of S_2 an ellipsoid of revolution [see the diagram on p. 66 above]. Thereupon we put the question—What is the reason for this difference in the two bodies? No answer can be admitted as epistemologically satisfactory, unless the reason given is an *observable fact of experience*. The law of causality has not the significance of a statement as to the world of experience, except when *observable facts* ultimately appear as causes and effects.

Newtonian mechanics does not give a satisfactory answer to this question. It pronounces as follows:—The laws of mechanics apply to the space R_1, in respect to which the body S_1 is at rest, but not to the space R_2, in respect to which the body S_2 is at rest. But the privileged space R_1 of Galileo [that is, an inertial frame]

205

thus introduced, is a merely *factitious* cause, and not a thing that can be observed. It is therefore clear that Newton's mechanics does not really satisfy the requirement of causality in the case under consideration, but only apparently does so, since it makes the factitious cause R_1 responsible for the observable difference in the bodies S_1 and S_2. ([28], 112-113).

It is evident that Einstein intended his theory of 1916 to overcome this "epistemological defect."

How exactly was general relativity to solve our problem? *A priori* there are two possible strategies. First, we could try to *extend* the relativity principle for inertial motions to the class of all motions. We could reformulate our laws so that no differential effects would arise between S_1 and S_2, just as no differential effects arise between two bodies moving with a relative constant velocity. Such a reformulated theory would satisfy a general relativity principle precisely analogous to the special principle of relativity. But this kind of general relativity principle is obviously false, since differential effects of the kind under consideration actually do occur in nature. A theory reformulated along the lines of this first strategy would be, as Einstein says in a footnote to the above passage, "satisfactory from the point of view of epistemology, and yet unsound physically" ([28], 113). Second, we could try the strategy of *Machianization*. We could admit the differential effects between S_1 and S_2 but offer an alternative explanation for these effects. Instead of invoking the absolute motion of S_2 with respect to an inertial frame R_1, we could invoke the relative motion of S_2 with respect to some third physical object or class of physical objects: for example, the relative motion of S_2 with respect to the fixed stars. Since the first strategy—a truly general relativity principle—is quite hopeless, it is extremely important to keep the two strategies distinct. They agree only when S_1 and S_2 are *alone* in the universe, in which case both strategies predict that no differential effects will occur—contrary to the prediction of classical physics, special relativity, and all "absolutist" theories.

Confusion begins to set in when Einstein mixes the two strategies:

The only satisfactory answer must be that the physical system consisting of S_1 and S_2 reveals within itself no imaginable cause to which the differing behaviour of S_1 and S_2 can be referred. The cause must therefore lie *outside* this system. We have to take it that the general laws of motion, which in particular determine the shapes of S_1 and S_2, must be such that the mechanical behaviour

206

of S_1 and S_2 is partly conditioned, in quite essential respects, by distant masses which we have not included in the system under consideration. These distant masses and their motions relative to S_1 and S_2 must then be regarded as the seat of the causes (which must be susceptible to observation) of the different behaviour of our two bodies S_1 and S_2. They take over the role of the factitious cause R_1. Of all imaginable spaces R_1, R_2, etc., in any kind of motion relatively to one another, there is none which we may look upon as privileged *a priori* without reviving the above-mentioned epistemological objection. *The laws of physics must be of such a nature that they apply to systems of reference in any kind of motion.* Along this road we arrive at an extension of the postulate of relativity. ([28], 113)

Thus, Einstein slides from a sketch of Mach's solution to the problem of rotation to a call for an extension of the principle of relativity.

However, it is Einstein's next confusion that is really decisive, for he identifies the desired extended relativity principle with the principle of general covariance. The requirement that the laws of nature "apply to systems of reference in any kind of motion" is identified with the requirement that the laws of nature be generally covariant:

> *The general laws of nature are to be expressed by equations which hold good for all systems of co-ordinates, that is, are co-variant with respect to any substitutions whatever (generally co-variant).*
>
> It is clear that a physical theory which satisfies this postulate will also be suitable for the general postulate of relativity. For the sum of *all* substitutions in any case includes those which correspond to all relative motions of three-dimensional systems of co-ordinates. ([28], 117)

This same identification is repeated in numerous later expositions of the theory—for example, by Adler, Bazin, and Schiffer: "accelerated systems must be considered to be quite as respectable as inertial systems, so we demand that physical laws do not distinguish between the two. This will clearly be so if the laws are tensor laws, for then the system of coordinates does not enter the equations at all" ([1], 132). The general principle of relativity requires that any pair of reference frames in any kind of relative motion be equivalent or indistinguishable. Two frames of reference are equivalent or indistinguishable if the laws of nature take the same form in both. Generally covariant laws take the same form in all reference frames.

207

Hence, generally covariant laws satisfy the general principle of relativity.

Of course, this line of thought is quite mistaken. "Sameness of form" is much too weak to guarantee physical equivalence and therefore much too weak to express a relativity principle. The notions of "sameness of form" and covariance correspond to the notions of physical equivalence and relativity only in the context of flat space-time theories in which there exists a privileged class of inertial coordinate systems. Such theories possess a standard formulation $\mathcal{D}(\Phi^\alpha, \Theta^\beta)$ that takes the same form in all inertial systems and whose covariance group is just the group of all transformations from one inertial frame to another. All inertial frames are indistinguishable, and so the indistinguishability group = the covariance group of the standard formulation $\mathcal{D}(\Phi^\alpha, \Theta^\beta)$. But in nonflat space-time theories like general relativity these "nice" connections between indistinguishability and covariance break down. Instead, we have no privileged subclass of inertial frames and no standard formulation $\mathcal{D}(\Phi^\alpha, \Theta^\beta)$ holding in all and only the inertial frames. The only "standard formulation" available is the generally covariant formulation holding in all coordinate systems, and this fact does not imply a generalized equivalence of reference frames.

To see this explicitly, consider a local inertial frame F and an arbitrarily accelerating normal frame F' intersecting at a point p. Let h map F' onto F. The law of motion at p relative to our original connection D in F is just (12):

$$\frac{d^2 y_i}{dy_0^2} = 0$$

and the law of motion at p relative to hD in F is given by (11):

$$\frac{d^2 y_i}{dy_0^2} + a^i + 2\Omega^i_j \frac{dy_j}{dy_0} - 2a^j \frac{dy_j}{dy_0} \frac{dy_i}{dy_0} = 0$$

(since the components of hD in F = the components of D in F'), where a^i is the acceleration and Ω^i_j the rotation of F' relative to D. Hence, absolutely accelerating and rotating frames are just as distinguishable from inertially moving frames in general relativity as in all previous theories. This is as it should be, since such frames are in fact physically distinct.

So Einstein's vision of a Leibnizian, or relationalist, theory rests on a double confusion: a confusion of the strategy of Machianization with the strategy of relativization and a confusion of the notions of

208

physical equivalence and relativity with the notions of sameness of form and covariance. This double confusion is reinforced to the point of irresistibility by the principle of equivalence.[17] On the one hand, the considerations motivating the principle of equivalence look very much like an argument for a generalized relativity principle for acceleration (§V.4). Moreover, although the principle does not actually implement such a relativity principle, it does lead to a nonflat, dynamical space-time in which there are no inertial frames. It leads to a theory with no standard formulation besides the generally covariant formulation. Thus, relativity and covariance are conflated. On the other hand, a dynamical affine structure goes some of the way toward Machianization. The affine connection depends on the total distribution of mass-energy; hence, absolute acceleration, absolute rotation, and (therefore) the differential effects experienced by noninertially moving bodies are in fact "partly conditioned, in quite essential respects, by distant masses which we have not included in the system under consideration." Thus, relativity, covariance, and Machianization are conflated.

But is general relativity a Machian theory? In such a theory absolute acceleration and rotation would not just be "partly conditioned" by distant masses, they would be *wholly determined* by distant masses. Moreover, in order to realize Einstein's relationalist ambitions, acceleration and rotation would have to be determined by distant masses in a quite specific way. An object should be absolutely accelerating just in case it is relatively accelerating with respect to the average matter of the universe (the fixed stars); an object should be absolutely rotating just in case it is relatively rotating with respect to the average matter of the universe. Neither condition will be met, obviously, if the average matter of the universe can be itself either absolutely accelerating or absolutely rotating. Now, in any theory in which $F = ma$, the average matter of the universe cannot itself be absolutely accelerating, because there is no external force available to produce such an acceleration. However, the situation is quite otherwise with absolute rotation. Absolute rotation persists in the absence of all external forces via conservation of angular momentum. In general, then, nothing prevents the average matter of the universe from itself absolutely rotating, so the proposed equivalence between absolute acceleration and rotation and relative

[17] Note that Einstein's discussion of this principle in his 1916 paper links the philosophical arguments about absolute motion and the mathematical treatment of general covariance ([28], 113–115).

acceleration and rotation with respect to the average matter will not hold.

This proposed equivalence does not hold in general relativity either. In two well-known models of the theory, one due to Gödel [48] and the other to Ozsváth and Schücking [77], the average matter of the universe rotates.[18] In both models the stress-energy tensor takes the form of a so-called pressureless field of "dust"—$T = \rho \mathbf{u} \otimes \mathbf{u}$, where $\mathbf{u}$ is the normalized velocity field of the world-lines of matter and ρ is the rest mass density—and in both models a frame of reference in which matter is on average at rest is *not* a local inertial frame. Rather, a frame of reference in which matter (the velocity field $\mathbf{u}$) is on average at rest is itself absolutely rotating; so one cannot equate absolute acceleration and rotation with relative acceleration and rotation with respect to such a frame. The existence of these explicitly non-Machian models should not be too surprising, in view of the close structural similarities between general relativity, the Newtonian gravitation theory of §III.4, and the conventional Newtonian gravitation theory of §III.3. The Newtonian theories of §III.3 and §III.4 treat absolute *rotation* exactly alike.[19] So if absolute rotation is independent of surrounding matter in §III.3—which it certainly is—then absolute rotation is independent of surrounding matter in §III.4 as well. But general relativity, as we saw in §V.2, is essentially just the conjunction of the theory of §III.4 with the kinematics of special relativity. So we should expect absolute rotation to have much the same status in general relativity as it does in our two Newtonian theories, and the Gödel and Ozsváth-Schücking models confirm this expectation.

Clearly, then, general relativity realizes neither of our two strategies for relativizing motion. It does not satisfy a generalized or extended relativity principle, since inertially moving frames are just as distinguishable from accelerating and rotating frames as in all previous theories; and it does not conform to the strategy of Ma-

---

[18] Both models are solutions of the generalized field equation (6) with nonvanishing cosmological term. Note also that, whereas the Gödel model is spatially infinite (noncompact), the Ozsváth-Schücking model is spatially finite (topology $S^3 \times R$, where S^3 is the three-sphere). This shows that Machianization cannot be achieved simply by adding a global finiteness requirement to our theory, a suggestion made, for example, in Einstein [29], 99–108; Misner, Thorne, and Wheeler [74], 543–549; Adler, Bazin, and Schiffer [1], 377. Adler, Bazin, and Schiffer (pp. 367–377) contains a very clear discussion of the Gödel model and its non-Machian character.

[19] It is evident from (III.51) that $\Gamma^i_{j0} = \mathring{\Gamma}^i_{j0}$ $(i,j = 1,2,3)$. The difference between D and $\mathring{D}$ concerns absolute acceleration only: $\Gamma^i_{00} \neq \mathring{\Gamma}^i_{00}$ $(i = 1,2,3)$.

chianization, since absolute rotation has essentially the same status—namely, independence from external masses—as in all previous theories. In the end, therefore, general relativity does not solve Einstein's problem of the rotating globes (or, equivalently, Newton's problem of the rotating bucket). General relativity, like all previous "absolutist" theories, predicts that if S_1 and S_2 were alone in the universe, it would still be possible for one and only one of them to experience distorting differential effects. The reasons that led Einstein and numerous later writers on the theory to think the general theory avoids this "absolutist" problem can be summarized under three main headings.

(1) Confusion of the Two Strategies for Relativizing Motion.

His failure to distinguish Machianization from generalized relativity, along with the fact that the general theory goes some of the way toward Machianization, prevented Einstein from seeing how hopeless a truly general relativity principle actually is. This confusion is intensified by the principle of equivalence, which is apparently connected with both strategies. The principle of equivalence, via the dynamical character of the new affine connection, leads both toward Machianization and, apparently, toward a relativization of acceleration. Confusion of our two stragegies only obscures the fact that the principle of equivalence does not in practice fulfill either one.

(2) Confusion over the Different Senses of "Absolute."

This confusion is especially evident in the slide from *acceleration and rotation are partly determined by distant masses* (true) to *acceleration and rotation are wholly determined by—reducible to—relations to distant masses* (false). The affine structure of space-time is indeed nonabsolute in sense (iii) of §II.3, but it remains absolute in sense (i). This is why general relativity goes only *some* of the way toward Machianization. Thus, consider the following passage from Einstein:

it is contrary to the mode of thinking in science to conceive of a thing (the space-time continuum) which acts itself, but which cannot be acted upon. This is the reason why E. Mach was led to make the attempt to eliminate space as an active cause in the system of mechanics. According to him, a material particle does not move in unaccelerated motion relatively to space, but relatively to the centre of all the other masses in the universe; in this

211

way the series of causes of mechanical phenomena was closed, in contrast to the mechanics of Newton and Galileo. ([29], 55-56)

In the finished theory, of course, we do not eliminate space-time as an active cause; rather, space-time is *both* an active cause and a passive receiver of effects—space-time is dynamical. Note also that the different senses of "absolute" are associated with different philosophical motivations. If one objects to the absoluteness of space-time in sense (iii), one is objecting to its independence or "fixity." If one objects to the absoluteness of space-time in sense (i), one is objecting to its unobservability. The second objection is much more extreme and, therefore, much more difficult to overcome; in the finished theory it is *not* overcome.

(3) The Late Invention of the Generally Covariant Formalism.

This formalism was developed in 1901 by Ricci and Levi-Civita [91] and first assimilated into physics through Einstein's work on general relativity. Given the connection between Lorentz covariance (of the standard formulation) and the special principle of relativity, it was entirely natural for Einstein to assume a similar connection between general covariance and the hoped-for general principle of relativity. The late invention and assimilation of generally covariant techniques reinforced this confusion, since it appeared that general covariance *uniquely* characterized Einstein's new theory. We now know that all space-time theories can be given a generally covariant formulation: general covariance is only a new mathematical technique, not an expression of new physical content. What is new about general relativity is only the necessity for a generally covariant formulation, a necessity due to the use of nonflat space-times in which no inertial coordinate systems exist. Of course, it is a commonplace in the history of science that the development of new physical theories stimulates the invention of new mathematical techniques. In this sense, the late invention of generally covariant techniques was no accident: before general relativity, physics had no need for them. The present case is unique because of the extent to which philosophical motivations became entangled with both the new mathematics and the new physics—so thoroughly as to obscure the true relationship between the mathematics and the physics for half a century.

Looking back on the development of relativity theory from our present point of view, we can see that there are three distinct notions

212

that have been inadvertently conflated: symmetry, indistinguishability, and covariance. The symmetry group of a space-time theory characterizes the *objects* of that theory: it tells us which objects are absolute and which dynamical, and the size of the symmetry group is inversely proportional to the number of absolute objects. The indistinguishability group of a space-time theory characterizes the *laws* of that theory: it determines which reference frames (states of motion) are distinguishable (by a "mechanical experiment") relative to those laws, and in well-behaved theories the indistinguishability group is contained in the symmetry group (indistinguishable models are identical). Covariance, on the other hand, is really a property of *formulations* of space-time theories rather than space-time theories themselves: it characterizes systems of differential equations $\mathscr{D}(\Phi^\alpha, \Theta^\beta)$ representing the intrinsic laws of a space-time theory relative to some particular coordinatization in R^4. The covariance group of such a formulation reflects the range of coordinate systems in which that particular system of equations holds good.

In pre-general-relativistic physics these three distinct notions happen to coincide.[20] In well-behaved theories with inertial coordinate systems (flat space-time) the symmetry group $=$ the indistinguishability group $=$ the covariance group of the standard formulation (with respect to a subclass of inertial coordinate systems). Thus, for example, in classical electrodynamics the symmetry group $= O^3 \times T =$ the indistinguishability group $=$ the covariance group of the standard formulation (III.83), (III.85), (III.88). In relativistic electrodynamics, on the other hand, the symmetry group $=$ the Lorentz group $=$ the indistinguishability group $=$ the covariance group of the standard formulation (IV.41), (IV.43), (IV.45). Moreover, the progressive widening of all three groups reflects an evolving relativization of motion. The indistinguishability group is the primary indicator of such relativization, but in pre-general-relativistic theories our three notions are interchangeable. Se we can express our relativity principles indifferently as symmetry requirements, indistinguishability requirements, or covariance requirements.

In general relativity, however, our three notions are *not* interchangeable. As we have seen, we cannot interpret the general principle of relativity as an indistinguishability requirement, for the indistinguishability group of the general theory is just the restricted

---

[20] They coincide for the most part anyway. As we have seen, Newtonian gravitation theory is an exception. The breakdown of this coincidence in Newtonian gravitation theory prefigures its breakdown in general relativity.

group of transformations from one local inertial frame to another. Nor can we interpret it as a covariance requirement, for the general theory has no standard formulation in the usual sense, and the covariance group of the *theory* is the same as the covariance group of every other space-time *theory*. Hence, in neither of these interpretations is the general principle of relativity any kind of *generalization* of the special principle of relativity. As Anderson was the first to realize,[21] the only way to interpret the general principle as such is to make it a symmetry requirement. That is, we interpret the general principle of relativity as the requirement that the symmetry group of our theory include all differentiable transformations: in effect, that it be just the group $\mathcal{M}$. This requirement means that our theory can have no absolute objects, for the only geometrical objects invariant under all differentiable transformations are constant-valued scalars.

This interpretation of the general principle of relativity therefore implements *some* of Einstein's motivations. It has the consequence that all geometrical objects of our theory are dynamical objects, that the space-time structure of our theory is not "fixed" but depends on the physical processes occurring within it. Furthermore, unlike the principle of general covariance, this construal distinguishes general relativity from all other space-time theories and, in particular, allows the general principle to be seen as a natural generalization of the special principle (viewed as a symmetry requirement, of course). However, this interpretation has nothing at all to do with indistinguishability or with the relativity of motion. This is as it should be, for the general theory has nothing at all to do with generalized relativity of motion either. Just as "general relativity" is not really an appropriate name for our theory,[22] "the general principle of relativity" is not an appropriate name for our principle.

Finally, let us observe that this construal of the general principle has an interesting connection with the principle of equivalence. The principle of equivalence (version (2) of §V.4) says that the gravitational connection D is unique. Hence, if we apply the principle of equivalence to a relativistic Lorentzian manifold, our generalized symmetry requirement follows. If the dynamical connection D is *the* connection of space-time, then the Lorentzian metric g must be compatible with D. The metric must therefore be dynamical also, and we are left with no absolute objects: hence, the group $\mathcal{M}$. In

[21] Anderson [3], 338–343.
[22] See Fock [35].

this context the principle of equivalence and the implied principle of generalized symmetry have further important consequences. The gravitational field affects not only mechanical and electrodynamic phenomena; it has metrical effects as well. In particular, the rates of clocks and the three-dimensional spatial geometry (behavior of "rigid rods") are both influenced by gravitation.

On the other hand, if we apply the principle of equivalence to a Newtonian manifold, the situation is different. Since a Newtonian manifold is not semi-Riemannian, the transition from the flat connection $\overset{\circ}{D}$ to the nonflat connection D has no effect whatever on the underlying metrical structure. Both the flat connection $\overset{\circ}{D}$ and the nonflat gravitational connection D are compatible with the very same temporal metric t and spatial metric h. Consequently, in Newtonian space-time gravitation has mechanical and electro-dynamic effects, but no metrical effects. Moreover, in the Newtonian theory of §III.4, which satisfies the principle of equivalence just as much as does general relativity, t and h remain absolute objects, so the symmetry group of this theory is a proper subgroup of the group $\mathcal{M}$. Thus, the principle of equivalence does not by itself yield the general principle of relativity construed as a symmetry require-ment. We need in addition the assumption of a semi-Riemannian Lorentzian manifold.

VI
Relationalism

In the preceding chapters I have presented a markedly (some would say naively) literal picture of our space-time theories. Their subject matter is taken to be a highly theoretical entity, the space-time manifold; they attribute various types of geometrical structure to this entity; and they relate such geometrical structure to the physical processes and events occurring within space-time. I think there can be little doubt that this picture has a number of attractive features. It gives a relatively simple and precise account of the content of our different space-time theories, thereby making their comparison especially perspicuous, and it provides a fruitful framework for the discussion of traditional problems relating to their foundations—notably, the status of relativity principles and laws of motion.

Must we really take this picture literally? A long and impressive tradition, the relationalist tradition, says we should not. We should regard the use that physical theory makes of space-time and its geometrical structure merely as a convenient way of saying something about the spatio-temporal properties and relations of concrete physical objects. In particular, we should not regard the system of concrete physical bodies as literally embedded in space-time as in some great "container." I shall try to come to terms with this relationalist tradition in the present chapter, especially with the following questions. What is the precise difference between the absolutist picture of space-time and the relationalist picture? What is really at stake in the traditional debate? How does the development of relativity theory merge with this debate and, in particular, with the relationalist strategies of Leibniz and Mach? What can we learn about the *general* conditions that decide in favor of a literal or nonliteral conception of a given piece of theoretical structure?

1. Space-Time: Representation or Explanation?

Our first problem is to give some clear content to the traditional absolutist/relationalist debate: we should explain the difference between the two pictures and why it matters. Let us begin by recalling the distinction from §II.3 between Leibnizean and Reichenbachian relationalism. Leibnizean relationalism places constraints on the *ontology* of our space-time theories. It wishes to limit the domain over which the quantifiers of our theories range to the set of physical events, that is, the set of space-time points that are actually occupied by material objects or processes. Any claim about the spatio-temporal structure of a given region of space-time is to be construed as a claim about the spatio-temporal properties and relations of the concrete physical objects and events that are "in" that region. Reichenbachian relationalism, on the other hand, places constraints on the *ideology*[1] of our space-time theories. It wishes to limit the vocabulary of our theories to some stock of preferred, not obviously spatio-temporal, predicates: for example, predicates definable in terms of "causal" relations. Any claim about spatio-temporal relations (whether between occupied or unoccupied space-time points) is to be construed as a claim about, say, "causal" relations. As the labels I have used suggest, I think the first issue represents what is actually at stake in the Newton-Leibniz debate, while the second represents what is at stake in the more recent Reichenbach-Grünbaum tradition. However, I shall not try to argue such historical claims here. What is important for our present purposes is that we clearly distinguish the two.

Consider special relativity, for example. In our absolutist formulation (Chapter IV) the theory postulates a four-dimensional manifold M of space-time points. Defined on M is a real-valued function $I(a,b)$ of pairs of space-time points: the square of the space-time interval (defined via the Minkowski metric) between them. $I(a,b)$ induces various relations on M: the relation $\tau(a,b)$ of timelike separation ($\tau(a,b)$ iff $I(a,b) > 0$); the relation $\sigma(a,b)$ of spacelike separation ($\sigma(a,b)$ iff $I(a,b) < 0$); and the relation $\lambda(a,b)$ of null or lightlike separation ($\lambda(a,b)$ iff $I(a,b) = 0$). Now, let $\mathscr{P}$ be the subset of M consisting of space-time points occupied by concrete physical events. The ontological relationalist wishes to postulate only the subset $\mathscr{P}$ and, accordingly, to consider only the *restrictions* of τ, σ,

---

[1] This distinction and terminology are from Quine. See essays 17 and 19 in [81].

and λ to $\mathscr{P}$. On his view, we can say everything we want to say by talking only about the set $\mathscr{P}$ and the restricted relations $\tau|_\mathscr{P}$, $\sigma|_\mathscr{P}$, and $\lambda|_\mathscr{P}$. It is not necessary to view these as embedded (respectively) in the more inclusive M, τ, σ, and λ. The ideological relationalist, on the other hand, is not primarily concerned with the domain of our theory, with the choice between M and $\mathscr{P}$. Rather, he is concerned with how our relations are *defined*—whether they are taken to be τ, σ, λ or $\tau|_\mathscr{P}$, $\sigma|_\mathscr{P}$, and $\lambda|_\mathscr{P}$. Thus, the ideological relationalist wishes to define τ by some such definition as

$\tau(a,b)$ iff a (massive) particle can be sent from a to b.

Here, the presence of a and b in $\mathscr{P}$ is not at issue. Conversely, for the ontological relationalist, whether $\tau|_\mathscr{P}$ can be defined in the above manner is not at issue. His primary concern is to do without $(\tau - \tau|_\mathscr{P})$.

In this chapter I shall be concerned almost exclusively with the ontological form of relationalism, with the advantages and disadvantages of representing the set $\mathscr{P}$ of concrete events as literally embedded in the inclusive manifold M of space-time points. To see what some of the *disadvantages* might be, and to bring out the essential differences between absolutism and (ontological) relationalism most clearly, consider Leibniz's famous "indiscernibility" argument.

> Space is something absolutely uniform; and, without the things placed in it, one point of space does not absolutely differ in any respect whatsoever from another point of space. Now from hence it follows (supposing space to be something in itself, besides the order of bodies among themselves) that 'tis impossible there should be a reason, why God, preserving the same situations of bodies among themselves, should have placed them in space after one certain particular manner, and not otherwise; why every thing was not placed the quite contrary way, for instance, by changing East into West. But if space is nothing else, but the possibility of placing them; then those two states, the one such as it now is, the other supposed to be the quite contrary way, would not at all differ from one another. Their difference therefore is only to be found in our chimerical supposition of the reality of space in itself. But in truth the one would be the same thing as the other, they being absolutely indiscernible; and consequently there is no room to enquire after a reason of the preference of the one to the other. ([2], 26)

Freed of theology, Leibniz's argument is this.[2] According to the Newtonian absolutist theory, there are an infinite number of distinct states of affairs consisting of the world (the set of concrete physical bodies) occupying one or another of the infinite number of distinct spatial locations. But on Newtonian theory itself there is no way to detect any differences among these states of affairs. Therefore, it is absurd to regard the world as located in an absolute space.

There are several important things to notice about this argument. First, although Leibniz phrases it as an objection to absolute *space*, it can easily be applied to the four-dimensional *space-time* manifold as well. To see this, let $\langle M,D,g,T_{\hat{a}} \rangle$ be a model for special relativity, where $T_{\hat{a}}$ fits the world-lines of matter. If h is in the indistinguishability group of our theory (that is, if h is a Lorentz transformation), then h generates an indistinguishable but nonidentical model $\langle M,D,g,\mathrm{h}T_{\hat{a}} \rangle$. The latter model differs from the former model over which space-time points are occupied by which material events, but, according to special relativity itself, there is no way to distinguish the two experimentally. Second, Leibniz's argument is not a simple application of verificationism. The problem is not that the absolutist postulates unobservable states of affairs; rather, it is that he commits himself to distinct states of affairs that are not distinguishable *even given his own theoretical apparatus* (the chosen h is in the indistinguishability group of the theory). In this respect, Leibniz's argument is far more subtle and sophisticated than typical arguments against unobservable entities. Recall, for example, that Einstein objected to inertial frames (§V.5) because they function as unobservable causes of observable effects. Leibniz's objection to Newtonian absolute space (and *mutatis mutandis* to special relativistic space-time) appeals not to unobservability *per se* but rather to *uniformity*, to the fact that Newtonian absolute space (and special relativistic space-time) admits nontrivial symmetries.

How does the relationalist solve this problem? How does his conception of the role of space or space-time avoid Leibniz's indiscernibility argument? Instead of asserting that the set $\mathscr{P}$ of concrete physical events is actually embedded in an inclusive space-time M as one particular *subset* of space-time points, he asserts only that the set $\mathscr{P}$ is *isomorphically embeddable* into a space-time

---

[2] Of course, to free the argument from theology is to free it also from the context of Leibniz's metaphysics. In general, the Leibniz I speak of here is not the actual historical Leibniz, but, rather, the Leibniz of the positivists.

M. What does this mean? Consider again a model $\langle M,\tau,\sigma,\lambda\rangle$ for special relativity. Let $\langle \mathscr{P},\tau',\sigma',\lambda'\rangle$ be the set of concrete physical events with the relations τ, σ, λ restricted to $\mathscr{P}$. The absolutist thinks of the latter structure as a *submodel* of the former. That is, $\mathscr{P} \subseteq M$ and

$$\tau' = \tau\big|_{\mathscr{P}} \qquad \sigma' = \sigma\big|_{\mathscr{P}} \qquad \lambda' = \lambda\big|_{\mathscr{P}}.$$

By contrast, the relationalist asserts only that the latter structure is *embeddable into* the former. That is, there is a one-one mapping $\phi:\mathscr{P} \to M$ such that

$$\phi(\tau') = \tau\big|_{\phi(\mathscr{P})} \qquad \phi(\sigma') = \sigma\big|_{\phi(\mathscr{P})} \qquad \phi(\lambda') = \lambda\big|_{\phi(\mathscr{P})}.$$

The absolutist regards the larger structure $\langle M,\tau,\sigma,\lambda\rangle$ as an *explanation* or *reduction* of the properties of the smaller structure $\langle \mathscr{P},\tau',\sigma',\lambda'\rangle$: elements of the latter are literally identified with elements of the former. The relationalist regards the larger structure $\langle M,\tau,\sigma,\lambda\rangle$ as only a *representation* of the properties of the smaller structure $\langle \mathscr{P},\tau',\sigma',\lambda'\rangle$: elements of the latter are only correlated with or mapped onto elements of the former.[3]

Let us now note two crucial differences between the absolutist conception and the relationalist conception. First, since the absolutist asserts that $\mathscr{P}$ is one particular subset of the inclusive manifold M, a mapping h onto a different subset h($\mathscr{P}$) generates a distinct state of affairs. Consequently, a mapping h onto a different subset h($\mathscr{P}$) that preserves the relations τ, σ, and λ (a nontrivial symmetry of relativistic space-time) will generate a distinct but indistinguishable state of affairs. By contrast, the relationalist asserts only that $\mathscr{P}$ is mapped onto a subset $\phi(\mathscr{P})$ of M by *some* function ϕ—no particular function is mentioned. Consequently, a nontrivial symmetry h does not generate a distinct state of affairs: we simply map $\mathscr{P}$ into M by h $\circ$ ϕ rather than ϕ. This makes no difference at all to the structure $\langle \mathscr{P},\tau',\sigma',\lambda'\rangle$. Second, since the relationalist deals only with an embedding of $\langle \mathscr{P},\tau',\sigma',\lambda'\rangle$ into $\langle M,\tau,\sigma,\lambda\rangle$, he can regard the latter structure as a purely mathematical model for the behavior of $\langle \mathscr{P},\tau',\sigma',\lambda'\rangle$. That is, he can make do, for example, with a numerical model $\langle R^4,\underline{\tau},\underline{\sigma},\underline{\lambda}\rangle$, where $\underline{\tau}$, $\underline{\sigma}$, and $\underline{\lambda}$ are defined in the obvious way from the pseudo-metric $(a_0 - b_0)^2 - (a_1 - b_1)^2 -$

---

[3] This interpretation of relationalism is suggested by van Fraassen's view of space and time as "logical spaces"; see van Fraassen [112] and especially [113], 90. Van Fraassen extends his view to a general interpretation of instrumentalism, in the manner of §VI.3, in [114].

$(a_2 - b_2)^2 - (a_3 - b_3)^2$ on R^4. The absolutist, however, asserts that $\langle \mathscr{P}, \tau', \sigma', \lambda' \rangle$ is literally a submodel of $\langle M, \tau, \sigma, \lambda \rangle$; so the latter model must have just as much "physical reality" as the former. The existence of a model $\langle M, \tau, \sigma, \lambda \rangle$ meeting the absolutist's needs is not guaranteed by the consistency of his theory alone, and this, of course, is how it should be.

Hence, the submodel/embedding distinction nicely captures the essential difference between the absolutist and the (ontological) relationalist. The notion of a submodel corresponds to the absolutist metaphor of space-time as a "container," and the notion of an embedding corresponds to the relationalist metaphor of space-time as a "representation of possibilities." Moreover, the submodel/embedding distinction allows us to see precisely why the absolutist conception, but not the relationalist conception, is subject to the Leibnizean indiscernibility argument. We should observe, however, that the Leibnizean argument applies only to space-times, like the space-time of classical kinematics or the space-time of special relativity, that are sufficiently uniform to admit nontrivial symmetries. In space-times of variable curvature, like the space-time of general relativity, there may exist *no* nontrivial symmetries. So the Leibnizean argument is an objection not to space-time or absolutism *per se* but only to sufficiently uniform space-times.[4] Of course, a uniform space-time—in fact, $E^3 \times R$—was the only space-time available for Newton and Leibniz.

The debate between the absolutist and the (ontological) relationist depends on a distinction between matter on the one hand and space-time on the other, between the set $\mathscr{P}$ of concrete physical events and the manifold M of all actual and possible events, between points in M that are "occupied" and points in M that are "unoccupied." Before proceeding, it is worth seeing how this distinction itself becomes problematic with the development of relativity theory.[5]

The distinction between matter and space-time, between occupied and unoccupied points, is clearest in the case of classical particle mechanics. Here the occupied points are just the points with non-vanishing mass density, and the set $\mathscr{P}$ is just the union of all trajectories of actual massive particles. Hence, $\mathscr{P}$ is clearly a *proper* subset

---

[4] See Earman [20], 303.

[5] I am indebted to David Malament for emphasizing this problem to me.

of M. The relationalist version of classical particle mechanics asserts that there is a mathematical Newtonian space-time $\langle M,D,dt,h,\rho \rangle$ and a mapping ϕ from $\mathscr{P}$ onto the set of all points in M such that $\rho > 0$. ϕ maps temporal intervals between elements a and b of $\mathscr{P}$ onto equal temporal intervals (induced by dt) between $\phi(a)$ and $\phi(b)$, spatial intervals between simultaneous elements a and b of $\mathscr{P}$ onto equal spatial intervals (induced by h) between $\phi(a)$ and $\phi(b)$, and the value of the mass density of a in $\mathscr{P}$ is equal to $\rho(\phi(a))$. This conception of the relationship between matter and space-time underlies the traditional debate in Newton, Leibniz, and Mach.

Developments in physical theory upset this simple picture in several ways. First, electrodynamics introduces a new entity, the electromagnetic field, whose status is initially unclear. This entity is not concrete and "observable" in the way massive particles are, but, like massive particles, it carries energy. In that case, are points in M at which the electromagnetic field is nonvanishing occupied or unoccupied? Classical electrodynamics leaves this largely as a matter of choice, but in relativistic electrodynamics it becomes quite arbitrary to distinguish between mass-density on the one hand and the electromagnetic field on the other. This is because relativity theory combines mass and energy into a single object, the energy-momentum vector (§IV.3), and the electromagnetic field possesses an energy-momentum vector—the so-called Poynting vector[6]—just as much as massive particles do. In a relativistic context the proper representative of matter is not the mass-density ρ but the total stress-energy density $\mathbf{T}$. The occupied points are just the points with nonvanishing stress-energy, so points at which there is nonvanishing electromagnetic energy should count as occupied. The simple picture of $\mathscr{P}$ as a union of discrete particle trajectories breaks down.

Moreover, when we move from a particle-theoretic to a field-theoretic conception of $\mathscr{P}$, it is no longer clear that $\mathscr{P}$ is a *proper* subset of M. In fact, as far as we can tell, the real universe appears to be completely filled with electromagnetic background radiation. Hence, all space-time points turn out to be occupied. This means that the relationalist's ontology is just as rich as the absolutist's: $\mathscr{P}$ is not just embeddable into M, it is actually isomorphic to M. Hence, the traditional debate threatens to dissolve completely. Worse still, in general relativity the space-time metric g itself carries energy,

---

[6] See, e.g., Misner, Thorne, and Wheeler [74], 140–141.

although this gravitational energy is *not* included in the stress-energy tensor **T**. Space-time is just as dynamical as matter, and this remains true even if **T** vanishes everywhere. There are, for example, empty space-time solutions of Einstein's equations in which energy-carrying gravitational waves exist.[7] But if we say that a space-time point p is occupied when g is nonvanishing at p, then *all* space-time points will automatically be occupied whether or not the universe is filled with background radiation. This last move removes what sense is left in the traditional debate.

Can the debate no longer continue then? I think it can. First, we should dig in our heels and use the nonvanishing of **T**, not the nonvanishing of g, as our criterion for an occupied space-time point. This criterion is not arbitrary, for it accords with the general-relativistic practice of not counting the gravitational energy induced by g as a component of the total energy, and it allows us to preserve a measure of continuity between general relativity and our previous space-time theories. Second, we should not simply compare $\mathscr{P}$ with M; rather, we should compare the relationalist *theory* of $\mathscr{P}$ with the corresponding absolutist *theory*. The point is that, although $\mathscr{P}$ may be isomorphic to M as a matter of fact, the relationalist theory of $\mathscr{P}$ only asserts that $\mathscr{P}$ is representable by *some* physically possible mass-energy distribution: for example, that there is a (mathematical) model $\langle M,D,g,\mathbf{T} \rangle$ of general relativity and a mapping ϕ of $\mathscr{P}$ onto the set of points in M at which **T** does not vanish. The laws that hold of $\mathscr{P}$ imply only that $\phi(\mathscr{P})$ is a subset of M; there is no theoretical guarantee that $\mathscr{P}$ is isomorphic to M. By contrast, the absolutist theory does imply the existence of all space-time points; its ontology is guaranteed to be isomorphic to M. Hence, whether or not the universe happens to be filled with nonvanishing stress-energy as a matter of fact, the laws of the relationalist theory and the laws of the absolutist theory still have very different theoretical implications. As we shall see, this makes a large difference indeed.

2. The Problem of Motion and the Development of Relativity Theory

We have seen some of the disadvantages of an absolutist position: the postulation of the entire space-time manifold M leaves us vulnerable to the Leibnizean indiscernibility argument. What are

---

[7] See Hawking and Ellis [57], 178–179.

some of the problems for a relationalist? As our example of special relativity shows, the relationalist has no difficulty at all in describing the spatio-temporal relations between events: his assertion that the relations τ', σ', and λ' are embeddable into a special relativistic manifold has exactly the same consequences as the absolutist's assertion that $\langle \mathscr{P}, \tau', \sigma', \lambda' \rangle$ is a submodel of such a manifold. Similarly, the relationalist can easily describe the spatio-temporal relations between events postulated by Newtonian theory. He can suppose that there are two-place functions $t(a,b)$ and $d_S(a,b)$ defined on events in $\mathscr{P}$, where $t(a,b)$ is the temporal interval between a and b and $d_S(a,b)$ is the spatial distance between simultaneous ($t(a,b) = 0$) a and b, and he can assert that $t(a,b)$ and $d_S(a,b)$ are isomorphically embeddable into a Newtonian space-time $\langle M, D, dt, h \rangle$. That is, there is a one-one mapping $\phi : \mathscr{P} \to M$ such that

$$t(a,b) = |t'(\phi(a)) - t'(\phi(b))| \text{ for some time-function } t' \text{ of } dt$$
$$d_S(a,b) = d_{\tilde{S}}^*(\phi(a), \phi(b)), \text{ where } d_{\tilde{S}}^* \text{ is the spatial metric on the} \tag{1}$$
$$\text{planes of simultaneity induced by } h.$$

This theory about $t(a,b)$ and $d_S(a,b)$ generates all the same consequences as the absolutist theory.

Problems begin to arise for the relationalist when we move from spatio-temporal geometry to the theory of motion. All known theories of motion involve *absolute* motion: motion with respect to a three-dimensional absolute space (in Newton's original theory) or motion with respect to a privileged class of four-dimensional inertial trajectories (in all relativistic theories). Further, no known theory supplies any guarantee that the privileged entities used to define absolute motion are occupied by concrete physical objects or events. In other words, all absolutist theories of motion proceed by postulating a class of curves or trajectories (the curves that fit absolute space in Newton's original theory, the inertial trajectories in all relativistic theories) that are probably not to be found in $\mathscr{P}$. These curves are therefore not available to the relationalist, and he is faced with the problem of giving an alternative description of motion.

To illustrate, let us see what happens when we try to make do with *relative* motion—motion with respect to a concrete physical object— in a Newtonian setting. Suppose we start with a particular three-dimensional rigid body B as our concrete reference object. We can construct a rigid Euclidean frame centered on B as follows. Let the center of mass of B be the point p. Choose three points a_1, a_2, a_3 on the surface of B such that the spatial angle $\alpha(a_i,p,a_j)$ is orthogonal for $i \neq j$. Finally, arbitrarily pick an event e_p at p. Any concrete event e is assigned the coordinates

$$t(e) = \pm t(e,e_p)$$
$$x_i^B(e) = \cos(\alpha(e,p,a_i))d_S(e,p). \tag{2}$$

Equation (2) assigns an element of R^4 to each physical event and a continuous curve in R^4 to each physical trajectory. Each physical trajectory has a well-defined velocity and acceleration with respect to B:

$$v_B^i = \frac{dx_i^B}{dt}, \qquad a_B^i = \frac{d^2 x_i^B}{dt^2}. \tag{3}$$

Can the relationalist state a law of motion in terms of the relative quantities (3)?

If our reference object B happens to be inertially moving (neither absolutely accelerating nor rotating), there is no problem. The rigid Euclidean frame generated by our assignment of spatio-temporal coordinates (2) is an inertial frame, and the law of motion is just

$$F^i = ma_B^i. \tag{4}$$

However, there is no guarantee in general that any such inertially moving reference object exists. First, it is quite unlikely that any concrete physical object has vanishing net external force acting on it. No object is completely isolated from external forces, and a precise balancing of external forces is not to be expected. Second, even if B happens to have vanishing net external force acting on it, and hence vanishing absolute acceleration, there is still no guarantee that it has vanishing absolute rotation. Uniform rotation, unlike acceleration, requires no external force. Thus, even if p is the center of mass of the entire universe, a frame of reference in which B is at rest might very well be rotating and therefore noninertial.

By contrast, the absolutist does have a guarantee that inertial reference frames exist. Think of an inertial frame f as given by four inertial trajectories $\sigma_0, \sigma_1, \sigma_2, \sigma_3$ such that $\sigma_0, \sigma_1, \sigma_2, \sigma_3$ are all at rest relative to one another and $\alpha(\sigma_i,\sigma_0,\sigma_j)$ is orthogonal for $i \neq j$. Although such trajectories may not be found in $\mathscr{P}$, we know they exist in M. We can generate the spatio-temporal coordinates of f by arbitrarily picking a point e_{σ_0} on σ_0 and mimicking (2):

$$t(e) = \pm t(e,e_{\sigma_0})$$
$$x_i^f(e) = \cos(\alpha(e,\sigma_0,\sigma_i))d_S(e,\sigma_0). \tag{5}$$

Any physical trajectory has a well-defined velocity and acceleration with respect to f:

$$v_f^i = \frac{dx_i^f}{dt}, \qquad a_f^i = \frac{d^2 x_i^f}{dt^2}. \tag{6}$$

But this time we know that the law of motion can indeed be correctly stated as

$$F^i = ma_f^i. \tag{7}$$

Given the functions $t(a,b)$, $d_S(a,b)$, and $\alpha(a,b,c)$, together with the absolutist's inclusive ontology, we *can* state a correct law of motion. Given the same functions and the relationalist's restricted ontology, we cannot. Restricting these functions to the set $\mathscr{P}$ leads to a real loss of expressive power.

Further, the absolutist version of Newtonian theory explains the deviations from (4) that are actually observed. Newtonian theory asserts that in an arbitrary rigid Euclidean reference frame generated by some concrete object B as in (2), the correct law of motion is not (4) but

$$\frac{F^i}{m} = a_B^i + a^i + 2\dot{a}_{ji} v_B^j + \ddot{a}_{ki} x_k^B \tag{8}$$

where $a^i(t)$ is the acceleration of B with respect to an inertial frame coinciding with the frame defined by (2) at t, and $\dot{a}_{ij}(t)$ is the rotation of B with respect to such a frame (III.18). To formulate an accurate law of motion, we must refer to B's acceleration and rotation with respect to some nonconcrete (probably unoccupied) reference frame. Of course, for most practical purposes we can be satisfied by an approximately true law of motion and make do with concrete reference frames in which (4) holds to a high degree of approximation. Nevertheless, a deeper and more accurate theory is clearly preferable, and the Newtonian law of motion in its absolutist form (7) constitutes just such a deeper theory.

In sum, then, the basic problem is that a correct theory of motion appears to require two essentially different kinds of motion: relative motion, which can be interpreted in terms of changing spatial relations between concrete physical objects; and absolute motion, which is understood in terms of changing spatial relations between a concrete physical object and some nonconcrete (probably unoccupied) curve or trajectory. How can the relationalist respond to this difficulty? We can distinguish two traditional strategies. The first attempts to eliminate the absolute quantities from (8) and make do with relative quantities *à la* (4). This strategy corresponds to the search for *relativity principles*. The second tries to define the absolute quantities in (8) in relationalistically unobjectionable terms—to show that, contrary to appearances, they are relative quantities after all. This second strategy corresponds to what I have called *Machianization*. Historically, the first strategy is associated with the views of Leibniz, the second with the views of Mach.

Leibniz's belief in, or hope for, a general relativity principle is expressed in passages like the following.

> In order to prove that space, without bodies, is an absolute reality; the author [Clarke] objected, that a finite material universe might move forward in space. I answered, it does not appear reasonable that the material universe should be finite; and, though we should suppose it to be finite, yet 'tis unreasonable it should have motion any otherwise, than as its parts change their situation among themselves; because such a motion would produce no change that could be observed, and would be without design. ([2], 74)

This passage attempts to extend Leibniz's indiscernibility argument from absolute position to absolute motion. Just as differences in absolute position have no observable effects, so (it is claimed) differences in absolute motion have no observable effects either. If this kind of general relativity principle were true, we would never have to appeal to anything but the relative quantities (3): we could state an empirically adequate law of motion by an equation like (4).

The trouble with this strategy, of course, is that it fails to make the crucial distinction between velocity and acceleration. We do have a relativity principle for velocity: differences in absolute velocity have no observable effects according to Newtonian theory. The law of motion (7) holds equally in all inertial frames, and, consequently, the velocity of an inertial frame plays no role at all in the Newtonian theory of motion. But we do not have a relativity principle for acceleration: differences in absolute acceleration have observable effects in Newtonian theory. The law of motion (7) does not hold in *arbitrary* frames, and this is precisely why (4) is inadequate. The correct law of motion in an arbitrary reference frame is (8), which explicitly contains the absolute quantities $a^i(t)$ and $\dot{a}_{ij}(t)$. Thus, we can simply eliminate absolute velocity from our theory by moving

227

from the space-time $\langle M,D,dt,h,V \rangle$ to the space-time $\langle M,D,dt,h \rangle$. But we cannot simply drop absolute acceleration from our theory.

Nevertheless, Leibniz can hardly be faulted for failing to distinguish the problem of velocity from the problem of acceleration, for a clear appreciation of this distinction requires the four-dimensional point of view. From the three-dimensional point of view, velocity is change of absolute spatial position with respect to time, and acceleration is change of velocity with respect to time. If one type of absolute motion is meaningful, then so is the other. Hence, from the three-dimensional point of view, the inference from the Leibnizean indiscernibility argument to a general relativity principle looks irresistible: if absolute position has no observable significance, then changes in absolute position can have no observable significance either.[8] From the four-dimensional point of view, on the other hand, the original indiscernibility argument applies to *spatio-temporal* location, not to spatial position. It tells us that certain variations in the way in which the four-dimensional material universe is embedded in space-time (those generated by nontrivial symmetries) make no observable difference, while other variations in the way in which the material universe is embedded in space-time do make a difference. The former correspond precisely to variations in velocity (elements of the Galilean group or the Lorentz group), the latter to variations in acceleration. Note also that Leibniz was not the only one to be misled by the three-dimensional point of view. Newton, in effect, made the opposite inference: absolute acceleration is meaningful; therefore, so is absolute spatial position.

In any case, it is clear that our first strategy is a failure: we cannot simply drop the terms involving absolute quantities from (8). The next best relationalist strategy is to try to explain these quantities in acceptable terms, to define absolute acceleration without referring to unoccupied inertial trajectories. Some passages from Leibniz suggest this second possibility.

> I grant that there is a difference between an absolute true motion of a body, and a mere relative change of its situation with respect to another body. For when the immediate cause of the change is in the body, that body is truly in motion; and then the situation of other bodies, with respect to it, will be changed consequently, though the case of that change be not in them. ([2], 74)

[8] Instances of this line of thought can be found in Reichenbach [84], 7–9, and Schlick [97], 2–4.

This passage is somewhat obscure, but Leibniz seems to have something like the following in mind. We know that every absolute acceleration is precisely correlated with a force. We might hope, therefore, to replace the reference to absolute acceleration in (8) with a reference to the actual physical processes generating the forces acting on B. Suppose, for example, that there are two rockets accelerating relative to one another in empty space. If one is firing its engines and the other is not, we know that the first is absolutely accelerating and the second is not. Moreover, we can make this distinction on the basis of physical features of the two rockets, on the basis of where the "cause of change" occurs. We do not need to refer to absolute space or unoccupied inertial trajectories.

The approach that Leibniz hints at here will obviously not work in general. First, it assumes an exclusively action-by-contact model of force: accelerating objects can be distinguished from nonaccelerating objects by "causal" processes (like collisions) occurring in the immediate neighborhood of the accelerating object in question. This is not generally true, of course, for objects can be absolutely accelerating (under the influence of gravitational force, for example) even though no distinguishable "causal" processes are occurring in their neighborhood, that is, even though the "cause of change" is not "in them." Second, as we have emphasized many times already, uniform rotation is *not* correlated with external forces in Newtonian theory. Even if we could replace the reference to absolute acceleration in (8) with a reference to force-generating "causal" processes, Newtonian theory does not allow us to replace the reference to absolute rotation in a similar manner.

It appears, then, that Leibniz in fact had no coherent response to the problem of motion. Still, we can learn some valuable lessons from his attempts to solve it. First, the relationalist cannot just drop absolute motion from his theory: his only hope for elimination is a definition in nonabsolutist terms. Second, absolute rotation is of overriding importance: the real problems for the relationalist are created by the terms involving $\dot{a}_{ij}(t)$ in (8). As a consequence, a satisfactory definition of absolute motion must be formulated in the context of a substantial *revision* of Newtonian theory, for in Newtonian theory absolute rotation is completely independent of external objects and processes. Now, all these points were clearly grasped by Mach. What he proposes is to revise Newtonian theory in such a way that the rotational terms in (8) arise from B's relative rotation with respect to the rest of the objects in the universe, *not* from *B*'s absolute rotation with respect to some unoccupied inertial frame. Thus, for example, Mach agrees that in Newton's rotating bucket experiment we need to consider more than the relative rotation of the water and the bucket. Why, however, should we appeal to the rotation of the water relative to some nonphysical space? Why not simply invoke the rotation of the water relative to *other* physical objects?

> The principles of mechanics can, indeed, be so conceived, that even for relative rotations centrifugal forces arise.
>
> Newton's experiment with the rotating vessel of water simply informs us, that the relative rotation of the water with respect to the sides of the vessel produces *no* noticeable centrifugal forces, but that such forces *are* produced by its relative rotation with respect to the mass of the earth and other celestial bodies. ([68], 284)

Hence, a Machian theory will agree with the Newtonian equation (8) only when there are sufficient surrounding masses to generate the pseudo force terms. As the number of surrounding masses is decreased, we should expect these terms to converge to zero. If there are no surrounding masses, (8) should necessarily take the form (4).

In terms of the problem of defining absolute motion we can understand Mach's proposal as follows. Absolute motion is definable in the structure $\langle M,D,dt,h \rangle$ from $t(a,b)$, $d_S(a,b)$, and $\alpha(a,b,c)$, because we have a guarantee that inertial frames (quadruples of inertial trajectories as in (5)) exist in M. An arbitrary trajectory is absolutely accelerating (rotating) just in case it is accelerating (rotating) with respect to some inertial frame. Absolute motion will not in general be definable in the substructure $\langle \mathscr{P},t(a,b),d_S(a,b),\alpha(a,b,c) \rangle$, because Newtonian theory supplies no guarantee that inertial frames exist in $\mathscr{P}$. Mach proposes to revise Newtonian theory in such a way that inertial frames *are* guaranteed to exist in $\mathscr{P}$. Thus, if we could carry out the interpretation of pseudo forces sketched above, we would know that a frame of reference in which the average matter of the entire universe is at rest must be an inertial frame. We could then define absolute acceleration (rotation) in terms of relative acceleration (rotation) with respect to such a physically definable frame.

These considerations make it clear that a Machian theory of absolute motion needs to be *created*. We cannot know whether "the principles of mechanics can indeed be so conceived, that even for relative rotations centrifugal forces arise," until a theory with this

230

consequence is produced. In the absence of a positive theory, Mach's proposal is not a real alternative to the Newtonian account of absolute motion; it is merely a pious hope for such an alternative. As we have seen, it was at first hoped, most notably by Einstein himself, that the general theory of relativity would be precisely this kind of Machian theory, that the general relativistic dependence of inertia on distant masses would amount to a full-fledged definition of inertia in terms of distant masses. But, as we have also seen, this hope was in vain. General relativity supplies no more guarantee that (local) inertial frames exist in $\mathscr{P}$ than does Newtonian theory. In particular, a frame of reference in which the average matter of the entire universe is at rest has just as much chance of absolutely rotating in general relativistic space-time as in Newtonian space-time.[9]

In short, relativity theory accords no better with relationalism than does Newtonian theory. All known theories of motion— Newtonian, special relativistic, and general relativistic—imply the existence of a class of probably unoccupied inertial trajectories and require the existence of curves that are not guaranteed to be found in $\mathscr{P}$ (according to these theories themselves). It is true that the development of relativity theory has been intimately associated with our two traditional strategies. The development of special relativity goes some of the way toward the Leibnizean idea of a general relativity principle, for it finally dispenses with absolute velocity and institutes a throughgoing equivalence (both mechanical and electrodynamic) between all inertial frames. The development of general relativity goes some of the way toward the Machian idea of a reinterpretation of absolute motion, for it exhibits the absolute acceleration and rotation of objects as dependent on the total distribution of matter in the universe. Nevertheless, neither traditional strategy is completely realized. Special relativity preserves the notions of absolute acceleration and rotation. General relativity preserves the

[9] As a matter of fact, there is one respect in which general relativity makes a Machian redefinition of absolute acceleration (rotation) even more difficult than in Newtonian theory. In the nonflat space-time of general relativity there are no true inertial frames, just *local* inertial frames; and in a local inertial frame (7) holds only along the inertial trajectory that defines the frame. The absolute acceleration of an arbitrary trajectory σ at a time τ does not equal its relative acceleration with respect to just any local inertial frame, but only with respect to a local inertial frame that is *tangent* to σ at τ. Therefore, to define absolute acceleration *à la* Mach in a nonflat space-time we need many *more* (local) inertial frames: we need (local) inertial frames tangent to every trajectory σ at every point $\sigma(\tau)$.

231

possibility that even the average matter of the universe may be absolutely rotating. The confusions catalogued in earlier chapters—confusion of the two traditional strategies, confusion of indistinguishability with covariance, confusion of the different senses of "absolute," and so forth—have stood in the way of a clear appreciation of the problem of motion and its status in relativity theory.

What does the development of relativity theory teach us about the problem of motion from our present point of view? First, we learn that the only forms of absolute motion—motion that cannot be interpreted in terms of changing spatial relations to some concrete physical reference object—required are absolute acceleration and rotation. We do not need to appeal to absolute velocity and absolute rest. Second, we learn that acceleration and rotation are essentially four-dimensional notions. They can be defined in terms of a class of inertial or geodesic trajectories in a four-dimensional space-time or, what amounts to the same thing, in terms of an affine structure on such a space-time. In particular, we do not need to invoke a three-dimensional absolute space in order to make sense of absolute acceleration and rotation. But the four-dimensional character of absolute acceleration and rotation was not clearly understood, even in the case of special relativity, until at least Minkowski's 1908 paper [73]. In the case of Newtonian theory, it was not understood until Weyl in 1918 [116] and Cartan in 1923–1924 [13]. So it is not at all surprising that the elimination of Newtonian absolute space was thought to vindicate Leibnizean relationalism. Only considerable hindsight allows us to realize that we have replaced a three-dimensional absolute entity with a four-dimensional one.

Is the story over then? Has relationalism suffered a decisive defeat? Not at all. For a third approach to the problem of motion has been ignored by the tradition. Why should the relationalist not accept absolute acceleration and rotation—not, to be sure, as acceleration and rotation with respect to an absolute space or space-time, but as *primitive properties of concrete physical trajectories*? The absolutist thinks of absolute acceleration and rotation as species of relative acceleration and rotation: they are quantities with two argument places, one of which is filled by a privileged absolute entity (an unoccupied reference frame). The relationalist, on the other hand, can simply deny that absolute acceleration and rotation are species of relative acceleration and rotation. He can think of them as additional primitive quantities with only one argument place: they characterize a concrete physical trajectory (at a time), but they make no reference to the state of motion of that trajectory with respect to any other entity, concrete or otherwise.[10] It is appropriate to remind ourselves here that the present dispute is ontological rather than ideological. The relationalist is free to adopt any primitive vocabulary he likes, so long as his quantifiers range only over concrete physical events. He is not restricted to predicates that literally express *relations* between physical events. (What possible motivations could such a restriction have?)

Nevertheless, this third strategy is liable to appear rather pointless at first. The neglect it has suffered may seem eminently justified. For what is the point of replacing a mysterious entity like space or space-time with mysterious quantities like absolute acceleration and rotation? After all, the relationalist's postulated quantities are no more observational than the absolutist's postulated entities. Just as motion with respect to space or space-time is only detectable through its distorting effects (the pseudo forces), so the absolute accelerations and rotations of physical bodies are only detectable through these same effects. There is clearly no advantage to be gained in general by substituting unobservable properties for unobservable entities. Thus, for example, the phenomenalist cannot solve his problems by accepting "is veridical" or even "is of a chair" as primitive predicates of sense-data. His suspicions about material objects extend naturally to such predicates as "is veridical." Why should the relationalist's suspicions about absolute space or space-time not extend likewise to such predicates as "is absolutely accelerated"?

To answer this worry we need only recall the Leibnizean indiscernibility argument. The objection raised by that argument is not just that absolute space or space-time is unobservable, but rather that absolute space or space-time generates distinct states of affairs that are indistinguishable *according to the absolutist's own theory*. Leibniz objects not to unobservable entities *per se* but to the very special

---

[10] As far as I know, Lawrence Sklar was the first author to articulate this possibility clearly; see [100], 229–232; [101], 15. However, such a view would fit in very well with the "monadism" of the historical Leibniz, and, in fact, it may be possible to interpret passages like the one quoted on p. 228 above in this fashion. For Leibniz did not have our concept of force as something "impressed" and "external." Rather, it was an internal power ("living force") by which a body determined its own future state. It is true that Leibniz tended to identify "force" with *vis viva* or (twice) our kinetic energy, and he lacked our (Newtonian) conception of the general relationship between (our) force and absolute acceleration. But our absolute acceleration, interpreted as a monadic quantity *à la* Sklar, could perhaps fulfill the role of Leibniz's "force."

233

(uniform) entities postulated by the absolutist. Does the postulation of absolute acceleration and rotation as primitive quantities give rise to the same objection? Clearly not: differences in absolute acceleration and rotation, although not directly observable to be sure, do generate observational effects in the context of the appropriate theory. Thus, for example, the relationalist can simply add a vector-valued function $a^i(\sigma)$ of concrete trajectories to the functions $t(a,b)$, $d_S(a,b)$, and $\alpha(a,b,c)$, and add the condition

$$a^i(\sigma)_{\langle y_i \circ \phi \rangle} = \left[D_{T_{\phi(\sigma)}} T_{\phi(\sigma)} \right]^i_{\langle y_i \rangle} \tag{9}$$

(where $\langle y_i \rangle$ is an admissible coordinate system on M) to the conditions (1) on the mapping $\phi : \mathscr{P} \to \langle M,D,dt,h \rangle$. This theory will have all the same consequences about $a^i(\sigma)$ that the absolutist theory does. In particular, it implies the correct law of motion (8) and *a fortiori* the distorting effects distinguishing differences in absolute acceleration and rotation.

In other words, the assertion that $\langle \mathscr{P},t(a,b),d_S(a,b),a^i(\sigma) \rangle$ is embeddable into $\langle M,D,dt,h \rangle$ by a mapping ϕ satisfying (1) and (9) does not generate distinct but indistinguishable states of affairs. By contrast, the assertion that $\langle \mathscr{P},t(a,b),d_S(a,b),a^i(\sigma) \rangle$ is a submodel of $\langle M,D,dt,h \rangle$ does. Therefore, from the relationalist's own point of view, there *is* an advantage in dispensing with absolute space or space-time in favor of absolute acceleration as a primitive quantity. A theory ascribing the primitive property of absolute acceleration to concrete physical trajectories avoids just the problems that the Leibnizean indiscernibility argument raises for theories embedding concrete trajectories in an absolute space or space-time. On his own principles, then, the relationalist has good reason to prefer the former kind of theory to the latter. Here we see one important difference between the present dispute and other traditional ontological disputes. In other disputes, such as the dispute over phenomenalism, unobservable properties are just as bad as unobservable entities. There is no point in general in taking an expression like "is veridical," which under its standard interpretation indicates a relation to a problematic entity, and turning it into a primitive one-place predicate. But in the present dispute it makes good sense to do precisely this with the expression "is absolutely accelerated": the Leibnizean indiscernibility argument supplies just the necessary difference.

These considerations also help to explain why no historical relationalist entertained the possibility of taking absolute acceleration as primitive. To appreciate this possibility it is essential to

234

isolate the problem of absolute acceleration from the problem of absolute motion in general. This, as we have emphasized repeatedly, requires the four-dimensional point of view, whereas the historical debate about relationalism is dominated by the three-dimensional point of view. Thus, from the three-dimensional point of view, there is no absolute acceleration without absolute velocity (and even absolute spatial position) as well, but if we adopt absolute velocity (or worse, absolute spatial position) as a primitive property of concrete trajectories, we are again vulnerable to the Leibnizean argument. There will again be an infinite number of distinct states of affairs, consisting of the center of mass of the material universe having one or another absolute velocity, that are completely indistinguishable according to our theory. In the case of absolute velocity, then, unlike the case of absolute acceleration, it is pointless to substitute a primitive absolute quantity for an absolute entity. The proper course for the relationalist here is simply to drop absolute velocity completely. Acceleration and velocity must be effectively separated, therefore, before our third strategy makes good relationalist sense. (Another consequence of these considerations is that our statement that the relationalist is free to adopt any primitive vocabulary he likes must be modified. The relationalist is free to adopt any primitive vocabulary *that is not vulnerable to the Leibnizean argument.*)

It is necessary, then, to consider seriously a relationalist theory adopting absolute acceleration as a primitive property of concrete trajectories. Such a theory is not a mere trivial variant of an absolutist theory: the relationalist can embrace it with a clear Leibnizean conscience. Hence, the traditional relationalist attempts to eliminate or reinterpret absolute acceleration—attempts that play such a central role in the development of relativity theory—are in an important way misconceived, because they have focused our attention on the structure $\langle \mathscr{P}, t(a,b), d_S(a,b) \rangle$ and the question of the definability of $a^i(\sigma)$ within that structure. From this point of view, the relationalist case ends up looking very weak indeed, for the absolutist structure $\langle M, D, dt, h \rangle$ (or a relativistic analogue) is obviously much richer in definable properties and relations. But $\langle \mathscr{P}, t(a,b), d_S(a,b), a^i(\sigma) \rangle$ (or a relativistic analogue) is itself a perfectly good (Leibnizean) relationalist structure. Given this latter structure, the relationalist can formulate a theory of motion (including the correct law of motion (8), for example) that is precisely as accurate as the absolutist's theory. Why, then, should we prefer the structure $\langle M, D, dt, h \rangle$ (or a relativistic analogue) to the smaller structure $\langle \mathscr{P}, t(a,b), d_S(a,b), a^i(\sigma) \rangle$

235

(or a relativistic analogue)? Is there any advantage to be gained by claiming that the latter structure is actually a submodel of the former? Why not rest content with the much weaker claim that asserts only the existence of an embedding and is therefore not vulnerable to the Leibnizean argument? As we shall see, these questions lead us to problems of fundamental importance.

3. Theoretical Structure and Theoretical Unification

The claim that the absolutist wishes to make—that the smaller structure $\langle \mathcal{P}, t(a,b), d_S(a,b), a^i(\sigma) \rangle$ is embedded in, or is a submodel of, the larger structure $\langle M, D, dt, h \rangle$—is actually typical of the explanations found in mathematical physics. These explanations characteristically proceed by deriving the properties of some relatively concrete and observable phenomenon via an embedding of that phenomenon in some larger, relatively abstract and unobservable, theoretical structure. We explain the properties of gases by embedding them in the world of molecular theory: we identify gases with configurations of molecules and properties of gases, like temperature, with properties of molecular configurations, like mean kinetic energy. We explain the properties of chemical compounds by embedding them in the world of atomic theory: we identify chemical elements with particular types of atoms and properties of chemical elements, like valence, with properties of atoms, like excesses and defects in the number of electrons in the outermost shell. Similar examples are found in the nonphysical sciences: for instance, the contemporary embedding of genetics in the world of molecular biology.

These explanations follow a common pattern. They postulate a theoretical structure $\mathcal{A} = \langle A, R_1, \ldots, R_n \rangle$ with complex and interesting mathematical properties, and they construe some observational structure $\mathcal{B} = \langle B, R'_1, \ldots, R'_m \rangle$ $(m \leq n)$ as a substructure of $\mathcal{A}$. Under this construal $\mathcal{A}$ functions as a genuine explanation or *reduction* of the properties of $\mathcal{B}$, for elements of $\mathcal{B}$ are literally identified with elements of $\mathcal{A}$. That is, we assert that $B \subseteq A$ and $R'_i = R_i|_B$ $(i \leq m)$. Yet it is also possible to construe our use of $\mathcal{A}$ as a mere *representation* of the properties of $\mathcal{B}$. Instead of asserting that $\mathcal{B}$ is a submodel of $\mathcal{A}$, we can assert only that $\mathcal{B}$ is embeddable into $\mathcal{A}$, that there is a one-one map $\phi : B \rightarrow A$ such that $\phi(R'_i) = R_i|_{\phi(B)}$ $(i \leq m)$. Under this latter construal, $\mathcal{A}$ can be thought of as a purely mathematical object. It is no longer necessary to attribute

236

3. *Theoretical Structure and Unification*

physical reality to $\mathscr{A}$, since we no longer literally identify elements of the concrete structure $\mathscr{B}$ with elements of $\mathscr{A}$. But now, we are faced with an obvious question. Why should we *ever* regard theoretical structure as something more than a mathematical representation of the observable phenomena? Why should we ever move from the weaker claim of representation to the stronger claim of an actual reduction? After all, the two claims have precisely the same consequences for the observational structure $\mathscr{B}$. What advantage do we gain by making the stronger claim?

This question is made more pressing by the fact that the representation/reduction distinction is not a mere philosopher's invention; it plays a genuine role in scientific practice. To see this, one has only to reflect on the use of coordinates in our space-time theories. When we use a spatio-temporal coordinate system, we assert the existence of a one-one mapping $\phi : A \to R^4$, where A is an open region in space-time, and it is crystal clear that R^4 is functioning only as a representation here. We do not identify space-time with R^4, we merely *correlate* it with R^4 for the purpose of mathematical representation. But why do we not identify space-time with R^4? It is not simply because R^4 is a mathematical object; more important, it is because R^4 is a very rich mathematical structure. There are numerous properties and relations definable in R^4 (or in any structure isomorphic to R^4) that are not thought to reflect physically meaningful properties and relations in space-time.

Consider, for example, a Newtonian space-time M with the relations $t(a,b)$ and $d_S(a,b)$. One of the basic results of Newtonian theory is the existence of rigid Euclidean coordinates—one-one mappings $\phi : A \to R^4$ such that

$$t(a,b) = |\phi(a)_0 - \phi(b)_0|$$
$$d_S(a,b)^2 = (\phi(a)_1 - \phi(b)_1)^2 + (\phi(a)_2 - \phi(b)_2)^2 \qquad (10)$$
$$+ (\phi(a)_3 - \phi(b)_3)^2, \quad if \quad \phi(a)_0 = \phi(b)_0.$$

(Note that $\phi(a)_0 = \phi(b)_0$ represents *simultaneity*.) Nevertheless, an infinite number of other quantities and relations can be defined in R^4 besides the two displayed on the right-hand side of (10), and not all of these represent physically real properties and relations. Thus, from a modern point of view, the condition $\phi(a)_i = \phi(b)_i$ $(i = 1,2,3)$ (sameness of spatial position) represents no physically real relation, although from Newton's original point of view it does: it is just the relation of being at rest relative to absolute space. In a

237

modern treatment we drop this latter (two-place) relation; instead we postulate a three-place relation $R(a,b,c)$ and add the condition

$$R(a,b,c) \text{ iff } \phi(a), \phi(b), \phi(c) \text{ are } \textit{collinear} \qquad (11)$$

to (10). (Thus, $R(a,b,c)$ is just the relation of lying on an inertial trajectory or space-time *geodesic*.)

The difference between Newtonian and contemporary physics comes to this. There is an aspect of the mathematical structure of R^4—the relation $\phi(a)_i = \phi(b)_i$ ($i = 1,2,3$)—that Newton took seriously and we do not. According to Newton, this relation functions as a genuine explanation or reduction, for he uses it to define the relation $R(a,b,c)$ and thereby to state his laws of motion. We, on the other hand, take $R(a,b,c)$ as primitive and state the laws of motion directly in terms of this latter notion. According to us, the relation $\phi(a)_i = \phi(b)_i$ ($i = 1,2,3$) functions only as a representation: it is simply another piece of mathematical structure in R^4 with which the real relation $R(a,b,c)$ is correlated via (11). This difference between Newtonian and contemporary physics is reflected in the degree of arbitrariness attributed to the representation $\phi: A \rightarrow R^4$. According to Newton, ϕ is unique up to an element of $O^3 \times T$; according to us, ϕ is unique only up to a Galilean transformation.

In any case, physicists themselves distinguish between aspects of mathematical structure that are meant to be taken literally—that really correspond to pieces of the physical world—and aspects of mathematical structure that are purely representative. What is the rationale for this distinction? Why not regard all theoretical structure as equally representative? Why not identify the physically significant part of the theoretical world with precisely the observable part of this world? We are free to use any theoretical structure we like to derive consequences about the observable world, so long as we regard this theoretical structure as a mere mathematical representation. So again, what advantage do we gain by claiming an actual reduction?

In order to get a handle on these questions, let us begin by considering a relatively uncontroversial physical theory: our old friend the kinetic theory of gases. This theory attempts to explain the behavior of gases by assuming (i) that gases are large configurations of tiny molecules; (ii) that the pressure p of a gas is generated by the collisions of molecules with the container in which the gas is enclosed and the temperature T of a gas is proportional to the mean kinetic

238

energy of the molecules; and (iii) that molecules interact according to the laws of Newtonian mechanics. Here we see the typical pattern of a theoretical embedding. Assumptions (i) and (ii) identify gases and their properties with elements of a higher-level theoretical structure, and (iii) supplies the laws governing this higher-level structure. On the basis of this embedding we derive the laws governing the lower-level structure. For example, in the presence of the idealizing assumption (iv) that molecules have negligible size and that there are negligible intermolecular forces, we can derive the Boyle-Charles law for an ideal gas:

$$pV = nRT. \tag{12}$$

Our question now becomes: Why should we prefer the complicated description of a gas given by kinetic theory to the phenomenological law (12)? Or, rather, why should we regard the description supplied by kinetic theory as anything more than a representation of the law (12)?

One obvious problem is that (12) is false. No ideal gases actually exist, and the behavior of any real gas diverges from (12) to a greater or lesser degree. In terms of the kinetic theory, (12) holds only under the idealizing assumption (iv), which is never more than approximately true. To obtain a more accurate law we have to add terms to (12) that take account of both the finite sizes of real molecules and the real intermolecular forces. One such law is the van der Waals law:

$$\left(p + \frac{n^2a}{V^2}\right)(V - nb) = nRT \tag{13}$$

where b depends on the size of the molecules constituting the gas in question and a depends on the intermolecular forces. Given an appropriate theory of molecular structure and intermolecular forces, we can explicitly define a and b and then go on to derive (13) from the kinetic theory.[11] Thus, our higher-level structure supplies properties and relations—the quantities a and b—that are not available in our lower-level structure, and these properties and relations are necessary for the formulation of accurate laws. By contrast, if we stay on the level of our original phenomenological structure—with just the quantities p, V, and T—we cannot formulate an accurate gas law.

[11] See, e.g., Castellan [15], 477–479.

The relationship between (12) and (13) is precisely analogous to the relationship between (4) and (8). In both cases we start with a phenomenological law formulated in terms of more or less observational quantities: relative acceleration in (4); p, V, and T in (12). It turns out that this phenomenological law is false: no actual concrete reference frame is inertial; no actual gas is ideal. In both cases we can see just what is missing from the point of view of a higher-level theory: Newtonian theory tells us that we have to take account of B's absolute acceleration and rotation via (8); kinetic theory tells us that we have to take account of the sizes and interactions of the molecules constituting our gas via (13). In both cases we correct an initial phenomenological law by relating relatively observable entities to higher-level theoretical entities: we correct (4) by relating B to the Newtonian's unoccupied inertial trajectories; we correct (12) by relating our gas to the kinetic-theorist's unobservable molecules.[12]

Here we see one clear virtue in literally embedding a phenomenological structure $\mathscr{B} = \langle B, R'_1, \ldots, R'_m \rangle$ in a theoretical structure $\mathscr{A} = \langle A, R_1, \ldots, R_n \rangle$. If $\mathscr{B}$ is literally a submodel of $\mathscr{A}$, then $\mathscr{A}$ induces theoretical properties and relations on objects in B, properties that are in general necessary for stating accurate laws about these objects. On the other hand, the assertion that $\mathscr{B}$ is only embeddable into $\mathscr{A}$ will not induce the necessary theoretical properties and relations. Let R_j be a theoretical property that corresponds to no property in $\mathscr{B}$ (that is, $j > m$). Unless R_j is definable from $R_1, \ldots, R_m$, there will be automorphisms of A that preserve these latter but do not preserve R_j. Therefore, there will be two different embeddings, ϕ and ψ, of $\mathscr{B}$ into $\mathscr{A}$ such that, for some object b in B, $R_j(\phi(b))$ and $\neg R_j(\psi(b))$. Hence, unless the theoretical property R_j is actually definable from the observational properties $R_1, \ldots, R_m$, it will not induce a well-defined theoretical property on objects in B. (This is why the relation $\phi(a)_i = \phi(b)_i$ ($i = 1,2,3$) on R^4 does not in general induce a relation of sameness-of-spatial-position on spacetime.) Here we see one way in which the *uniqueness* of the mapping (the identity mapping) involved in the submodel relation pays off.

Note, however, that this kind of defect in a purely representational interpretation of $\mathscr{A}$ can always be solved by the addition of new

---

[12] This "corrective" function of theoretical structure is emphasized by Sellars in "The Language of Theories" ([99], chap. 4).

primitive properties and relations to $\mathscr{B}$. If we add the restricted relations $R'_{m+1}, \ldots, R'_n$ to $\mathscr{B}$ and the conditions $\phi(R'_{m+j}) = R_{m+j}|_{\phi(B)}$ ($j \leq (n - m)$) on our mapping ϕ, the assertion that $\mathscr{B}$ is embeddable into $\mathscr{A}$ will generate precisely the same laws about objects in B as the assertion that $\mathscr{B}$ is a submodel of $\mathscr{A}$. This, of course, is just what happens when we decide to postulate absolute acceleration as an additional primitive quantity: we expand $\langle \mathscr{P}, t(a,b), d_S(a,b) \rangle$ to $\langle \mathscr{P}, t(a,b), d_S(a,b), a^i(\sigma) \rangle$. We can follow a similar strategy in the case of our phenomenological gas law: we add new primitive quantities (a and b) and additional appropriate constraints on our embedding of gases into the world of kinetic theory; from this expanded embedding the more accurate gas law (13) will follow. Hence, although we cannot eliminate theoretical *ideology* without sacrificing accuracy of laws, we can always eliminate theoretical *ontology* with no such sacrifice.[13] So why not give theoretical ontology a purely representative interpretation?

To answer this question we must turn our attention from the problem of stating accurate laws to the problem of confirming such laws. Moreover, I believe that this is where the truly decisive difference between representations and reductions emerges: representations do not behave the way reductions do in the process of confirmation. In general, a purely representative construal of theoretical structure (ontological or ideological) will be less well-confirmed than a literal interpretation. Thus, for example, the ordinary literal interpretation of kinetic theory is much more plausible—much less *ad hoc*—than the crazy theory we get by taking a and b in (13) as primitive properties of actual gases. Why is this so?

The key to understanding these matters is provided by a property of theoretical structure that has been often noticed in the philosophical literature, although not, I think, with the proper emphasis:

---

[13] Again, one may question the point of such ontological parsimony if we have to postulate nonobservational primitive properties and relations instead. In the case of space-time, we saw that the Leibnizean indiscernibility argument supplies the difference. Yet something very much like the Leibnizean argument applies in the kinetic theory example as well. Consider a uniform gas composed of qualitatively identical molecules: for example, a hydrogen gas. Now exchange the positions in phase space of any two of these molecules: the result is a distinct but indistinguishable state of affairs. So Leibniz's argument applies to *any* theoretically uniform ontology. (In Leibniz's original argument the theoretically uniform property is curvature.) Here I am indebted to Paul Horwich [61], 409.

namely, its *unifying power*.[14] A good or fruitful theoretical structure does not serve only to provide a model for the particular phenomenon it was designed to explain; rather, in conjunction with other pieces of theoretical structure, it plays a role in the explanation

[14] Thus, for example, Hempel in one place writes: "a worthwhile scientific theory explains an empirical law by exhibiting it as one aspect of more comprehensive underlying regularities, which have a variety of other testable aspects as well, i.e., which also imply various other empirical laws. Such a theory thus provides a systematically unified account of many different empirical laws" ([58], 444). Several authors have identified the unifying power of scientific theories with the goal of scientific understanding. William Kneale wrote in 1949: "When we explain a given proposition we show that it follows logically from some other proposition or propositions. But this can scarcely be a complete account of the matter. . . . An explanation must in some sense simplify what we have to accept. Now the explanation of laws by showing that they follow from other laws is a simplification of what we have to accept because it reduces the number of untransparent necessitations we need to assume. . . . What we can achieve . . . is a reduction of the number of independent laws we need to assume for a complete description of nature" ([63], 91–92). Schlick said essentially the same thing in 1918: "if we proceed in the fashion described, the number of phenomena explained by one and the same principle becomes ever greater, and hence the number of principles needed to explain the totality of phenomena becomes ever smaller. For since one thing is continually being reduced to something else, the set of things not yet reduced . . . steadily diminishes. Consequently, the number of explanatory principles used may serve as a measure of the level of knowledge attained, the highest level being that which gets along with the fewest explanatory principles that are not themselves susceptible of further explanation. Thus the ultimate task of knowing is to make this minimum as small as possible" ([98], 13). The present author tried to elaborate such an account of scientific understanding in [38]. Our present discussion goes beyond these ideas by connecting theoretical unification with theoretical *confirmation*. However, as several people (especially William Harper and Mary Hesse) have pointed out to me, precisely this connection was made by William Whewell in 1840 in his *Philosophy of the Inductive Sciences* [117]. Whewell introduces the idea in the following striking passage: "the evidence in favour of our induction is of a much higher and more forcible character when it enables us to explain and determine cases of a *kind different* from those which were contemplated in the formation of our hypothesis. . . . Accordingly the cases in which inductions from classes of facts altogether different have thus *jumped together*, belong only to the best established theories which the history of science contains. . . . I will take the liberty of describing [this feature] by a particular phrase; and will term it the *Consilience of Inductions*" ([117], vol. 2, 230; more generally, see pp. 212–259 and the rather wonderful "Inductive Tables" following p. 282). If the present treatment adds anything to Whewell's notion of consilience it is perhaps the thought (itself borrowed from Boyd and Putnam; see note 16 below) that the whole point of *theoretical* structure is to facilitate this process of inductive "jumping together." (Thus, Whewell does not apply his notion of consilience where, from the present point of view, one would most expect it: namely, in the explanation of theoretical postulation, or what Whewell calls "the Induction of Causes" (ibid., 260–271, 577–586).)

242

of many other phenomena as well. Thus, by assuming that gases are literally composed of tiny molecules subject to the laws of Newtonian mechanics, we can explain the van der Waals law (13), but this is only a small fraction of our total gain. First, we can also explain numerous other laws governing the behavior of gases, such as Graham's law of diffusion and Dalton's law of partial pressures—though of course these laws will also follow from a nonliteral interpretation of kinetic theory. Second, however, we can also integrate the behavior of gases with the behavior of other kinds of objects. The hypothesis of molecular constitution, in conjunction with atomic theory, for example, helps us to explain chemical bonding, thermal and electrical conduction, atomic energy, and literally hundreds of other phenomena. Similarly, Newton's laws of mechanics (another central hypothesis of kinetic theory) figure in the explanation of planetary motion, projectile motion, the tides, and so on. In the absence of the theoretical structure supplied by our molecular model—in the absence of a literal molecular world—the behavior of gases simply has no connection at all with these other phenomena, and our picture of the world is much less unified.

Now, I claim that the point of this kind of theoretical unification is not merely aesthetic; it also results in our picture of the world being much better confirmed than it otherwise would be. A theoretical structure that plays an explanatory role in many diverse areas picks up confirmation from all these areas. The hypotheses that collectively describe the molecular model of a gas of course receive confirmation via their explanation of the behavior of gases, but they also receive confirmation from all the other areas in which they are applied: from chemical phenomena, thermal and electrical phenomena, and so on. By contrast, the purely phenomenological description of a gas—the nonliteral embeddability claim—receives confirmation from one area only: from the behavior of gases themselves. Hence, the theoretical description, in virtue of its far greater unifying power, is actually capable of acquiring more confirmation than is the phenomenological description.

This may seem paradoxical. Since the phenomenological description of a gas is a logical consequence of the theoretical description, the degree of confirmation of the former must be at least as great as the degree of confirmation of the latter. My claim, however, is not that the phenomenological description is less well-confirmed than the theoretical description after the former is derived from the latter—this, of course, is impossible. Rather, the phenomenological

243

description is less well-confirmed than it would be if it were *not* derived from the theoretical description but instead taken as primitive. The phenomenological description is better confirmed in the context of a total theory that includes the theoretical description than it is in the context of a total theory that excludes that description. This is because the theoretical description receives confirmation from indirect evidence—from chemical phenomena, thermal and electrical phenomena, and the like—which it then "transfers" to the phenomenological description.[15] If the phenomenological description is removed from the context of higher-level theory, on the other hand, it receives confirmation only from direct evidence—from the behavior of gases themselves.

In other words, a total theory rich in higher-level structure is likely to be better confirmed than a total theory staying on the phenomenological level, even though the latter theory may have precisely the same observational consequences as the former. It is not that phenomenological laws cannot themselves have unifying power. Newton's law of gravitation, for example, even in the absence of theoretical hypotheses about molecules and so on, unifies planetary motions, projectile motions, the behavior of the tides, and so on. The point is that theoretical structure dramatically increases the unifying power of our total picture of the world, thereby dramatically increasing its potential for confirmation.

Can we give this intuitive notion of unifying power more content? Can we describe in more detail the alleged difference in confirmation between the literal interpretation of $\mathscr{A}$ as a real reduction and the phenomenological interpretation of $\mathscr{A}$ as a mere representation? The way to do this, I think, is to look at the differences in how reductions and representations interact *over time*. Let $\mathscr{A}$ be the theoretical structure provided by molecular theory, for example. On a literal construal $\mathscr{A}$ is thought of as a real molecular world, and we suppose that all observable things—objects in B—are literally constituted by molecules. When we give a theory of gases,

---

[15] I am not claiming that every increase in the degree of confirmation of the theoretical description leads to an increase in the degree of confirmation of the phenomenological description. I do not endorse the dubious "consequence condition" (see Hempel [58], 31; Carnap [11], 471ff.). I am only pointing out that if the degree of confirmation of the theoretical description is *sufficiently* increased (for example, if it actually exceeds the prior probability of the phenomenological description), the degree of confirmation of the phenomenological description will be correspondingly increased as well. (Here I am indebted to Adolf Grünbaum.)

244

we make certain assumptions about $\mathscr{A}$—that molecules obey the laws of Newtonian mechanics, are affected by such-and-such inter-molecular forces, and so on—and certain assumptions about properties and relations in $\mathscr{B}$—that the property of being a gas is identical with the property of having such-and-such molecular constitution, the property of temperature is identical with mean kinetic energy, and so forth. Given these assumptions, we derive laws about the behavior of gases. Now, when we apply the molecular model to chemical phenomena, we take the same structure $\mathscr{A}$ and make further assumptions about it. We suppose that molecules are themselves composed of tiny atoms and that atoms obey the valence-theoretic rules of combination represented in the periodic table. We then make further assumptions about the relationship between the observable world $\mathscr{B}$ and the structure $\mathscr{A}$: that each chemical substance is a configuration of molecules with a particular atomic structure, that properties of chemical elements are properties of atomic structures, and so on. Given these assumptions, we derive the phenomenological laws of chemical combination.

The important point here is that, on a literal construal of $\mathscr{A}$, our theory *evolves by conjunction*. Certain assumptions about $\mathscr{A}$ play a role in the explanation of the gas laws. These same assumptions, in conjunction with further assumptions about $\mathscr{A}$, play a role in the later explanation of chemical combination. As a result, our initial assumptions about $\mathscr{A}$ receive confirmation at two different times. This is just what does not happen on a nonliteral construal of $\mathscr{A}$. Under this latter interpretation, of course, there is no real molecular world $\mathscr{A}$, only the phenomenological world $\mathscr{B}$. We start by asserting the existence of one representation ϕ into a mathematical molecular model $\mathscr{A}$: ϕ is appropriately constrained on the property of being a gas, of having such-and-such a temperature, and so forth. Given this representation, we derive laws about the behavior of gases. When we extend our theory to cover chemical phenomena, we assert the existence of a second representation ψ into a mathematical molecular-atomic model $\mathscr{A}'$: ψ is appropriately constrained on the property of being a chemical substance, of having such-and-such a valence, and so on. Given this second representation, we derive the laws of chemical combination. Note, however, that under this interpretation our theory does not evolve by conjunction: there is no common hypothesis that receives confirmation at two different times. On the contrary, all initial assumptions about $\mathscr{A}$ have been "trapped" within the scope of the mapping ϕ.

245

To put the matter more precisely, there is a central type of theoretical inference that is valid on the hypothesis of a genuine reduction but not on the hypothesis of a mere representation. This inference has the form:

$$\frac{\langle B,R_1 \rangle \subseteq \mathscr{A} \quad \text{and} \quad \mathscr{A} \in \Delta_1}{\langle B,R_1,R_2 \rangle \subseteq \mathscr{A} \quad \text{and} \quad \mathscr{A} \in \Delta_1 \cap \Delta_2} \tag{14}$$

where Δ_1 and Δ_2 are classes of models. Under a nonliteral interpretation the corresponding inference is:

$$\frac{\exists \mathscr{A} \exists \phi : \langle B,R_1 \rangle \to \mathscr{A} \quad \text{and} \quad \mathscr{A} \in \Delta_1}{\exists \mathscr{A}'' \exists \chi : \langle B,R_1,R_2 \rangle \to \mathscr{A}'' \quad \text{and} \quad \mathscr{A}'' \in \Delta_1 \cap \Delta_2.} \tag{15}$$

But, of course, (15) is not valid. The problem is twofold. First, we have different models $\mathscr{A}$ and $\mathscr{A}'$ in the two premises. This reflects the fact that, under a nonliteral interpretation, $\mathscr{A}$ is simply an abstract mathematical object; it is not a real particular "world" in its own right. Second, even if we do hold $\mathscr{A}$ constant in the premises (15), the inference still does not go through, for we have no guarantee that the class of mappings ϕ conforming to the first premise and the class of mappings ψ conforming to the second premise have a mapping χ in common. In (14) we have a single mapping, the identity map, in both cases. Here is another way in which the uniqueness of the mapping involved in the submodel relation pays off.

In short, the conjunction of two reductions implies a single joint reduction; the conjunction of two representations does not in general imply a single joint representation.[16] As a result, reductions and representations interact differently over time and receive correspondingly different boosts in confirmation. Suppose we have two reductions, A and B, that receive individual boosts in confirmation at t_1 and t_2 respectively. Then, working in conjunction, they imply

[16] This point was first made by Richard Boyd and Hilary Putnam, who also used it in support of a "realist" interpretation of scientific theories; see Boyd [6] and Putnam [80], chap. 11 of vol. 2. The Boyd-Putnam idea was that only a "realist" (literalist) attitude toward theoretical structure explains the presumably common scientific practice of conjoining two independently supported theories. The point of the present discussion is that this practice plays a central role in the *confirmation* of theories; if we give up this practice, we give up an important source of confirmation.

246

a new observational prediction at t_3 that does not follow from either A or B separately. Diagramatically:

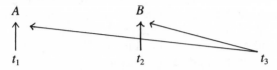

A receives *repeated* boosts in confirmation: one at t_1 and one at t_3. B also receives repeated boosts: one at t_2 and one at t_3. Now, consider the associated representations $\exists\phi(A)$ and $\exists\psi(B)$. Since they have the same observational consequences as A and B respectively, they receive at least the same boosts at t_1 and t_2 respectively. However, we cannot derive the same observational prediction at t_3 from the conjunction $\ulcorner \exists\phi(A) \ \& \ \exists\psi(B) \urcorner$; instead, we need the new joint representation $\exists\chi(A \ \& \ B)$. Diagramatically:

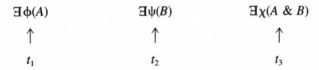

Here there is no hypothesis receiving repeated boosts in confirmation. This is why representations are in general less well-confirmed than reductions.

Note the central importance of time in our picture. Confirmation accrues to an hypothesis by virtue of the derivation of an observational consequence at a particular time, not by virtue of the timeless fact of derivability *simpliciter*. Thus, consider the reduction A in our schematic example above. Let us suppose that A is formulated before t_1, before we have any idea about conjoining A with the further hypothesis B. When we do conjoin A with B, therefore, we are subjecting A to a new and more severe test, one whose outcome cannot be anticipated at t_1. If A passes this test (at t_3), it picks up a substantial increase in confirmation. By contrast, consider the representation $\exists\chi(A \ \& \ B)$. Although $\exists\chi(A \ \& \ B)$ certainly implies timelessly all the observational consequences of A and B, including the observational consequences verified at t_3, it is not formulated until t_3. So $\exists\chi(A \ \& \ B)$ is not subject to the same kind of test as A and B at t_3, and it therefore picks up substantially less confirmation. In fact, the only way in which the representation $\exists\chi(A \ \& \ B)$ could be as well-confirmed as the reduction $\ulcorner A \ \& \ B \urcorner$ is if some genius

247

had already formulated $\exists\chi(A \ \& \ B)$ before t_1, that is, only if temporal evolution is completely removed from our picture.

These are the considerations, it seems to me, that explain our preference for a literal interpretation of kinetic theory over a non-literal, purely representative interpretation. This is why the latter seems so completely *ad hoc*. On a literal interpretation the temporal evolution of our theory displays a characteristic conjunctive pattern:

On a nonliteral interpretation there is no conjunctive pattern:

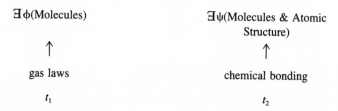

In particular, the representation $\exists\psi$(Molecules & Atomic Structure) is not implied by the conjunction $\ulcorner\exists\phi$(Molecules) & $\exists\chi$(Atomic Structure)$\urcorner$. As a consequence, the hypotheses involved in the nonliteral version of molecular theory receive fewer boosts in confirmation. Moreover, as times goes on, as our molecular hypotheses play a role in more and more derivations, this disparity in confirmation becomes greater and greater. The difference in *logical form* between reductions and representations, as reflected in the difference between (14) and (15), makes all the difference in the world.

These considerations gain further support from our earlier example of Newtonian kinematics and the relation $\phi(a)_i = \phi(b)_i$ ($i = 1,2,3$) of absolute rest. The point of that example, it will be recalled, was that Newton took the relation of absolute rest literally, whereas contemporary physics regards this relation as having a purely representative significance. How can we explain our current preference for a nonliteral interpretation? One way, of course, is to emphasize that the relation of absolute rest is not needed for for-

248

mulating an accurate law of motion: the relation $R(a,b,c)$ of collinearity (11) suffices. However, there is a second and, I think, more illuminating explanation: the relation of absolute rest has no unifying power in the context of Newtonian theory. It plays no role, for example, in unifying Newtonian space-time geometry and Newtonian gravitation theory. To see this, observe that the Newtonian theory of space-time geometry is equivalent to the assertion:

$$\exists \phi : M \to R^4 \text{ satisfying (10).} \tag{16}$$

On the other hand, Newtonian gravitation theory (§III.3 version) requires:

$\exists \psi : M \to R^4$ such that freely falling trajectories satisfy

$$\frac{d^2\psi_i}{dt^2} = -\frac{\partial \Phi}{\partial \psi_i} \qquad (i = 1,2,3) \tag{17}$$

where Φ is the gravitational potential of §III.3. The unification of (16) and (17) is just

$$\exists \chi : M \to R^4 \text{ satisfying (10) and (17).} \tag{18}$$

But (18) will be indeed forthcoming just so long as the relation $R(a,b,c)$ of collinearity (11) is defined on M. The class of mappings ϕ satisfying (16) and the class of mappings ψ satisfying (17) will have a common mapping χ just so long as *inertial* coordinate systems exist on M, and precisely this is guaranteed by the condition (11) on the relation $R(a,b,c)$. There is no need to invoke the relation of absolute rest—no need, that is, for the existence of rest systems.

This example also vividly illustrates how the notion of unifying power is a relative notion: an aspect of theoretical structure has unifying power (or not) only in the context of some second theory with which our initial theory is to be unified. Thus, we have just seen that absolute rest has no unifying power in the context of Newtonian gravitation theory. However, in the context of electrodynamics absolute rest does have unifying power; the relation $R(a,b,c)$ will not suffice. To see this, observe that electrodynamics implies the assertion:

$$\exists \psi : M \to R^4 \text{ such that } (g^{ij}) = \text{diag}(1,-1,-1,-1) \tag{19}$$

where g^{ij} is the inverse Minkowski metric. The unification of (16) and (19) requires

$$\exists \chi : M \to R^4 \text{ satisfying (10) and (19).} \tag{20}$$

But (20) will not follow unless the relation of absolute rest is defined on M and g^{ij} is defined via (III.72a). Unlike (18), (20) requires the existence of rest systems, not simply the existence of inertial systems (see p. 157). In this context, then, absolute rest does have unifying power and therefore should be taken literally.

More generally, we can put the matter as follows. Physical theories postulate a structure $\mathscr{A} = \langle A, R_1, \ldots, R_n \rangle$ that is intended to be taken literally, this is supposed to have physical reality. Physical theories also typically invoke various representative elements—pieces of mathematical structure that are not intended to be taken literally. In this latter category we find such things as choices of coordinates, units, and so on;[17] in our present context, we are always dealing with a representation $\phi: \mathscr{A} \to R^4$. Our problem is to find a rationale for this practice. On what basis do we assign some pieces of structure to the "world" of physical reality $\mathscr{A}$ and other pieces to the "world" of mathematical representation R^4? My answer is that this practice is based on the unifying power of theoretical structure. A particular piece of structure postulated by an initial theory of the form

$$\exists \phi \in \Phi : \mathscr{A} \to R^4$$

(where Φ is a class of mappings) has unifying power in the context of a second theory of the form

$$\exists \psi \in \Psi : \mathscr{A} \to R^4$$

(where Ψ is a second class of mappings) just in case it facilitates the inference to

$$\exists \chi \in \Phi \cap \Psi : \mathscr{A} \to R^4.$$

If this inference goes through even without the structure in question, as in the example of absolute rest and gravitation theory, it has no unifying power and can be safely dropped from $\mathscr{A}$. If, on the other hand, the structure in question plays a necessary role in many such inferences, we have no choice but to take it literally, to assign it a rightful place in the "world" of physical reality $\mathscr{A}$. For otherwise our total theory of $\mathscr{A}$ is much less well-confirmed.

[17] See Hartry Field [33] for an illuminating discussion of these matters in the context of a general account of applied mathematics.

4. Realism about Space-Time

We have now seen enough, I hope, to give us some confidence in the account of theoretical unification sketched above. It appears to capture faithfully our practice of assigning some bits of structure to the "world" of physical reality and others to the "world" of mathematical representation; moreover, it supplies a plausible rationale for this practice. It explains why we do not limit the "world" of physical reality to the observable "world"—this would lead to a drastic loss of unifying power—and it also explains why we do not go so far as to attribute physical reality to coordinate systems, choices of units, and so on—for these latter can be treated as purely representative with no loss of unifying power whatever. Furthermore, the account gives us just the results we want in the case of our paradigmatically "bad" theoretical entity: Newtonian absolute space. Of course, this case is an instance of ideological parsimony; it is a matter of dropping an illegitimate relation from the structure $\mathscr{A}$. Our account should apply equally well, however, to issues about ontological parsimony, that is, questions concerning the proper scope of the domain A. So let us now apply our account of theoretical unification to the ontological dispute underlying the absolutist/relationalist debate. Do considerations of unifying power favor the structure $\langle M,D,dt,h \rangle$ (or a relativistic analogue) over the structure $\langle \mathscr{P},t(a,b),d_S(a,b),a^i(\sigma) \rangle$ (or a relativistic analogue)?

To answer this question we must compare the temporal evolution of our theories under the absolutist interpretation with their temporal evolution under the relationalist interpretation. We want to know if the former exhibits an important conjunctive pattern that the latter does not. For simplicity, let us examine only a segment of this evolution: the evolution of Newtonian space-time geometry and Newtonian kinematics, and the transition to special relativistic space-time geometry and special relativistic kinematics. For the present, we shall ignore gravitation theory, electrodynamics, and general relativity. Nevertheless, an examination of this small segment should be sufficient to capture the essential differences (if any) in the inference patterns involved in the two interpretations.

What does this segment of theoretical evolution look like on the absolutist conception? The underlying assumption of this conception is that there exists a class of inertial curves or trajectories, a class of space-time geodesics, that is completely independent of

251

occupation by material processes and events and serves as a background framework for all spatio-temporal properties and relations. We begin, then, with

$$\text{There exists an affine space-time } \mathscr{A} = \langle M,D\rangle. \quad (21)$$

Of course, Newton himself began with the class of temporally extended spatial positions, rather than the class of space-time geodesics: that is, with the much stronger assumption that $\mathscr{A} = E^3 \times R$. Here, and elsewhere, clarity is served by employing weaker assumptions than Newton's.

To construct a Newtonian theory of space-time geometry within this framework we add the objects dt and h to $\mathscr{A}$ and the corresponding functions $t(a,b)$ and $d_S(a,b)$ to $\mathscr{B} = \langle \mathscr{P},t(a,b),d_S(a,b)\rangle$. The basic result we want is the existence of rigid Euclidean coordinates (10), and, as we have seen, this can be secured by equations (III.57)–(III.61). Let us call a space-time $\langle M,D,dt,h\rangle$ *rigid Euclidean* if (III.57)–(III.61) hold. We assert

$$\mathscr{A} \text{ is rigid Euclidean \& } \mathscr{P} \subseteq M. \quad (22)$$

(Again, of course, Newton himself made the much stronger assumption that $\mathscr{A}$ is flat; but it will better serve our purposes to separate space-time geometry and kinematics.) We also assert

$$t(a,b) = |t'(a) - t'(b)| \text{ for some time-function } t' \text{ of } dt$$
$$d_S(a,b) = d_S^*(a,b), \text{ where } d_S^* \text{ is the spatial metric on planes of} \quad (23)$$
$$\text{simultaneity induced by } h.$$

In other words, we assert that $\langle \mathscr{P},t(a,b),d_S(a,b)\rangle \subseteq \langle M,D,dt,h\rangle$. From (22) and (23) it follows that

$$\exists\phi:\langle \mathscr{P},t(a,b),d_S(a,b)\rangle \to R^4 \text{ satisfying (10)} \quad (24)$$

and this is all we need for our theory of Newtonian space-time geometry.

In formulating Newtonian kinematics, the basic result we want is Newton's Law of Inertia. We can easily secure this result by asserting

$$\mathscr{A} \text{ is flat,} \quad (25)$$

by adding the quantity $a^i(\sigma)$ to $\mathscr{B}$, and by identifying $a^i(\sigma)$ with the flat affine structure D. That is,

$$a^i(\sigma)_{\langle x_i\rangle} = (D_{T_\sigma}T_\sigma)^i_{\langle x_i\rangle} \quad (26)$$

or $\langle \mathscr{P}, a^i(\sigma) \rangle \subseteq \langle M, D \rangle$. From (25) and (26) we obtain the desired representation

$$\exists \psi : \langle \mathscr{P}, a^i(\sigma) \rangle \rightarrow R^4 \text{ such that } a^i = \frac{d^2 \psi_i}{du^2}. \tag{27}$$

Moreover, from (22), (23), (25), and (26) we obtain the joint representation

$$\exists \chi : \langle \mathscr{P}, t(a,b), d_S(a,b), a^i(\sigma) \rangle \rightarrow R^4 \text{ satisfying (10) and (27)}. \tag{28}$$

In other words, we can prove the existence of Newtonian *inertial* coordinates. Assumption (28) fully represents our combined Newtonian theory of space-time geometry and kinematics.

When we move to special relativity, on the other hand, we change our conception of space-time geometry. We replace the objects dt and h by the Minkowski metric g, and we replace the pair of functions $t(a,b)$ and $d_S(a,b)$ by the single function $I(a,b)$. We replace the first conjunct of (22) by

$$\mathscr{A} \text{ is Lorentzian}. \tag{29}$$

So $\mathscr{A} = \langle M, D, g \rangle$, where g has signature $\langle 1,3 \rangle$ and is compatible with D. We replace (23) by

$$I(a,b) = \left[\int_a^b |T_\sigma| \, d\tau \right]^2 \quad \begin{array}{l} \text{where } \sigma \text{ is a geodesic} \\ \text{connecting } a \text{ and } b. \end{array} \tag{30}$$

That is, we assert that $\langle \mathscr{P}, I(a,b) \rangle \subseteq \langle M, D, g \rangle$. We need no further changes to express special relativistic kinematics, and we can retain both (25) and (26). From (25), (26), (29), and (30) we obtain the joint representation

$$\exists \xi : \langle \mathscr{P}, I(a,b), a^i(\sigma) \rangle \rightarrow R^4$$

such that

$$a^i = \frac{d^2 \xi_i}{du^2}$$

and

$$I(a,b) = (\xi(a)_0 - \xi(b)_0)^2 - (\xi(a)_1 - \xi(b)_1)^2 \\ - (\xi(a)_2 - \xi(b)_2)^2 - (\xi(a)_3 - \xi(b)_3)^2. \tag{31}$$

In other words, we can prove the existence of special relativistic inertial coordinates. Equation (31) fully represents our combined special relativistic theory of space-time geometry and kinematics.

253

What does this same segment of theoretical evolution look like on the relationalist conception? On this conception, of course, we do not postulate a class of inertial trajectories or space-time geodesics. No underlying affine space-time $\langle M,D \rangle$ serves as a background framework for spatio-temporal properties and relations. Instead, we are always dealing with the set $\mathscr{P}$ of concrete physical events and the spatio-temporal properties and relations defined on $\mathscr{P}$: we are always dealing with the structure $\mathscr{B}$ rather than the structure $\mathscr{A}$. Thus, in constructing a Newtonian theory of space-time geometry on this basis, we do not assert (21), (22), or (23). We postulate the functions $t(a,b)$ and $d_S(a,b)$ on $\mathscr{P}$ and assert

$\langle \mathscr{P}, t(a,b), d_S(a,b) \rangle$ is embeddable into a rigid Euclidean
manifold $\langle M,D,dt,h \rangle$. (32)

Assumption (32) is equivalent to the representation (24). Similarly, in formulating Newtonian kinematics, we do not assert (25) and (26). We postulate the primitive quantity $a^i(\sigma)$ on $\mathscr{P}$ and assert

$\langle \mathscr{P}, a^i(\sigma) \rangle$ is embeddable into a flat manifold $\langle M,D \rangle$. (33)

Assumption (33) is equivalent to the representation (27). Note, however, that (32) and (33) do not imply the joint representation

$\langle \mathscr{P}, t(a,b), d_S(a,b), a^i(\sigma) \rangle$ is embeddable into a flat, rigid
Euclidean manifold $\langle M,D,dt,h \rangle$. (34)

This is because (24) and (27) do not by themselves imply (28). In the absence of the common affine structure D on M, there is no guarantee that the class of rigid Euclidean coordinates (24) and the class of affine coordinates (27) overlap.

The same thing happens when we move to special relativity. We do not, of course, assert (29). We replace the pair of functions $t(a,b)$ and $d_S(a,b)$ with the single function $I(a,b)$ on $\mathscr{P}$ and assert

$\langle \mathscr{P}, I(a,b) \rangle$ is embeddable into a Lorentzian
manifold $\langle M,D,g \rangle$. (35)

Again, however, we cannot derive the joint representation (31) from the conjunction of (35) with (33). In the absence of the common affine structure D on M, there is no guarantee that the class of affine coordinates (27) and the class of Minkowskian coordinates—where $(g_{ij}) = \text{diag}(1,-1,-1,-1)$—overlap. Hence, to obtain (31) we must

254

postulate the necessary joint representation independently, namely,

$\langle \mathscr{P}, I(a,b), a^i(\sigma) \rangle$ is embeddable into a flat, Lorentzian
manifold $\langle M, D, g \rangle$. (36)

Only (36) itself implies the existence of special relativistic inertial coordinates; the conjunction of (33) and (35) does not.

Thus, the evolution (21)–(31) exhibits the conjunctive pattern (14); the evolution (32)–(36) does not. Since there is no common affine structure D on $\mathscr{P}$, we cannot "tie together" the representations (24) and (27) as in (28); nor can we "tie together" the representations (33) and (35) as in (31). Instead of the sequence (22), (23), *therefore* (24); (25), (26), *therefore* (27) and (28); and (29), (30), *therefore* (in conjunction with (25) and (26)) (31)—we have only the sequence (32), (34), (35), (36). The important point here is that the sequence of hypotheses (21)–(31) contains a set of common hypotheses $\{(21),(25),(26)\}$ that remains constant throughout our theoretical evolution. These hypotheses receive the *repeated* boosts of confirmation described above. By contrast, the sequence of hypotheses (32), (34), (35), (36) contains no common set; we simply move from one joint representation to another, and the hypotheses receive no repeated boosts of confirmation. As a result, the hypotheses involved in the absolutist sequence (21)–(31) are better confirmed that the hypotheses involved in the relationalist sequence (32)–(36).

Although this line of thought is correct as far as it goes, it does not really go far enough, for it gives us no explanation for why the *ontological* difference between $\mathscr{A}$ and $\mathscr{B}$—the difference between M and $\mathscr{P}$—should be decisive. We have seen, to be sure, that the affine connection D on M plays an essential role in tying together individual representations like (24) and (27) into joint representations like (28) and (31). The affine connection D indeed has unifying power, but this looks like an *ideological* point: we need the connection D—or, what amounts to much the same thing, the relation $R(a,b,c)$ of collinearity (11)—defined on our underlying set M. Since D (or $R(a,b,c)$) is a constant element of our theoretical structure $\mathscr{A}$, we are able to tie together our evolving theoretical hypotheses in the above manner. But the relationalist has a component of ideology precisely parallel to D: the quantity $a^i(\sigma)$—or, what amounts to much the same thing, the relation $R(a,b,c)$ restricted to $\mathscr{P}$. What happens if we make $a^i(\sigma)$ (or $R(a,b,c)$) a constant element of our relationalist structure $\mathscr{B}$? This would give our evolving relationalist theory a structure precisely parallel to our evolving absolutist theory:

$\langle \mathscr{P}, a^i(\sigma) \rangle$ is embeddable into an affine manifold $\langle M, D \rangle$ (37)

$\langle \mathscr{P}, a^i(\sigma), t(a,b), d_S(a,b) \rangle$ is embeddable into a rigid Euclidean manifold $\langle M, D, dt, h \rangle$ (38)

$\langle \mathscr{P}, a^i(\sigma) \rangle$ is embeddable into a flat manifold $\langle M, D \rangle$ (39)

$\langle \mathscr{P}, a^i(\sigma), I(a,b) \rangle$ is embeddable into a Lorentzian manifold $\langle M, D, g \rangle$. (40)

In (37)–(40), unlike in (32), (33), (35), and (36), the quantity $a^i(\sigma)$ is a constant element parallel to D. Can the constant quantity $a^i(\sigma)$ on $\mathscr{P}$ tie together our representations in the style of the constant object D on M?

It cannot. Even if the quantity $a^i(\sigma)$ or the relation $R(a,b,c)$ occurs in every single one of our evolving relationalist structures, the conjunctive inference (15) still does not go through. For example, we still cannot proceed from (39) and (40) to the joint representation (36). To see this, consider a local inertial frame adapted to a freely falling trajectory σ in a *nonflat* general relativistic space-time. Let Σ

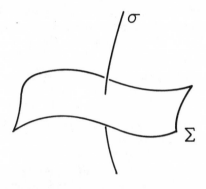

be a simultaneity surface orthogonal to σ, and suppose that Σ has non-Euclidean spatial geometry: that is, the spatial metric $d_\Sigma(a,b) = \sqrt{-I(a,b)}|_\Sigma$ is non-Euclidean. Let $\mathscr{P} = \sigma \cup \Sigma$. Clearly,

$$\langle \mathscr{P}, a^i(\sigma), I(a,b) \rangle$$

satisfies (40): by hypothesis, it is a submodel of a Lorentzian manifold. Moreover, $\langle \mathscr{P}, a^i(\sigma) \rangle$ satisfies (39): $a^i = d^2\phi_i/d\phi_0^2$ along σ, where ϕ is our local inertial coordinate system (V.12). But

$\langle \mathscr{P}, a^i(\sigma), I(a,b) \rangle$ is not embeddable into a flat, Lorentzian manifold: if it were, Σ would have Euclidean spatial geometry via (31), contrary to hypothesis.

The problem here is that the domain of definition for a^i contains just the single trajectory σ; and we can *always* map a single trajectory into a flat manifold via (27). In the context of such an impoverished domain for a^i, therefore, the embedding (27) does not force the rest of our domain to be flat as well. In particular, the plane of simultaneity Σ can have any geometry whatsoever. By contrast, if we expand the domain of definition of a^i to all the trajectories in M, the embedding (27) forces M to be flat; in this inclusive context Σ will be forced to be Euclidean as well. Hence, if our domain of definition for a^i is M rather than $\mathscr{P}$, the conjunctive inference from (27) (or, equivalently, (39)) and (40) to (36) does go through. *Here* is where the ontological richness of M pays off.[18]

In other words, the decisive advantage of the absolutist structure $\mathscr{A}$ over the relationalist structure $\mathscr{B}$ does not lie solely in the fact that $\mathscr{A}$ contains an affine connection D, for $\mathscr{B}$ contains a counterpart of affine structure, namely, the quantity $a^i(\sigma)$. The point, rather, is that the abolutist's affine structure is pervasive: its domain of definition is the entire manifold M, and, consequently, it serves as a general background framework for all other spatio-temporal properties and relations. So the basic assertions we make about this affine structure—that it is flat, or rigid Euclidean, or Einsteinian (obeying the field equation (V.2))—have important consequences for all other spatio-temporal properties and relations. We can make conjunctive inferences such as the one from (25), (26), (29), and (30) to (31). By contrast, the relationalist's primitive quantity $a^i(\sigma)$ will not serve this function. The corresponding assertions the relationalist might make about $a^i(\sigma)$—that it is embeddable into a flat, or rigid Euclidean, or Einsteinian manifold—will not have corresponding consequences. If we want to ensure the joint representation (31), for example, we have no choice but to lay down (36) as a basic primitive law: we cannot derive it from (39) and (40). As a result,

---

[18] An awkward question arises: is the set $\mathscr{P} = \sigma \cup \Sigma$ a *physically possible* distribution of mass-energy? Offhand it would seem not, for the stress-energy tensor $\mathbf{T}$ is normally assumed to be sufficiently *continuous*, and the distribution $\mathscr{P} = \sigma \cup \Sigma$ is clearly not continuous. On the other hand, I have no idea what happens to $\mathbf{T}$ when quantum mechanics is taken into account. I leave it to the reader to investigate the questions of whether $\mathscr{P} = \sigma \cup \Sigma$ is physically possible, and, if not, whether there is a physically possible distribution that makes the same point.

the relationalist's assertions about $a^i(\sigma)$ do not receive repeated boosts in confirmation via the conjunctive pattern (14).

Thus far our discussion has been rather abstract. Let us return to the problem of motion, and, in particular, to Newton's rotating bucket. Newton used his explanation of the pseudo forces generated by the rotating bucket—his derivation of the law (8)—as a central argument for the reality of both absolute motion and absolute space. We know now, of course, that Newton's argument goes too far. The derivation of (8) does not require absolute motion *per se*, just absolute acceleration and rotation. Similarly, we do not require three-dimensional absolute space, just the four-dimensional manifold of space-time geodesics. Nevertheless, the contemporary absolutist inference from (8) to the affine structure $\langle M, D \rangle$ has precisely the same *form* as Newton's original inference from (8) to absolute space. This inference has been characterized as an "inference to the best explanation":[19] Newton's theory (our current absolutist theory) provides the best available explanation for phenomena like (8); therefore Newton's postulation of absolute space (our current postulation of $\langle M, D \rangle$) is inductively warranted and scientifically legitimate.

It seems to me, however, that the notion of "inference to the best explanation" is an unfortunate one. First, theoretical explanations of the kind under consideration—a derivation of the properties of some observational structure $\mathscr{B}$ by means of an embedding of $\mathscr{B}$ in some theoretical structure $\mathscr{A}$—are all too easy to come by. Given sufficient mathematical ingenuity, we can construct such a theoretical model for any phenomenon whatsoever. Suppose, then, that we start with some observational structure $\mathscr{B}$ and that we provide a theoretical model $\mathscr{A}_1$. Suppose that $\mathscr{A}_1$ is the best available model, that it is the *only* available model, for example. According to our principle of inference, we accept $\mathscr{A}_1$, at least provisionally. But now, with sufficient ingenuity, we can construct a stronger theoretical structure, $\mathscr{A}_2$, which models $\mathscr{A}_1$ in the same way that $\mathscr{A}_1$ models $\mathscr{B}$. Should we accept $\mathscr{A}_2$? If so, what about $\mathscr{A}_3 \ldots$? Thus, a consistent application of "inference to the best explanation" leads to an infinite hierarchy of explanations of ever increasing strength. But this is absurd: virtually *any* theoretical structure will

[19] Earman [20], 299–300.

be eventually accepted in this way. More realistic principles of scientific inference should tell us when it is rational to ascend the infinite hierarchy of potential explanations *and* when it is rational to stop.

A second, related point is that the notion of "best available explanation" carries insufficient critical force. The best available explanation may nevertheless be a bad explanation, even in the absence of better alternatives. Newton's absolute space is a "bad" theoretical entity even though it provided the best available explanation for (8), and this can be seen from the point of view of Newton's theory itself (see §III.6). So Newton ascended too far in the abstract hierarchy of potential explanations when he postulated the full product structure $E^3 \times R$; he should have stopped, according to us, with the affine structure $\langle M,D \rangle$. The notion of best available explanation provides us with no tools for making this kind of criticism and, therefore, no tools for criticizing our "best available" *current* theories.

Finally, cavalier appeals to the best available explanation miss the point of the reduction/representation distinction. Given a successful derivation of the properties of some observational structure $\mathscr{B}$ via a theoretical structure $\mathscr{A}$, we can still opt for either a literal or a nonliteral interpretation of $\mathscr{A}$. We can construe $\mathscr{A}$ as a genuine reduction and suppose that $\mathscr{B}$ is literally a submodel of $\mathscr{A}$; or we can construe $\mathscr{A}$ as a mere representation and suppose that $\mathscr{B}$ is only embeddable into $\mathscr{A}$. Which construal should we adopt? The principle of "inference to the best explanation" gives us no guidance. Why is not the weakest assumption—that $\mathscr{B}$ is only embeddable into $\mathscr{A}$—the best available explanation? In other words, once we see that all theoretical structure can be given a purely representative interpretation, it is no longer clear that "inference to the best explanation" will warrant any theoretical inference. Our principle fails to explain why theoretical structure should ever be taken literally.

From the present point of view, theoretical inference is based on the notion of unifying power, not on the notion of "best available explanation." A theoretical structure should be taken literally when, and only when, it has sufficient unifying power. In rough outline the inference looks like this. At t_1 we successfully derive the properties of some observational structure $\mathscr{B}_1$ via a theoretical structure $\mathscr{A}$. As a result, the hypotheses that collectively describe the literal interpretation of $\mathscr{A}$ acquire some confirmation at t_1. Yet the amount

259

of confirmation these hypotheses acquire from this single derivation will in general be quite small, at any rate, smaller than the confirmation acquired by the nonliteral interpretation of $\mathcal{B}_1$ as merely embeddable into $\mathcal{A}$. Suppose now, however, that we successfully derive the properties of some second observational structure $\mathcal{B}_2$ via the same theoretical structure $\mathcal{A}$ at t_2. The hypotheses that collectively describe the literal interpretation of $\mathcal{A}$ acquire a second boost of confirmation at t_2. By contrast, the assertion that $\mathcal{B}_1$ is only embeddable into $\mathcal{A}$ acquires no direct confirmation at t_2—only the assertion that $\mathcal{B}_2$ is embeddable into $\mathcal{A}$ acquires such direct confirmation. If our theoretical structure $\mathcal{A}$ has sufficient unifying power, therefore, the literal interpretation of $\mathcal{A}$ will eventually acquire more direct confirmation than does the nonliteral interpretation, for the former will receive *repeated* boosts in confirmation where the latter does not. Then, and only then, will the literal interpretation of $\mathcal{A}$ be preferable.

Moreover, from our present point of view it is precisely this process of theoretical unification, and not any simple "inference to the best explanation," that justifies the absolutist's postulation of the affine structure $\langle M,D \rangle$. If we concentrate solely on Newton's original derivation of pseudo forces (8), we do not find sufficient grounds for postulating $\langle M,D \rangle$. Only as our literal interpretation puts its extra strength to work in the derivation of *further* phenomena will sufficient grounds for postulating $\langle M,D \rangle$ emerge.[20] Thus, for example, just as absolute acceleration and rotation produce dynamical effects via (8), they also produce electromagnetic effects. Just as the correct law of motion relative to a concrete reference object B is not $F = ma$ but (8), which we can abbreviate as

$$T(a,\Omega)[Newton], \tag{41}$$

the correct laws for the electromagnetic field are not Maxwell's equations but rather

$$T(a,\Omega)[Maxwell]. \tag{42}$$

The literal interpretation of $\langle M,D \rangle$ provides a unified explanation for both (41) and (42), so it receives boosts in confirmation from both.

---

[20] This no doubt *partly* explains the skepticism of historical relationalists like Leibniz and Mach. From the standpoint of "inference to the best explanation," on the other hand, this skepticism is liable to appear silly and uninteresting. See Earman [20], 300–301.

Hence, a realistic attitude toward the affine structure $\langle M,D \rangle$ is more warranted after the derivation of (42) than it was before. Nevertheless, the role of affine structure in classical electrodynamics is still quite limited. In prerelativistic space-time affine structure is largely independent of the spatial and temporal metrics: absolute acceleration and rotation have no metrical effects, and we have the geometry $E^3 \times R$ in all rigid Euclidean reference frames. In the semi-Riemannian space-time of relativity theory, on the other hand, affine structure is inseparable from metrical structure: absolute acceleration and rotation do have metrical effects. Thus, while inertial frames have the geometry $E^3 \times R$ in special relativity, non-inertial frames have the geometry

$$T(a,\Omega)[E^3 \times R] \tag{43}$$

(see §IV.1). After the special relativistic derivation of (43) our affine structure $\langle M,D \rangle$ looks even better. What happens in general relativity? We retain the affine structure $\langle M,D \rangle$, and it continues to play a role in the explanations of dynamical, electromagnetic, and metrical phenomena: we can derive analogues of (41), (42), and (43). The difference is that this affine structure is no longer absolute in sense (iii) of §II.3: it is no longer fixed independently of the material content of space-time but depends explicitly on the distribution of mass-energy. However, this last fact only increases the unifying power of $\langle M,D \rangle$, for it now figures in the explanations of many additional phenomena—from the bending of light rays to black holes. By using a space-time of variable, mass-energy dependent curvature, we dramatically increase the explanatory potential of our theory (a potential that contemporary astrophysics and cosmology are just beginning to realize). Hence, a realistic attitude toward the space-time structure $\langle M,D \rangle$ actually makes more sense after the development of relativity theory than it did before, since the unifying power of this structure has steadily increased.

In any case, an extensive theoretical unification is required before our postulation of $\langle M,D \rangle$ is warranted. The mere fact that (8) is derivable from this structure does not suffice. By the same token, the notion of theoretical unification clarifies the sense in which $\langle M,D \rangle$ provides a genuine *explanation* for (8). The claim that $\langle \mathscr{P},a^i(\sigma) \rangle$ is literally a submodel of $\langle M,D \rangle$ provides something more than just a derivation of (8)—after all, a closely analogous derivation is provided by the claim that $\langle \mathscr{P},a^i(\sigma) \rangle$ is embeddable

261

into $\langle M,D \rangle$. Rather, the former claim provides a derivation with unifying power: we deduce (8) from hypotheses that play a role in many further derivations, like the derivations of (42) and (43). These hypotheses pick up additional confirmation that, under favorable circumstances, can be "transferred" back to (8).[21] Evidence that supports (42), for example, provides indirect support for (8); and it is precisely this kind of indirect support that is missing from the derivation of (8) via the nonliteral embeddability claim. In other words, derivation from the literal interpretation of $\langle M,D \rangle$ allows the confirmation of (8) to be increased from "above"—from the additional confirmation that our absolutist theory acquires from indirect evidence—as well as from "below"—from the observational instances that (8) itself implies. Thus, we might say that to *explain* a phenomenological law is to provide a derivation for it with unifying power, to provide a derivation that can increase its confirmation from "above" via indirect evidence.

Finally, these last considerations help to account for the persistent feeling that a Machian relationalist strategy—if only it could be carried out—would result in theories that were more explanatory and less *ad hoc* than those provided by the primitive quantity strategy. For the whole point of the Machian strategy is to supply a concrete surrogate for the absolutist's affine structure, namely, a concrete inertial reference frame. A theory in which the existence of concrete inertial frames were provable would capture some of the unifying power, and therefore some of the explanatory power, of our absolutist theories. Suppose, for example, that we try to construct a unified explanation for (41) and (42) as follows. We postulate a transformation law of the form

For all frames x, the laws in x are $T(a(x),\Omega(x))$ [the laws in y], where $a(y) = \Omega(y) = 0$. (44)

We then try to derive (41) by conjoining (44) with

For all frames y, if $a(y) = \Omega(y) = 0$, then the law of motion in y is *Newton*. (45)

Similarly, we try to derive (42) by conjoining (44) with

For all frames y, if $a(y) = \Omega(y) = 0$, then the electromagnetic laws in y are *Maxwell*. (46)

[21] See p. 244 above, especially note 15.

The problem, of course, is that neither of these derivations goes through without the existential assumption.

$$\text{There } exists \text{ a frame } y \text{ such that } a(y) = \Omega(y) = 0. \qquad (47)$$

Moreover, although (44)–(46) are consequences of both strategies, (47) would be provable only in a Machian theory (if such a theory were constructed). So (47) represents the "extra" unifying power of the Machian strategy.

VII
Conventionalism

Conventionalism, like relationalism, raises skeptical doubts about the structures postulated by our space-time theories. Whereas ontological relationalism raises doubts about the domain of individuals of our theories—whether it is permissible to quantify over the entire manifold of space-time points—conventionalism raises doubts about the geometrical properties and relations, especially the metrical properties and relations, that are defined on this domain—whether or not it includes unoccupied points. Thus, conventionalism, as I understand it, is closely connected with *ideological* relationalism. We shall explore their interconnections in more detail as we proceed.

The basic conventionalist strategy is to argue that certain *prima facie* incompatible systems of description—Euclidean and non-Euclidean physical geometries, for example—are in reality "equivalent descriptions" of the same facts. Neither system is objectively or determinately true or false; rather, both can be held as true relative to different, arbitrarily chosen "coordinative definitions." Therefore, the properties and relations that occur in such systems—spatial and temporal congruence relations, for example—do not objectively and determinately exist: they can be properly said to hold or fail to hold only relative to one or another arbitrary choice of "coordinative definition."

Prominent among the many motivations for the conventionalist strategy are general epistemological worries about choosing between competing theoretical descriptions. We often encounter situations in which the grounds for making such theoretical choices—the choice between Euclidean and non-Euclidean geometries is paradigmatic here—are quite unclear. We find ourselves unable to specify the evidence or observations that could "in principle" decisively settle the issue. We are faced with a skeptical problem: the problem of theoretical underdetermination. The conventionalist

264

strategy attempts to solve this problem by maintaining that the alternative theoretical descriptions in question are not really incompatible: they are all equally correct—again, relative to different, arbitrarily chosen "coordinative definitions." So there is no more epistemological problem involved here than there is in choosing between different scales of measurement (for example, Fahrenheit and Centigrade) or different systems of coordinates (for example, polar and rectangular). To be sure, there can be pragmatic reasons of convenience for making such choices, but there is no question of truth.

These general epistemological worries will be the focus of §VII.1. From the present point of view, however, there are specific motivations deriving from the development of relativity theory that are even more pressing. Relativity theory itself appears to be based on a conception of "equivalent descriptions" straight out of the conventionalist strategy. Thus, for example, in special relativity we deny that an object has a single, true (absolute) velocity; rather, it has a variety of different, equally correct velocities relative to different, arbitrarily chosen inertial frames. Similarly, a rod no longer has a single, true (absolute) length; rather, it has a variety of different, equally correct lengths—again, relative to different, arbitrarily chosen inertial frames. In general relativity a similar relativization applies to the notions of gravitational force and gravitational acceleration. It no longer makes sense to talk about *the* gravitational acceleration of an object; rather, different gravitational accelerations will be assigned by different, arbitrarily chosen (Galilean) reference frames. Hence, relativity theory appears to exemplify conventionalism in its central principles, and this fact can furnish strong motivations—as it does for twentieth-century conventionalists like Reichenbach and Grünbaum—for a generalized conventionalist answer to the problem of theory choice (or, at least, a generalized conventionalism about *geometrical* theory).

In §VII.2 we shall examine the relativization of concepts characteristic of relativity theory and its connection with the conventionalist strategy of "equivalent descriptions." Further, we shall try to show how the relativizations effected by the development of relativity theory are connected with the epistemological problems of §VII.1. The two are not independent, of course, for relativity theory is itself motivated by skeptical (anti-"metaphysical") epistemology. Our conclusion will be that the relativizations characteristic of relativity theory have a very special structure, a structure that does not give rise to any *generalized* conventionalism or skepticism. In §§VII.3–4 we apply this lesson to two paradigmatic cases for twentieth-century conventionalism: the concepts of congruence and simultaneity. Finally, §§VII.5–6 provide an overview of the *positive* conception of scientific method that emerges from our investigation of the development of relativity theory.

1. Theoretical Underdetermination and Empirical Equivalence

The epistemological motivations for conventionalism can be best appreciated if we begin by drawing a familiar contrast between two different ways in which scientific theories fail to be constrained or determined by evidence. Lower-level, or inductive, underdetermination consists in the fact that at any given time we possess only a finite amount of evidence. An infinite number of mutually incompatible hypotheses will agree on all the evidence we actually possess while disagreeing on possible evidence we do not yet possess. There are an infinite number of distinct smooth curves through a finite set of data points, an infinite number of distinct generalizations compatible with a finite number of instances, and so on. This kind of underdetermination applies to any inductive inference from our present data, whether it involves higher-level theoretical structure or mere empirical generalizations. The second kind of underdetermination, on the other hand, applies only to higher-level theories. Such theories, we are told, are not only underdetermined by all the evidence we actually possess, but they are underdetermined by all *possible* evidence as well. Perhaps Quine characterizes this *theoretical* underdetermination most clearly:

> physical theory . . . is underdetermined by past evidence; a future observation can conflict with it. Naturally, it is underdetermined by past and future evidence combined, since some observable event that conflicts with it can happen to go unobserved. Moreover many people will agree, far beyond all this, that physical theory is underdetermined even by all *possible* observations. Not to make a mystery of this mode of possibility, what I mean is the following. Consider all observation sentences of the language: all the occasion sentences that are suited for use in reporting observable events in the external world. Apply dates and positions to them in all combinations, without regard to whether observers were at the place and time. Some of these

place-timed sentences will be true and the others false, by virtue simply of the observable though unobserved past and future events in the world. Now my point about physical theory is that physical theory is underdetermined even by all these truths. Theory can still vary though all possible observations be fixed. Physical theories can be at odds with each other and yet compatible with all possible data even in the broadest sense. In a word, they can be logically incompatible and empirically equivalent. ([83], 178–179)

Note that Quine's characterization rests on a prior distinction between two kinds of vocabulary: observational and theoretical. Given this distinction, the underdetermination of theory follows as a trivial truth of logic. There can obviously be two consistent but mutually incompatible sets of sentences agreeing on all sentences containing only observational vocabulary while disagreeing on some sentences containing theoretical vocabulary. For the present, I shall take the distinction between observational and theoretical vocabulary for granted. Later, however, we shall look at it more critically.

Theoretical underdetermination raises acute skeptical problems. What justifies us in selecting one particular alternative from a class of mutually incompatible but empirically equivalent theories as the *correct* theory? Since by hypothesis no empirical evidence can guide us either now or in the future, such a choice appears to be quite arbitrary and gratuitous. As we have already observed, conventionalism responds to this skeptical problem by accepting the apparent arbitrariness of theory choice while at the same time extending this arbitrariness to the truth-value of the theory chosen. There is no need to select one alternative theory as correct, for their seeming incompatibility is only a surface phenomenon: they are in fact "equivalent descriptions." Therefore, the selection we actually make from a class of empirically equivalent theories needs no epistemological justification after all. Considerations of pragmatic convenience—the same kinds of considerations that explain a choice of scale or coordinate system, say—are quite sufficient.

What happens if we reject the conventionalist strategy and maintain that the alternative, empirically equivalent theories in question should be taken at face value, that they are in no way "equivalent" but are just what they appear to be, namely, conflicting descriptions of higher-level theoretical entities? To avoid skepticism we must

267

hold that, in addition to conformity to evidence, there are methodological principles or criteria—simplicity, parsimony, non-ad-hocness, explanatory power, and so on—that are capable of singling out one theory from a class of empirically equivalent theories (or at least capable of narrowing the choice considerably). Moreover, we must hold that these methodological principles are themselves non-arbitrary: they are rational or justifiable because they are likely to result in true, or at least approximately true, theories. The problem for such an anticonventionalist view, then, is to describe these methodological principles more precisely and to give some reason to think they in fact do lead to true theories. Of course, no one has come close to doing this, and that is one reason conventionalism is an attractive position. On the other hand, conventionalists have their problems, too: namely, giving a clear sense to the notion of "equivalent descriptions" and showing that in problematic cases— notably, the choice between alternative physical geometries—we in fact have "equivalent descriptions."

In the absence of specific proposals the anticonventionalist's appeal to methodological principles is liable to appear to be lame wishful thinking. It is tempting, therefore, to adopt a tough-minded attitude that rejects all appeals to vague methodological criteria and relies solely on conformity to evidence. But this would be a serious mistake, for *all* inductive inference, not just inference to higher-level theories, depends on methodological principles that go beyond mere conformity to evidence. Even if we avoid the problem of theoretical underdetermination by means of the conventionalist strategy, we are still left with the problem of lower-level inductive underdetermination. Thus, for example, given any finite set of data points, we must choose one smooth curve from the infinite class of curves compatible with these points. In practice we choose the "simplest," most "economical," least "ad hoc" curve (and, we might add, the curve that best fits whatever *theoretical* beliefs we have about the subject matter in question). We invoke the same kind of vague methodological criteria to settle both kinds of underdetermination. So one can reasonably ask, I think, why such criteria should be so much more problematic in their application to higher-level theories than in their application to empirical generalizations.

Conventionalists, of course, have a ready answer to this last question. They distinguish between *inductive simplicity* and *descriptive simplicity*. Inductive simplicity encapsulates the methodological virtues that explain the choice of one empirical generalization (for

268

example, one particular smooth curve) over others equally compatible with the available evidence. *Descriptive simplicity* encapsulates the methodological properties that lead us to choose one higher-level theory (for example, one particular physical geometry) over other empirically equivalent theories. Conventionalists contend that inductive simplicity is a guide to truth, whereas descriptive simplicity is a matter of pragmatic convenience only. If this distinction is to amount to more than a simple begging of the question, however, we need some independent explanation of why inductive simplicity is a guide to truth in a way descriptive simplicity is not. Reichenbach, who invented the distinction between inductive and descriptive simplicity, thought he had just such an explanation:

> Actually in cases of inductive simplicity it is not economy which determines our choice. The regulative principle of the construction of scientific theories is the postulate of the best predictive character; all our decisions as to the choice between unequivalent theories are determined by this postulate. If in such cases the question of simplicity plays a role for our decision, it is because we make the assumption that the simplest theory furnishes the best predictions. This assumption cannot be justified by convenience; it has a truth-character and demands a justification within the theory of probability and induction.
>
> Our theory of induction enables us to give this justification. We justified the inductive inference by showing that it corresponds to a procedure the continued application of which must lead to success, if success is possible at all. The same idea holds for the principle of the simplest curve. ([87], 376–377)

Reichenbach's thought is this. Suppose we construct a smooth curve through a finite set of data points by the usual techniques of curve fitting: we connect the data points by straight lines and "round off" the cusps to form a smooth curve. The more data points we collect, the better our constructed curve approximates the true curve. If we collect enough data points, we are guaranteed of being as close to the true curve as we like. According to Reichenbach, "This procedure of preliminary drawing and later correction must lead to the true curve, if there is such a curve at all" ([87], 377).

In short, the methodological principles involved in choosing the "simplest" curve are justifiable *in the long run*. Eventually they will produce hypotheses that are arbitrarily close to the truth. Similar long-run justifications can be given for other types of lower-level

269

inductive inference: Reichenbach's famous justification of the "straight rule" for inferring probabilities from observed relative frequencies is paradigmatic here.[1] More generally, we can give a long-run justification for all methodological principles whose target hypotheses contain only observational vocabulary in the following sense: when all the observational evidence is in, when we are given all true observation sentences at all space-time points, the truth or falsity of all such hypotheses is completely determined. Therefore, when all the observational data is in, our methods are guaranteed of true results.[2] On the other hand, no such long-run justification can be given for methodological principles whose target hypotheses essentially contain theoretical vocabulary. Even when all the observational evidence is in, there is still no guarantee of true results. This is one clear difference between lower-level inductive underdetermination and higher-level theoretical underdetermination—one clear difference, therefore, between methods involving inductive simplicity and those involving descriptive simplicity.

Does this difference show that our standards of descriptive simplicity (methods for choosing between empirically equivalent theories) are problematic in a way in which our standards of inductive simplicity (methods for choosing between competing empirical generalizations) are not? I do not see how, for a long-run justification is no justification at all! We know that there will eventually come a point (either finite or infinite) when our chosen hypotheses are guaranteed of truth, but the hypotheses chosen before this point may be arbitrarily far from the truth. In particular, all the inferences we actually make, and even all the inferences that are physically possible for us to make, may have no chance whatever of producing true hypotheses. Moreover, infinitely many distinct methods of inference can be given the same kind of long-run justification. Thus, for example, in Reichenbach's curve-fitting case, imagine a method joining adjacent data points by arbitrary smooth curves rather than straight lines. So long as these curves converge toward straight lines as the distance between adjacent data points approaches zero, this method will converge to the true curve in the long run in just the same sense as our usual procedure of linear interpolation. But in the

[1] Reichenbach [87], chap. 5, §39.
[2] Note that this general long-run justification is a bit weaker than Reichenbach's, for Reichenbach shows that convergence to truth takes place when only a *finite* amount of evidence is in. However, since we can place no finite bound on Reichenbach's point of convergence, this extra strength does not help us.

short run the two convergent methods produce arbitrarily divergent conclusions.[3] Similarly, in the general case, imagine a method preferring empirical generalizations containing Goodmanesque grue-like predicates[4] to our usual "projectible" generalizations. When all the evidence is in, this method is guaranteed of truth just as much as our actual methods. In the meantime, the two are as different as could be.

Clearly, then, virtually *any* standard of inductive simplicity can be given a long-run justification. So long-run justifications cannot explain the methodological principles we in fact employ. When faced with any real case of lower-level inductive underdetermination—choosing between the usual linear interpolation and some crazy curvilinear interpolation, choosing between our usual "projectible" generalizations and crazy Goodmanesque generalizations—long-run justifications leave us just where we started. We still need some assurance that the methodological criteria that issue in such choices are reliable guides to truth in actual practice. What must be shown is not that our criteria are *guaranteed* of true results in the long run, but that they have a *good chance* of producing true results in the actual world. Furthermore, once we shift our attention from the long run to the question of reliability in the actual world, no relevant difference between inductive simplicity and descriptive simplicity remains. With respect to this latter kind of justification, there is no reason to think that our standards of descriptive simplicity (methods for choosing between empirically equivalent theories) are any more problematic than our standards of inductive simplicity.

But, one might object, is there not still an important difference between the two cases? Any mistake we make in choosing between empirical generalizations equally compatible with the available evidence can "in principle" be *corrected* in light of future evidence. On the other hand, if we make a similar mistake in the case of empirically equivalent theories, we are stuck, for no possible observation can help us to correct our error. Does this difference support the conventionalist's distinction between inductive and descriptive simplicity? I think not. In the first place, the possible observations that are always available to correct our inductive mistakes are just as evanescent as Reichenbach's point of convergence. We know that

---

[3] See [87], 353–355, for Reichenbach's honest, but unsuccessful, attempt to deal with this problem. In effect, he argues that our usual methods are "less risky" than alternative convergent methods. But how can this possibly be known *a priori*?

[4] Goodman [50], 72–81.

there are place-timed observation sentences conflicting with any mistaken generalization, but there is no guarantee they are physically accessible to us.[5] So the above "in principle" is a matter of mere *logical* possibility. In the second place, the possibility of correction does not affect the basic problem of inductive inference. If scientific method is to have any utility at all, our methods must supply reliable means for arriving at true hypotheses, not simply for rejecting false hypotheses. Suppose again that we have examined a number of data points and wish to anticipate a future point, and suppose some matter of great practical importance depends on our prediction. We need some assurance that the choice we actually make (via the "simplest" linear interpolation) is more likely to be correct than are alternative choices (via some crazy curvilinear interpolation). It is no help at all to be told that if we are mistaken we will find out, for this is true no matter what method of interpolation we choose.

The idea that the possibility of correction makes a difference here is encouraged, I think, by the following picture. As time progresses, we are successively discarding more and more false empirical generalizations and more and more unreliable methods of inductive inference. Therefore, we are gradually narrowing the class of generalizations available to us and decreasing the degree of arbitrariness in any inductive choice among the remaining alternatives. By contrast, no such process of progressive refinement affects the arbitrariness of theoretical choice: we are never in a position to discard one member of a class of empirically equivalent theories based on observation alone. Yet this picture is quite misleading. Although it is true that we are successively discarding more and more false empirical generalizations, we do not at all reduce the arbitrariness involved in any actual predictive inference. All generalizations consistent with the *then* available evidence are still open to us. Our choice is still arbitrary to an infinite degree, because the discarded generalizations and methods pertain only to the past, as it were; the future remains "infinitely open." So the degree of arbitrariness in any actual predictive inference is precisely the same as the degree of arbitrariness in any theoretical inference. In both cases *all* hy-

[5] The "indistinguishable space-times" introduced by Glymour [45; 46] provide a good illustration of this problem. In such space-times a kind of epistemological underdetermination exists, because some space-times points are *causally inaccessible*: there are space-time points that forever lie outside the light cone of any given "observer." See also Malament [70] for a clear exposition.

potheses consistent with the evidence remain live options, and in both cases this arbitrariness can be reduced only by invoking and justifying appropriate methodological principles.

Hence, there are not two different epistemological problems, one relating to our methodological criteria for making lower-level inductive inferences, the other relating to our criteria for choosing higher-level theories. Rather, there is only the single problem of *nondeductive* inference. We are given a finite amount of evidence, but an infinite number of incompatible hypotheses conform to this evidence (unlike the situation in deductive inference, where all possible conclusions are compatible—provided the premises are themselves consistent). Consequently, we need additional methodological criteria, standards of simplicity, economy, and so forth, that go beyond mere conformity to evidence. These criteria can be justified only by showing that they tend to produce true hypotheses in actual scientific practice in the real world.[6] That our procedures for making lower-level inductive inferences can be shown to lead to truth in the long run is beside the point. All our methodological procedures, whether they involve higher-level theories or lower-level empirical generalizations, stand equally in need of justification in actual practice. It is misleading, then, to divide the general problem of underdetermination into two radically different types: inductive and theoretical. Such a division only encourages an underestimation of the difficulties involved in justifying lower-level inductive procedures (based, perhaps, on the possibility of a Reichenbachian long run justification) and a consequent overestimation of the problems involved in justifying our theoretical methodological criteria. Of course, it is just this kind of overestimation that makes the conventionalist position seem so compelling.

These considerations gain further support if we now take a closer look at our assumed observational evidence. We have been assuming that there is a distinguished class of observational predicates and that our evidence consists of place-timed observation sentences— observational predicates applied to space-time points. These are assumed to be epistemologically unproblematic; the problems arise in making further inferences *from* these privileged sentences. It is

---

[6] Hence, any justification of our actual inductive methods must itself be empirical; it cannot be carried out *a priori* in the manner of Reichenbach's long-run justifications. This involves us in an inevitable, but not necessarily vicious, circularity. Cf. Friedman [40].

273

clear, however, that our procedures for arriving at observational beliefs possess no more logical guarantee of truth than do our inductive procedures, either lower-level or theoretical. This follows immediately if we concede that observational beliefs describe medium-sized objects in the physical world, if we reject the idea of a private sense-datum language. Such observational beliefs are no more uniquely determined by actual sensory input than our theoretical beliefs are uniquely determined by our observational beliefs. They can be revised in the light of future sensory stimulation, and even in the light of conflicting theoretical beliefs. So our procedures for arriving at observational beliefs require the same kind of epistemological justification as our other nondeductive procedures. There is no *guarantee* of truth, so the best we can do is to establish their *reliability*: we can attempt to show that they have a good chance of leading to true beliefs in the actual world.

Consider, for example, Quine's well-known characterization of the class of observation sentences. According to Quine, observation sentences have two salient properties: their acceptance or rejection depends directly on sensory stimulation; there is communitywide agreement on patterns of acceptance and rejection—all speakers react in the same way to the same sensory stimulations.[7] This characterization has some clear virtues: for example, it nicely captures the sense in which observation sentences can play a role in settling disputes between opposing scientific theorists. However, I doubt whether it can bear the epistemological weight Quine wants it to. Note first that it is far from clear that any fixed distinction between two types of vocabulary has in fact been drawn. Any item of vocabulary, no matter how theoretical, can figure in observation sentences in Quine's sense, provided there is sufficient community-wide agreement on the relevant theory. Think of "Here is a sample of H_2O" or even "There are electrons moving through this wire." Although at each point in time there will undoubtedly be items of vocabulary too recondite to figure in Quinean observation reports, the advance of science may eventually secure enough community-wide agreement on any given item. Every predicate may count as an observational predicate at some time. So Quine's characterization of the class of observation sentences, viewed as acting across time, may very well undermine his own statement of the underdetermination of theory. Second, communitywide agreement has no necessary

[7] See Quine [82], 85–89.

connection with truth. Suppose everyone becomes convinced (via political indoctrination, say) of the truth of Newtonian physics with absolute space. Further, everyone believes that the center of mass of the solar system is at absolute rest. In such a community "is absolutely at rest" could easily become a Quinean observation predicate. But, we want to protest, according to Newtonian physics itself, there is no way to determine, let alone observe, absolute velocity. States of absolute velocity just cannot be reliably detected in a Newtonian world; whether everyone agrees about them or not is quite irrelevant.

It seems to me, therefore, that the role of observational evidence cannot adequately be captured by any linguistic distinction between types of vocabulary. There is no fixed class of observational predicates or sentences. Rather, there exists a class of observational *procedures*: procedures for accepting sentences that (i) are based directly on sensory stimulation (as opposed to inductive procedures, which apply to sentences rather than to stimulations), (ii) command widespread agreement in the scientific community, and (iii) have a high degree of reliability (we hope). Our observational procedures do not differ in principle from our other nondeductive procedures for acquiring beliefs. In particular, any item of vocabulary can figure in beliefs acquired through observational procedures, given sufficient theoretical sophistication on the part of the observer. Moreover, our observational procedures are to be justified in the same manner as other nondeductive procedures—by establishing their reliability, showing they are in fact likely to result in true beliefs in actual practice. If we have more confidence in observational procedures than in other nondeductive procedures, it is because we can see roughly how to establish their reliability. We can see roughly how the empirical connections between sensory stimulation and physical objects in our immediate environment yield reliable beliefs about the latter.

Note, however, that if no fixed distinction exists between observational and theoretical vocabulary, the underdetermination of theory does not follow as a trivial truth of logic. The existence of empirically equivalent theories cannot be established by simple logical considerations about disjoint vocabularies. Rather, theoretical underdetermination is itself a broadly empirical matter: it depends on just what kinds of facts about the world can and cannot be reliably ascertained by our observational procedures. To show that two theories are empirically equivalent, we must show that, given our

275

general picture of what the world is like, the different states of affairs postulated by the two theories are not distinguishable by our observational procedures, given our general picture of how these procedures work.

On the other hand, I think there can be little doubt that there are empirically equivalent theories in just this sense. Consider once again the Newtonian kinematics with absolute space V of §III.1. Call this theory T_V. Now generate alternative theories $T_{V'}, T_{V''}$, and so on by choosing distinct fields V', V'', and so on to represent absolute space. These theories are all incompatible with T_V and with one another: they generate mutually inconsistent assignments of absolute velocity to any given object. Nevertheless, *according to T_V itself*, the theories $T_V, T_{V'}, T_{V''}$, and so on are all empirically equivalent. In a world in which T_V is true, the theories $T_V, T_{V'}, T_{V''}$ are all perfectly indistinguishable. Similarly, consider the Newtonian gravitation theory T_Φ of §III.3. *According to this theory itself*, the alternative theories T_Φ, T_Ψ, and the like are all mutually incompatible but empirically equivalent: variations in the gravitational potential Φ generate mutually inconsistent assignments of gravitational force and gravitational acceleration, but are indistinguishable in a matter-filled universe.

In neither of these cases is empirical equivalence established by logical considerations about vocabulary; rather, it follows from special facts about the structure of the theories in question, coupled with general empirical beliefs about what the world is like and what aspects of the world are accessible to our observational procedures. For example, we are assuming that only the trajectories T_σ of gravitationally affected particles and the distribution of mass ρ are accessible to observation; neither the absolute space V nor the gravitational potential Φ is itself observable. Moreover, both our examples involve theories violating the parsimonious methodological principle (R2). The significance of this point will emerge in what follows.

In sum, then, there *are* empirically equivalent theories. At any rate, we have good reason to believe that there are theories indistinguishable by our observational procedures as we understand them. It is also true that the existence of empirically equivalent theories presents an important epistemological problem. If we accept these theories at face value—that is, as being mutually incompatible descriptions of higher-level theoretical structure—we must formulate methodological criteria capable of guiding our choice of one alternative over others, and we must give some reason to think such criteria have a reliable connection with truth. Moreover, since this epistemological problem is a very difficult one, the conventionalist strategy of "equivalent descriptions" is *prima facie* attractive. The above considerations are intended to undermine this *prima facie* attractiveness. Conventionalists do not really avoid our epistemological problem—they face exactly the same problem. Even lower-level inductive inference to empirical generalizations depends on methodological criteria, like simplicity and economy, that necessarily go beyond mere conformity to evidence. In fact, the only discernible difference between the two cases is that our lower-level inductive criteria can be given a long-run justification of the Reichenbachian type while our theoretical criteria cannot. Yet this justification is doubly spurious. Not only does it fail to establish any connection at all between our inductive methods and truth in the real world, but it also depends on the assumption of an epistemologically unproblematic class of observation sentences.

Thus, the motivations for conventionalism based on general epistemological considerations are in the end quite weak. There is no *special* problem of theoretical underdetermination and no special need, therefore, for a strategy of "equivalent descriptions" to meet the threat of skepticism posed by empirically equivalent theories. Nevertheless, although conventionalism is not a necessary strategy, it is a possible one: it can solve at least some cases of underdetermined theory choice. Further, this possibility is not merely abstract, for the conventionalist strategy appears to be explicitly realized in the development of relativity theory: it appears to play a central role in the way relativity theory handles the two cases of empirical equivalence described above. From the present point of view, it is this application to real physical theories that gives the conventionalist strategy its great interest and importance. We must come to terms with conventionalism in order to understand fully the development of relativity theory and, of course, vice versa. I shall try to provide the foundations for this kind of mutual understanding in the next section.

2. *Relativization and "Equivalent Descriptions"*

The essence of relativity theory, of course, is a relativization of concepts that are absolute, or nonrelativized, classically. Special

277

relativity effects a relativization of the concepts of distance, time, and velocity. General relativity effects a relativization of the concepts of gravitational force and gravitational acceleration. For present purposes, it is helpful to think of these relativizations as falling into two distinct classes: the relativization of velocity effected by special relativity and the relativization of gravitational force (gravitational acceleration) effected by general relativity belonging in one class, the relativizations of distance and time effected by special relativity belonging in the other. The relativizations in the first class are primarily motivated by methodological or "conceptual" considerations, whereas those in the second are primarily motivated by "empirical" considerations—the experimental discovery of the invariance of the velocity of light. Thus, the relativizations in the first class have Newtonian counterparts: the Newtonian kinematics of §III.2 and the Newtonian gravitation theory of §III.4. These theories have the same "empirical content" as their corresponding non-relativized Newtonian theories (the kinematics of §III.1 and the gravitation theory of §III.3), so they allow us to study the relativizations in the first class in a pure ("conceptual") form.

Let us begin, then, by reviewing the relativization of velocity effected by the transition from the kinematics of §III.1 to the kinematics of §III.2. In the space-time of §III.1 alternative ascriptions of velocity to a given inertially moving object are not equally correct. Every such object has a single true velocity, namely, its velocity relative to absolute space. As a consequence, there is a privileged subclass of inertial frames—frames at rest relative to absolute space—in which relative motions correspond to true absolute motions. Unfortunately, however, motion relative to absolute space is not detectable: alternative, incompatible ascriptions of velocity turn out to be empirically equivalent. In the space-time of §III.2, on the other hand, we are no longer committed to such incompatible but empirically equivalent alternative descriptions. All inertial frames are equivalent, and all ascriptions of velocity are equally correct. An object has no single true velocity; rather, it has different velocities relative to different inertial frames. Moreover, the choice of one inertial frame over another is itself arbitrary: all inertial frames have an equal claim to represent the true motions of bodies. Hence, this case has all the elements of the conventionalist scenario. We resolve a problem about empirically equivalent but incompatible alternatives by a strategy of "equivalent descriptions": we declare the alternatives to be equally correct.

278

The relativization of gravitational acceleration effected by the transition from the theory of §III.3 to the theory §III.4 has a similar structure. In the space-time of §III.3 alternative ascriptions of gravitational acceleration to a given object are not equally correct. Every object has a single true gravitational acceleration $\ddot{b}_i$: the acceleration induced by the gravitational potential Φ—$\ddot{b}_i = -\Phi_{,i}$. As a consequence, there is a privileged subclass of reference frames—the inertial frames of the flat connection $\overset{\circ}{D}$—in which relative motions correspond to true absolute motions. Unfortunately, however, motion relative to such inertial frames is not detectable (at least in a matter-filled universe): alternative, incompatible ascriptions of gravitational acceleration turn out to be empirically equivalent. In the space-time of §III.4, on the other hand, we are no longer committed to such incompatible but empirically equivalent descriptions. We replace the inertial frames with the *Galilean* frames, and all ascriptions of gravitational acceleration induced by such frames (resulting from transforming Φ to Ψ by (III.47)) are equally correct. An object has no single true gravitational acceleration; rather, it has different gravitational accelerations relative to different Galilean frames. Moreover, the choice of one Galilean frame over another is itself arbitrary: all Galilean frames have an equal claim to represent gravitational acceleration.[8] So we again resolve a problem about empirically equivalent but incompatible alternatives by a strategy of "equivalent descriptions": we declare the alternatives to be equally correct.

Notice that the key to implementing a strategy of "equivalent descriptions" in both cases is ideological: we change our minds about what geometrical structures are defined on space-time. In moving from the space-time of §III.1 to the space-time of §III.2, we drop the indeterminate and indistinguishable vector field V. In moving from the space-time of §III.3 to the space-time of §III.4, we drop the indeterminate and indistinguishable flat connection $\overset{\circ}{D}$ and gravitational potential Φ. Thus, in the space-time of §III.1 alternative assignments of velocity are empirically equivalent but not fully equivalent: one particular assignment is true, but we cannot tell which. In §III.2 alternative assignments of velocity are both

---

[8] Note, however, that all Galilean frames are not equivalent representations of absolute acceleration relative to the nonflat connection D (see §V.4). The fact that Galilean frames are not fully equivalent in §III.4 reflects an important asymmetry between our two relativistic cases: §III.4 does not effect a full relativization of the concept of acceleration (see §§III.4,8).

empirically equivalent and fully equivalent: they are all equally true (relative to different inertial frames). Similarly, in the space-time of §III.3 alternative assignments of gravitational acceleration are empirically equivalent but not fully equivalent: again, one is true, but we cannot tell which. In §III.4 alternative assignments of gravitational acceleration are both empirically equivalent and fully equivalent: they are all equally true (relative to different Galilean frames). The point is that to move from empirical equivalence to full equivalence we need to change our minds about which theoretical properties and relations actually exist.

This point seems to have been missed in the standard literature on "equivalent descriptions." Thus, for example, positivist writers like Reichenbach and Carnap are constantly tempted to use empirical equivalence alone as the standard for full theoretical equivalence, to say that two theoretical descriptions are equivalent just in case they agree on all possible observations.[9] But this simple and easy conception of "equivalent descriptions" is obviously inadequate to our two cases from the development of relativity, for the alternative descriptions in question are *empirically* equivalent both before and after the crucial theoretical transitions. They are not *fully* equivalent, however, until after we revise the ideology of our theories. Only after we have rejected absolute space can we declare that alternative attributions of velocity are not only empirically equivalent but also equally true; only after we have rejected the connection $\overset{\circ}{D}$ and potential Φ can we assert the full theoretical equivalence of alternative attributions of gravitational acceleration. Of course, considerations of empirical equivalence can and do *motivate* these revisions of theoretical ideology, and this fact raises important issues about the methodology of theoretical revision. Nevertheless, empirical equivalence alone is clearly not sufficient to establish full theoretical equivalence.

The simple positivistic conception of "equivalent descriptions" would make sense if the positivists had succeeded in explicitly defining all theoretical vocabulary in observational terms. If we could actually carry out this project, it would certainly follow that observationally equivalent theories had the same truth value. Moreover, we could give clear content to the idea that apparently (syn-

---

[9] Reichenbach often characterizes empirically equivalent theories as "logically equivalent" ([87], 374), and as late as 1966 we find Carnap asserting that two theories agreeing on all observations are "equivalent descriptions" ([12], 150).

tactically) incompatible sentences S and $\ulcorner\neg S\urcorner$ are equally true by pointing out that the same theoretical term (for example, "congruent") has different observational definitions in the two sentences. By contrast, in the absence of term-by-term or sentence-by-sentence translations into the observation language, we have given no content at all to the idea of syntactic incompatibility but semantic equivalence. We can say, if we like, that an entire *theory* is true just in case its observational consequences are true, but this leaves us with no clear account of individual sentences like S and $\ulcorner\neg S\urcorner$.

In the absence of observational definability, then, the positivistic conception of "equivalent descriptions" is excessively holistic. It supplies no adequate semantics for individual theoretical terms or sentences and no term-by-term or sentence-by-sentence correlations between alternative theoretical descriptions. A natural next move is to require just such a correlation: to require that two theoretical descriptions are fully equivalent just in case they are *intertranslatable*. This conception of theoretical equivalence has been proposed by Glymour. He suggests that two empirically equivalent theoretical descriptions should be considered to be fully equivalent only if each is *relatively interpretable* in the other. That is, it should be possible to define the primitive terms of one theory by means of the primitive terms of the other in such a way that the theorems of the first theory appear as logical consequences of the second theory and vice versa.[10] The existence of such definitions gives us just the term-by-term and sentence-by-sentence correlations that simple empirical equivalence does not itself supply. Moreover, the account makes good sense of the claim that fully equivalent descriptions are equally true: given the definitions, we can show that one is true just in case the other is. So we can give content to the idea of apparent incompatibility but real equivalence in a natural way: an apparent conflict between S and $\ulcorner\neg S\urcorner$ is explained by pointing out that the same term has different definitions in the two sentences—this time, different definitions in the *theoretical* language.

Although this kind of intertranslatability standard for "equivalent descriptions" is a definite improvement, it certainly cannot serve as a *sufficient* condition for full theoretical equivalence (Glymour himself claims only *necessity*).[11] Once again, it fails to make sense of our cases from the development of relativity. To see this, consider

[10] Glymour [43], 279, actually imposes the slightly stronger condition that the two theories have a common definitional extension.
[11] Glymour [43], 281.

the alternative theories T_V, $T_{V'}$, which induce different assignments of absolute velocity. These theories are intertranslatable even before the crucial transition to the new space-time of §III.2: simply translate "absolute velocity" in one theory by "velocity relative to inertial frame $F_{V'}$" in the other. Nevertheless, they are not fully equivalent until after we change our minds about the structure of space-time. Or consider the alternative theories T_Φ, T_Ψ, which induce different assignments of gravitational acceleration. They are also intertranslatable before the crucial transition to §III.4: simply translate "gravitational acceleration" in one theory by "acceleration relative to Galilean frame F_Ψ" in the other. But they are not fully equivalent until we revise our theory as in §III.4.

These points can be developed from a somewhat different perspective if we make use of the notion of a *representation in R^4* from §VI.3. All our space-time theories assert the existence of a coordinatization by R^4: a one-one map $\phi : A \rightarrow R^4$, where A is an open subset of the space-time manifold M. Yet the theories differ over which aspects of the mathematical structure of R^4—which of the totality of functions, properties, and relations defined on R^4—represent physically real structures on M. Thus, for example, all Newtonian theories assert the existence of rigid Euclidean coordinates: they postulate the functions $t(a,b)$ and $d_S(a,b)$ on M and a one-one mapping $\phi : A \rightarrow R^4$ such that

$$t(a,b) = |\phi(a)_0 - \phi(b)_0|$$
$$d_S(a,b)^2 = (\phi(a)_1 - \phi(b)_1)^2 + (\phi(a)_2 - \phi(b)_2)^2 \qquad (1)$$
$$+ (\phi(a)_3 - \phi(b)_3)^2.$$

However, the theory of §III.1 and the theory of §III.2 differ over what further structure is defined on M. In particular, the theory of §III.1 postulates a two-place relation $R(a,b)$ of absolute rest and constrains the mapping ϕ by

$$R(a,b) \quad \text{iff} \quad \phi(a)_i = \phi(b)_i \qquad (i = 1,2,3). \qquad (2)$$

The theory of III.2 rejects $R(a,b)$ and instead postulates a three-place relation $R(a,b,c)$ of "inertiality." Instead of (2) we have

$$R(a,b,c) \text{ iff } \phi(a), \phi(b), \phi(c) \text{ are } collinear. \qquad (3)$$

According to §III.2, collinearity in R^4 has physical significance, but the relation $\phi(a)_i = \phi(b)_i$ $(i = 1,2,3)$ does not. According to §III.1, both relations have physical significance.

282

A central feature of this case is a difference in the degree of arbitrariness of the representation $\phi : A \to R^4$. The class of representations satisfying (1) and (3) is wider than the class of representations satisfying (1) and (2): elements of the former class are generated by the Galilean group; elements of the latter are generated by $O^3 \times T$. Thus, according to §III.2, all inertial coordinate systems are *equivalent representations* of the structure $\langle M, t(a,b), d_S(a,b), R(a,b,c) \rangle$, a structure that can be characterized *intrinsically* by the existential quantification

$$\exists \phi : A \to R^4 \text{ satisfying (1) and (3).} \tag{4}$$

By contrast, in §III.1 all inertial coordinate systems are *not* equivalent representations of the structure $\langle M, t(a,b), d_S(a,b), R(a,b) \rangle$; only the rest systems count as equivalent representations of this structure. The latter is characterized intrinsically, not by (4), but by

$$\exists \phi : A \to R^4 \text{ satisfying (1) and (2).} \tag{5}$$

The difference between (4) and (5) gives precise expression to the idea that the widening of the class of "equivalent descriptions" effected by the move from §III.1 to §III.2 goes hand in hand with a narrowing of theoretical ideology, that is, with the rejection of $R(a,b)$ for $R(a,b,c)$.

A similar story can be told about the transition from §III.3 to §III.4. The two theories agree, of course, about the functions $t(a,b)$, $d_S(a,b)$, and condition (1). They differ over the physical reality of the gravitational potential Ψ and, consequently, over the relation $R(a,b,c)$ of "inertiality": according to §III.3, $R(a,b,c)$ is induced by a flat connection $\overset{\circ}{D}$; according to §III.4, by a nonflat connection D. §III.3 attributes physical reality to Ψ and adds the condition

gravitational trajectories satisfy

$$\frac{d^2\phi_i}{d\phi_0^2} = -\frac{\partial(\Psi \circ \phi^{-1})}{\partial \phi_i} \qquad (i = 1,2,3) \tag{6}$$

to (1). §III.4 rejects Ψ and replaces (6) by

gravitational trajectories satisfy

$$\frac{d^2\phi_i}{d\phi_0^2} = -\frac{\partial\Psi}{\partial\phi_i} \quad (i = 1,2,3) \quad \text{for } some \text{ potential} \atop \text{function } \Psi \text{ on } R^4. \tag{7}$$

In (7) the function Ψ is dependent on the representation ϕ. In (6) Ψ is not so dependent: it is part of the intrinsic structure of M. As

283

a result, (6) puts constraints on the representation ϕ that are absent in (7). In (6) ϕ is arbitrary up to an element of the Galilean group; in (7) the degree of arbitrariness is much greater, namely, up to a transformation of the form (III.66).[12] Once again an increase in the arbitrariness of the representation is associated with a decrease in the ideology of our theory.

The relativization of distance and time effected by special relativity follows the same pattern, although, as we remarked above, this relativization does not have the epistemological motivations that the other two do. Classical electrodynamics postulates the structure $\langle M, t(a,b), d_S(a,b), R(a,b) \rangle$, within which the Minkowskian space-time interval $I(a,b)$ is definable:

$$I(a,b) = t(a,b)^2 - d_S(a,c)^2, \qquad \text{where } R(b,c) \text{ and } t(a,c) = 0.$$

It follows that in a rest system $\phi: A \to R^4$ we have

$$\begin{aligned} I(a,b) = (\phi(a)_0 - \phi(b)_0)^2 &- (\phi(a)_1 - \phi(b)_1)^2 \\ &- (\phi(a)_2 - \phi(b)_2)^2 - (\phi(a)_3 - \phi(b)_3)^2 \end{aligned} \tag{8}$$

So in classical electrodynamics we have the characterization

$$\exists \phi: A \to R^4 \text{ satisfying (1), (2), and (8)} \tag{9}$$

where ϕ is arbitrary up to an element of $O^3 \times T$. In special relativity, on the other hand, we drop $t(a,b)$, $d_S(a,b)$, and $R(a,b)$, retaining only the space-time interval $I(a,b)$. Accordingly, we replace (9) by

$$\exists \phi: A \to R^4 \text{ satisfying (8)} \tag{10}$$

where this time ϕ is arbitrary up to an element of the Lorentz group. The equivalent representations satisfying (10), unlike the equivalent representations satisfying (9)—or even the wider class of representations satisfying (1) only—disagree about times and distances, about the functions $|\phi(a)_0 - \phi(b)_0|$ and $(\phi(a)_1 - \phi(b)_1)^2 + (\phi(a)_2 - \phi(b)_2)^2 + (\phi(a)_3 - \phi(b)_3)^2$. So in special relativity these latter functions are extrinsic features of the representation ϕ, not intrinsic features of M.

In sum, then, our different space-time theories postulate various pieces of theoretical structure—various quantities, properties, and

---

[12] Again, however, this case is somewhat different from the relativization of velocity effected by the move from (5) to (4). The class of coordinate systems related by (III.66) does not generate fully equivalent representations of the new, nonflat space-time structure; see note 8 above. To obtain fully equivalent representations we have to restrict (III.66) further; see pp. 329–330 below.

relations—on the underlying manifold M. They make use of representations $\phi: M \to R^4$ to characterize this structure by asserting an *isomorphism* with some purely mathematical structure—some set of purely mathematical functions, properties, and relations—defined on R^4. This procedure gives rise to a class of *equivalent representations* whose scope is inversely proportional to the amount of theoretical structure we postulate on M. Equivalent representations agree on all aspects of mathematical structure on R^4 corresponding to physically real quantities, properties, and relations on M, but disagree on aspects of mathematical structure having no such physical significance. Thus, in the space-time §III.1 only the choice of origin and orientation is arbitrary and without physical significance; in §III.2 the choice of rest system is arbitrary as well; in §III.4 even the choice of inertial system is arbitrary or conventional.[13]

Note also that two representations count as fully equivalent in the context of a given space-time theory T only if the transformation connecting them is a *symmetry* of the fundamental spatio-temporal structures postulated by T. Rest systems are equivalent in §III.1 because $O^3 \times T$ preserves $t(a,b)$, $d_S(a,b)$, and $R(a,b)$; inertial systems are equivalent in §III.2 and §III.3 because the Galilean transformations preserve $t(a,b)$, $d_S(a,b)$, and $R(a,b,c)$. The mere *existence* of a transformation connecting two representations—of a *translation* from one to the other—is obviously not sufficient. Hence, the idea that one can use intertranslatability alone as a sufficient condition for full theoretical equivalence is the same kind of mistake as the traditional confusion of covariance and invariance; neither is sensitive to the important differences in theoretical ideology among our different space-time theories.[14]

Here is a good place to take another brief look at Reichenbach's early work, *The Theory of Relativity and A Priori Knowledge* [84].

---

[13] The way we have set things up does not reflect the arbitrariness or conventionality involved in a choice of *scale* or *units*. However, we can easily accommodate this kind of arbitrary choice as well if we always use properties and relations on M rather than quantities. Thus, for example, we can reflect the arbitrariness of spatial and temporal units in Newtonian theory by starting with spatial and temporal congruence relations on M instead of with the functions $d_S(a,b)$ and $t(a,b)$. Similarly, we can formulate special relativity via the relation $p\lambda q$ of lightlike connectibility as on p. 164 above. See H. Field [33], esp. pp. 41–54 and note 32 on p. 116, for an extended treatment.

[14] As we saw in §IV.7 above, this is just the mistake Winnie makes in his argument for the "kinematical equivalence" of standard ($\varepsilon = \frac{1}{2}$) and nonstandard ($\varepsilon \neq \frac{1}{2}$) formulations of special relativity.

As is well known, this book contains Reichenbach's famous "refutation" of Kant based on the theory of relativity. According to Reichenbach, relativity theory makes it impossible to consistently maintain all the principles Kant thought to be *a priori*: for example, one cannot consistently maintain both Euclidean geometry and "normal causality." However, as we pointed out in the Introduction, the book also contains something far more interesting: namely, the idea that relativity theory provides a *reinterpretation* of Kant's notion of *a prioricity*, a reinterpretation that shows us what is right about Kantian epistemology—and, we should add, what is wrong with empiricist epistemology.[15]

Reichenbach distinguishes two meanings of the Kantian *a priori*: (i) necessarily and unrevisably true; (ii) constitutive of the concept of object ([84], 48). By way of clarifying (ii), he introduces the notion of *axioms of coordination* ([84], 54). Axioms of coordination must be laid down before our theories have any empirical content, any relation to reality. No meaningful question of empirical truth or falsity can be raised antecedent to such axioms; they are necessary to determine what objects, properties, and so on we are theorizing about. Thus, axioms of coordination constitute a kind of "background framework" for empirical theorizing. Once we have axioms of coordination and, therefore, a determinate empirical subject matter, we can then go on to formulate *axioms of connection* that make specific empirical claims about this subject matter. By contrast, the axioms of coordination do not make empirical claims: they represent the "subjective form" of knowledge, the "contribution of reason" to the object of knowledge. Empiricism ignores this distinction between specific empirical laws and principles of coordination ([84], 93) and thereby misses the role of reason in constituting the world of experience ([84], 77).

Although Reichenbach agrees with Kant about the existence of *a priori* principles in sense (ii), he denies that axioms of coordination are *a priori* in sense (i). Axioms of coordination are not laid down once and for all; they change and develop with the progress of empirical science. This, in fact, is the lesson of relativity theory. In moving from classical physics to relativity, we change our axioms

---

[15] This idea can also be found in Cassirer [14], published a year after Reichenbach's book (1921). Each read the other's manuscript while his own was going to press, and each expressed general approval of the other; see [84], 114–115, [14], 460.

of coordination and, consequently, our concept of object ([84], 94).
So axioms of coordination are not unrevisable "fixed points" of
empirical inquiry. As the development of relativity theory shows,
they can themselves be changed under pressure of empirical findings:

> Although we have rejected Kant's analysis of reason, we do not
> want to deny that experience contains rational elements. Indeed,
> the principles of coordination are determined by the nature of
> reason; experience merely selects from among all possible prin-
> ciples. It is only denied that the rational components of knowledge
> *remain* independent of experience. The principles of coordination
> represent the rational components of empirical science at any
> given stage. ([84], 87)

Moreover, Reichenbach takes the radical position that all concepts
and principles, even his own philosophical results, are subject to
revision in the light of experience: nothing is *a priori* in sense (i)
([84], 77–87). Nevertheless, one can still distinguish, *in the context
of any particular theory*, those elements that have their origin in
reason from those elements that have their origin in experience.

But how do we distinguish axioms of coordination from axioms
of connection? How do we isolate the rational or subjective components of knowledge? The answer is that the rational elements are
distinguished not by their unrevisability or self-evidence but by
their *arbitrariness*:

> The idea that the concept of object has its origin in reason can
> manifest itself only in the fact that this concept contains elements
> for which *no* selection is prescribed, that is, elements that are
> independent of the nature of reality. The arbitrariness of these
> elements shows that they owe their occurrence in the concept of
> knowledge altogether to reason. *The contribution of reason is
> not expressed by the fact that the system of coordination contains
> unchanging elements, but in the fact that arbitrary elements occur
> in the system.* This interpretation represents an essential modi-
> fication compared to Kant's conception of the contribution of
> reason. The theory of relativity has given an adequate presenta-
> tion of this modification. ([84], 88–89)

Axioms of coordination have a special epistemological position—
not, however, because they are unrevisably and necessarily true, but

because their truth-value is determined by arbitrary choice: no question of empirical truth or falsity can arise.[16]

Our above discussion of the arbitrariness of representation captures the kernal of truth in Reichenbach's idea. Given any particular mathematical representation of space-time via a coordinatization in R^4, we can separate out the *intrinsic* and the *extrinsic* elements. In Reichenbach's words, we have a "method of distinguishing the objective significance of a physical statement from the subjective form of the description through transformation formulas" ([84], 91–92). The objective elements are those preserved by the fundamental symmetries of our theory, by transformations leaving the geometrical ideology of our theory invariant. These elements of our representation in R^4 correspond to physically real properties and relations on M. By contrast, the "contribution of reason" is expressed in the degree of arbitrariness of our representation. Those elements that vary from representation to representation belong to the "subjective form." No question of empirical truth or falsity applies to them, because they do not correspond to physically real properties and relations on M.

That Reichenbach has something like this in mind is shown by the fact that he agrees with us about which elements are "subjective" or arbitrary. The arbitrary elements are essentially just the choice of rest system (as implied by the principle of special relativity; [84], 7) and the choice of inertial system (as implied by the principle of equivalence; [84], 33). In marked contrast to his later views in *The Philosophy of Space and Time* [86], Reichenbach maintains that the choice of *metric* is not arbitrary ([84], 33, 90).[17] However, these

---

[16] Note the striking similarities between Reichenbach's views here and Carnap's generalized conventionalist position in *The Logical Syntax of Language* [9]. Carnap's *analytic* sentences play the role of Reichenbach's axioms of coordination. The difference is that Carnap tries to give a logical or linguistic interpretation of these "rational components" (hence *analyticity*), whereas Reichenbach gives an interpretation drawn from mathematical physics. Note also that, since both views explicitly recognize the way in which their "rational components" change with the progress of empirical science, neither is simply or straightforwardly refuted by Quinean considerations about revisability. See especially §82 of *Logical Syntax*, which reads just like a capsule summary of Quinean philosophy of science! Here I am indebted to Thomas Ricketts [92].

[17] Still, Reichenbach does not consistently follow this point of view. Thus, for example, he makes the claim that the metric is an axiom of coordination in pre-general-relativistic (flat space-time) physics but an axiom of connection in general relativity ([84], 100). It is clear, however, that the metric is invariant or nonarbitrary in both. Reichenbach is torn here between the intuitive idea that axioms of coordina-

examples also show us just how weak and "thin" the "relativistic reinterpretation of the *a priori*" really is. The only "rational components" are such uncontroversial elements as coordinate systems, units,[18] and so forth. No "thick" Kantian conception of constitutive principles emerges. No generalized conventionalism is forthcoming either, as will become clear in what follows.

We have seen that the relativizations effected by the development of relativity theory are best understood as changes of theoretical ideology. We change our minds about what properties and relations really exist on space-time and, therefore, about which elements of mathematical structure on R^4—a structure used by all our theories to *represent* space-time—are to be taken literally. It is now time to consider the motivations for these crucial changes in theoretical ideology, especially the primarily epistemological motivations involved in the relativizations of velocity and gravitational acceleration. These two relativizations, unlike the relativization of distance and time effected by special relativity, are direct responses to "conceptual" problems about empirical equivalence. Hence, they provide us with our best opportunity for seeing the conventionalist strategy put to work in actual scientific practice.

The first thing to notice is that both cases involve the rejection of theoretically indeterminate or indistinguishable entities. In the theory of §III.1 the vector field V is indeterminate: it can vary (up to a Galilean transformation) while all other objects of the theory remain fixed. In the theory of §III.3 the connection $\overset{\circ}{D}$ and potential Ψ are indeterminate: they too can vary (up to a transformation of the form (III.66)) while all other objects of this theory remain fixed. In moving from §III.1 to §III.2, we drop the indeterminate V; in moving from §III.3 to §III.4, we drop the indeterminate $\overset{\circ}{D}$ and Ψ. In both cases we move from a non-well-behaved theory that violates the methodological principle (R2) to a well-behaved theory that satisfies (R2). What is the epistemological import of these facts? What is wrong with indeterminate entities and non-well-behaved theories (besides their name)?

---

tion are part of our general "background framework" (absolute objects?) and the idea that they are arbitrary or noninvariant. Interestingly enough, Carnap makes the very same claim about the status of the metric in general relativity as opposed to earlier theories ([9], 178–179). Of course, Carnap does not use the notion of relativistic invariance to explain *his* conception of arbitrariness.

[18] See note 13 above.

We have already made an initial suggestion in §VI.3: indeterminate entities like the vector field V have no *unifying power*. Such entities play no role in the process of theoretical unification and, consequently, do not contribute to the confirmation of the (non-well-behaved) theories in which they occur. Thus, the vector field V (relation $R(a,b)$) plays no role in unifying Newtonian space-time geometry and Newtonian kinematics. The former theory requires rigid Euclidean coordinate systems, namely, mappings $\phi: A \to R^4$ satisfying (1). The latter, on the other hand, requires *affine* coordinate systems, namely, mappings $\psi: A \to R^4$ such that

$$\text{free particles satisfy } \frac{d^2\psi_i}{d\psi_0^2} = 0 \qquad (i = 1,2,3). \tag{11}$$

The unification of these two theories requires *inertial* coordinate systems:

$$\exists \chi: A \to R^4 \text{ satisfying (1) and (11)}. \tag{12}$$

We therefore need a guarantee that the class of rigid Euclidean systems and the class of affine systems overlap. But precisely this is ensured by the presence of a flat affine connection $\overset{\circ}{D}$ on M that is compatible with the objects dt and $h: \overset{\circ}{D}(dt) = \overset{\circ}{D}(h) = 0$. So V plays no role in the unification (12).

Similarly, as we have already observed in §VI.3, V plays no role in unifying Newtonian space-time geometry with the gravitation theory of §III.3 (still less, of course, with the gravitation theory of §III.4). This latter requires mappings $\phi: A \to R^4$ satisfying (6). So our desired unification again requires *inertial* coordinates in which we have both (1) and (6). Once again, the connection $\overset{\circ}{D}$ (relation $R(a,b,c)$) suffices: there is no need for the field V. However, as we also observed in §VI.3, this is not to say that V has no unifying power *simpliciter*, for V does play a crucial role in unifying Newtonian space-time geometry (or kinematics or gravitation theory) with classical electrodynamics. The problem here is not methodological or "conceptual"; it is just that classical electrodynamics is false. *This* is how special relativity contributes to the elimination of V.

What about the indeterminate entities of §III.3: the connection $\overset{\circ}{D}$ and potential Ψ? To avoid begging the question in favor of Ψ from the very beginning, let us represent the gravitational law of motion as (7) rather than (6), and let us ask whether the connection $\overset{\circ}{D}$ (and *therefore* the potential Ψ) plays a role in unifying the *Pois-*

290

sonian coordinates in which (7) is valid with, say, the rigid Euclidean coordinates in which (1) is valid. In other words, what entities do we require in order to guarantee the existence of a joint representation or unification $\chi: A \rightarrow R^4$ satisfying (1) and (7)? What entities do we require in order to ensure the existence of *Galilean* coordinates? The answer, of course, is that the *nonflat* connection D suffices: all we need are the conditions $\bar{D}(h) = 0 = \bar{D}(dt)$. So $\mathring{D}$ and Ψ play no role in this unification. What about the unification of gravitation theory with other theories? The principle of equivalence (version (2) of §V.4) says that D is the *unique* connection of space-time: $\mathring{D}$ does not figure in any (true) theory of interaction. So, for example, electrodynamics is correctly formulated with D rather than $\mathring{D}$, and $\mathring{D}$ plays no role in the unification of gravitation theory with electrodynamics. Hence, the principle of equivalence, which holds true as far as we know, is itself equivalent to the assertion that $\mathring{D}$ has no unifying power.

The notion of unifying power therefore makes good sense of the methodological motivations for our central changes in theoretical ideology. By adopting a slightly different point of view, we can also bring more traditional methodological criteria into the picture. Consider again the different absolute velocity theories T_V, $T_{V'}$, and so on. All these theories are of course empirically equivalent (according to these theories themselves), but, in addition, they have no discernible "structural" differences. So there is no way to apply methodological criteria like simplicity, parsimony, and economy to force a choice among them. The theories T_V, $T_{V'}$, and so on are not only empirically equivalent, they are also "methodologically equivalent." On the other hand, consider the Galilean space-time theory of §III.2. Let us call this theory T_-, since it can be obtained from any of the theories T_V by simply dropping the field V. T_- is empirically equivalent to all the T_V, and it also has all the methodological virtues of the latter theories. However, it does better with respect to the principle of parsimony than any of the T_V, since it dispenses with the explanatorily superfluous field V. In accordance with good scientific practice, then, we choose the theory T_- over all the theories T_V.

What makes this story work is that the empirical equivalence of T_V, $T_{V'}$, and so on is of a very special kind: it arises from the presence of an indeterminate field V that can vary arbitrarily while the rest of our theoretical ideology remains fixed. Thus, let S_v be a sentence ascribing the particular absolute velocity v to some given object

291

(the center of mass of the solar system, say). Each T_V is equivalent to $\ulcorner S_v \ \& \ T_- \urcorner$, and, therefore, $\ulcorner S_v \ \& \ T_- \urcorner$ is empirically equivalent to $\ulcorner S_{v'} \ \& \ T_- \urcorner$. This is the precise sense in which the S_v fail to have empirical import *even given the rest of the relevant theory*. Moreover, this is why the empirically equivalent theories in question have no "structural" differences either: they differ only with respect to S_v. By contrast, most theoretical sentences do have empirical import in conjunction with the rest of their relevant theory. If S is a theoretical sentence and T is the rest of the relevant theory, then $\ulcorner S \ \& \ T \urcorner$ is not in general empirically equivalent to $\ulcorner (\neg S) \ \& \ T \urcorner$. In general, if we want to obtain an empirically equivalent theory, we have to make *compensatory adjustments* that change T to T': $\ulcorner S \ \& \ T \urcorner$ will only be empirically equivalent to $\ulcorner (\neg S) \ \& \ T' \urcorner$. Further, such compensatory adjustments in T will typically lead to "structural" differences between $\ulcorner S \ \& \ T \urcorner$ and $\ulcorner (\neg S) \ \& \ T' \urcorner$ that invite the application of methodological criteria to settle the choice between them.

A good illustration of these points is furnished by the electrodynamic theory of §III.5. In this theory absolute space and absolute velocity are no longer indeterminate. In general, if $\langle M,D,dt,h,V,J,F \rangle$ is a model for the theory of §III.5, then $\langle M,D,dt,h,V',J,F \rangle$ is not a model. Therefore, the different S_v no longer generate empirically equivalent theories as they do in the context of kinematics or gravitation theory: if T_E is the theory of §III.5 then $\ulcorner S_v \ \& \ T_E \urcorner$ is not empirically equivalent to $\ulcorner S_{v'} \ \& \ T_E \urcorner$. We can obtain empirically equivalent theories all right,[19] but we have to make compensatory adjustments in T_E. In particular, we can add a second vector field U (the "aether") to §III.5 and let U have velocity u relative to V. We then formulate our theory using the tensor $(U \otimes U - h)$ rather than the tensor $(V \otimes V - h)$. Call the resulting theory T_E^U. Although $\ulcorner S_v \ \& \ T_E \urcorner$ is not empirically equivalent to $\ulcorner S_{v+u} \ \& \ T_E \urcorner$, it is empirically equivalent to $\ulcorner S_{v+u} \ \& \ T_E^U \urcorner$ for each u. But the important point is this: the two empirically equivalent theories $\ulcorner S_v \ \& \ T_E \urcorner$ and $\ulcorner S_{v+u} \ \& \ T_E^U \urcorner$ are now "structurally" different. The latter postulates two privileged vector fields in place of one, and, as a consequence, the latter theory contains an explanatorily superfluous entity with no unifying power: the "idle" field V. In this case, therefore, methodological criteria can and do dictate the choice of $\ulcorner S_v \ \& \ T_E \urcorner$ over

[19] Here I follow the suggestions in Sklar [101], 18–19, [100], 196–197, and van Fraassen [114], 47–50.

292

the empirically equivalent $\ulcorner S_{v+u}$ & $T_E^U \urcorner$. In fact, the very same criterion of parsimony that favors the Galilean space-time theory T_- over the empirically equivalent T_V likewise favors the former electrodynamic theory over the latter.

We find the same pattern in the transition from §III.3 to §III.4. The different theories T_Φ, T_Ψ are both empirically equivalent (according to these theories themselves) and "methodologically equivalent": there are no discernible "structural" differences between them. On the other hand, the theory T_D of §III.4 is empirically equivalent to all the T_Φ but not "methodologically equivalent" to any. Again, T_D does better with respect to the methodological virtue of parsimony, since it dispenses with the explanatorily superfluous entities $\overset{\circ}{D}$ and Φ. Similarly, the empirical equivalence of T_Φ, T_Ψ is of a very special kind. Let $S_{\ddot{b}_i}$ be a sentence ascribing the particular gravitational acceleration $\ddot{b}_i = -\Phi_{,i}$ to some given object. Each T_Φ is equivalent to $\ulcorner S_{\ddot{b}_i}$ & $T_D \urcorner$, and, therefore, $\ulcorner S_{\ddot{b}_i}$ & $T_D \urcorner$ is empirically equivalent to $\ulcorner S_{\ddot{b}'_i}$ & $T_D \urcorner$, where $\ddot{b}'_i = -\Psi_{,i}$. So $S_{\ddot{b}_i}$, like S_v, fails to have empirical import *even given the rest of the relevant theory*. It makes good sense, then, to drop $S_{\ddot{b}_i}$ from our theory and retain only the methodologically superior T_D.

Finally, let us briefly consider the transition from the classical electrodynamics of §III.5 to the relativistic electrodynamics of §IV.4. These theories are not, of course, empirically equivalent: they differ over "observable" properties of the velocity of light. So there is no need to invoke methodological criteria to explain our preference for the latter. However, there is a stage in the evolution of the theory of §IV.4 at which methodological criteria do play a role, namely, the elimination of the Lorentz-Fitzgerald-type "aether" theory. As we observed in §IV.5, this latter theory is indeed empirically equivalent to relativistic electrodynamics, but, like the theories of §III.1 and §III.3, it violates the relativity principle (R2) and is not well-behaved: its indistinguishability group (the Lorentz group) exceeds its symmetry group ($O^3 \times T$). (By contrast, there is nothing methodologically or "conceptually" wrong with the theory of §III.5!) So this theory contains indeterminate entities—V, dt, and h—that have no unifying power in our revised theoretical context.[20] Hence, the methodological principle of parsimony tells us to drop these

---

[20] For example, we no longer need the rest systems to formulate a Lorentzian law of motion. Our revised law of motion holds equally in all special relativistic inertial systems. See §IV.5.

entities in favor of the Minkowski metric g. We end up with special relativity.

Our moral is this. The relativizations we encounter in the development of relativity theory do involve something like the conventionalist strategy. Motivated by methodological problems with empirically equivalent theories, we revise our theoretical ideology. As a result, we are able to say that various theoretical descriptions that formerly were *merely* empirically equivalent are now fully equivalent as well: they are nothing but different equivalent representations of the same intrinsic facts. Yet these theoretical revisions have a very special structure. We eliminate only those entities that are indeterminate and have no unifying power in a given theoretical context. So the parsimony exhibited in the evolution of relativity does not generalize to skepticism or conventionalism about all theoretical structure—or even to conventionalism about specifically *geometrical* structure. In the next two sections we shall apply this moral to a pair of controversial cases.

3. Congruence and Physical Geometry

The idea that the geometry of physical space is arbitrary or conventional, that there is no fact of the matter whether space is Euclidean or non-Euclidean, is an old one. Indeed, Poincaré's forceful presentation of this idea in *Science and Hypothesis* [79] is one of the primary historical sources for the *general* problems about theoretical underdetermination, "equivalent descriptions," and so on. The basic line of thought underlying geometrical conventionalism goes as follows. Physical space can be meaningfully said to have one or another geometry only relative to some particular method for measuring length, some particular way of determining congruence relations between spatial intervals. Antecedent to a choice of method for measuring length (determining congruence relations) physical space simply has no geometry: it is "metrically amorphous." Thus, our methods for measuring length (determining congruence relations) are themselves arbitrary or conventional. There is no fact of the matter, for example, whether rigid measuring rods change their length when transported or retain their length. Since there is no preexisting quantity of length (relation of congruence) defined on physical space, this question can itself be settled only by arbitrary stipulation.

294

The way in which the geometry ascribed to physical space depends on our methods for measuring length, and, in particular, how rigid rods are assumed to behave under transport, is easily illustrated. Perhaps the clearest example is provided by Reichenbach ([86], 11–12). Imagine two surfaces, a hemisphere on top and a plane below, as in the figure. Imagine that two-dimensional creatures

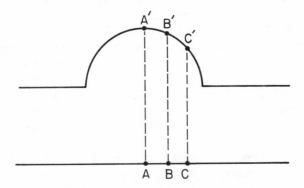

confined to each surface determine the geometries of their respective worlds by means of two-dimensional measuring rods. Creatures on the hemisphere measure length in the normal way, by assuming that rigid rods retain their length under transport. They find, for example, that $A'B'$ is congruent to $B'C'$. Moreover, they discover thereby that the geometry of their surface is non-Euclidean (for example, by comparing the radii and diameters of circles). If the creatures on the plane also measure length in this standard (length-retaining) way, they will of course discover that the geometry of their surface is Euclidean. On the other hand, suppose that the creatures on the plane change their methods of measurement in such a way that two intervals are judged congruent just in case they are the projections of congruent intervals on the hemisphere: this method would involve the assumption that rigid rods *expand* as they are transported toward the boundary of the hemisphere. The creatures "find," for example, that AB is congruent to BC, and they clearly "discover" thereby that the geometry of their surface is spherical or non-Euclidean.

That the geometry ascribed to physical space depends in this way on our methods for measuring length (determining congruence relations) is clear and incontestable. It is a simple reflection of the purely mathematical fact that the geometry of a space depends

295

completely on the metric of that space. What is controversial is the conventionalist's further claim that our methods for measuring length are themselves arbitrary, that there is no fact of the matter whether rigid rods retain their length under transport. On a naive view of physical geometry there is a preexisting, "true" quantity of length (relation of congruence) defined on physical space. So, for example, rigid rods either retain their length under transport (relative to the "true" congruence relation) or they do not: there is a fact of the matter. This is why we intuitively feel that physical space must be either Euclidean or non-Euclidean, but not both. Of course, the naive view of physical geometry may be wrong, just as the naive view of rest and velocity is wrong. The problem for the conventionalist is to give arguments to this effect. Why should we think that our methods for measuring length (our assumptions about the behavior of rigid rods) are arbitrary in the way the conventionalist suggests?

The most popular conventionalist argument, from Poincaré to the present, is epistemological. There is no way to verify assumptions about the behavior of rigid rods (notably, that they do or do not retain their length under transport) by direct observation. Such assumptions are unverifiable in principle. By some version of the principle of verifiability, then, descriptions of the behavior of rigid rods have no objective truth value. The clearest and most powerful presentation of this kind of argument is found in Reichenbach's *The Philosophy of Space and Time* [86]. On the other hand, the most influential post-Reichenbachian conventionalist, Adolf Grünbaum, attempts to avoid all such epistemological considerations. He tries to show on nonepistemic grounds that there just is no "true" quantity of length (relation of congruence). The reason assumptions about the behavior of rigid rods under transport have no determinate truth value is not that they are unverifiable but that there is no determinate, "intrinsic" quantity of length by virtue of which they could have an objective truth value. Thus, Grünbaum's approach is of interest as it attempts to free conventionalism from a positivist appeal to verifiability. Let us examine these two approaches in turn.

REICHENBACH. Consider again the example of the two-dimensional creatures confined to the surface of a plane. Why is the normal assumption of length retention unverifiable? Because there is an alternative description, a description postulating an "undetectable" expansive force affecting all bodies in the same way, that is empirically equivalent to the normal one:

> Assume two measuring rods which are equal in length. They are transported by different paths to a distant place; there again they are laid down side by side and found equal in length. Does this procedure prove that they did not change on the way? Such an assumption would be incorrect. The only observable fact is that the two measuring rods are always equal in length at the place where they are compared to each other. But it is impossible to know whether on the way the two rods expand or contract. An expansion that affects all bodies in the same way is not observable because a direct comparison of measuring rods at different places is impossible. ([86], 16)

In other words, we have two theories agreeing on all possible observations. One asserts that our surface is Euclidean and that there is no "undetectable" expansive force: rigid rods retain their length under transport. The other asserts that our surface is non-Euclidean and that there is a "universal" expansive force: rigid rods do not retain their length under transport. Since these two theories are empirically equivalent, they are also, by some version of the principle of verifiability, fully equivalent. Hence, there is no fact of the matter whether the surface is Euclidean or non-Euclidean, except relative to one or another "coordinative definition" (that is, specification of the behavior of rigid measuring rods):

> The problem does not concern a matter of *cognition* but of *definition*. There is no way of knowing whether a measuring rod retains its length when it is transported to another place; a statement of this kind can only be introduced by definition. For this purpose a coordinative definition is to be used, because two physical objects distant from each other are *defined* as equal in length. It is not the *concept* equality of length that is to be defined, but a *real object* corresponding to it is to be pointed out. A physical structure is coordinated to the concept equality of length, just as the standard meter is coordinated to the concept unit of length. ([86], 16)

Now, we already know, of course, that the positivist inference from empirical equivalence to full equivalence does not in general hold. We also know, however, that there are certain special cases—our paradigmatic relativizations of velocity and gravitational

Conventionalism

acceleration[21]—where precisely this move is made. So *our* question is whether Reichenbach's case of empirically equivalent geometrical theories has the same structure as these paradigmatic cases from the development of relativity theory. In particular, does it involve indeterminate theoretical entities with no unifying power? The answer is no.

Consider, for example, the spatial metric h of Newtonian kinematics. A non-Euclidean metric h' will not be compatible with the flat connection $\mathring{D}$; if it were, it would be forced to be Euclidean after all. So if $\langle M,\mathring{D},dt,h,(V),T_{\hat{a}}\rangle$ is a model for the theory of §III.2 (§III.1), $\langle M,\mathring{D},dt,h',(V),T_{\hat{a}}\rangle$ is not a model. On the other hand, suppose we introduce a nonflat connection D with which h' is compatible. $\langle M,D,dt,h',(V),T_{\hat{a}}\rangle$ is not a model for §III.2 (§III.1) either, because geodesics of $\mathring{D}$ will not be geodesics of D. If we want $\langle M,D,dt,h',(V),T_{\hat{a}}\rangle$ to be a model, we must revise our law of motion[22] from

$$\frac{d^2x_i}{dt^2} + \mathring{\Gamma}^i_{jk}\frac{dx_j}{dt}\frac{dx_k}{dt} = 0 \tag{11}$$

to

$$\frac{d^2x_i}{dt^2} + \Gamma^i_{jk}\frac{dx_j}{dt}\frac{dx_k}{dt} = F^i \tag{12}$$

where the Γ^i_{jk} are the components of D and F^i is a "universal" force given by

$$F^i = (\Gamma^i_{jk} - \mathring{\Gamma}^i_{jk})\frac{dx_j}{dt}\frac{dx_k}{dt}. \tag{13}$$

In this revised theory free particles no long traverse equal distances (with respect to the non-Euclidean metric h') in equal times. Their deviation from this inertial state of affairs is explained by a "universal" force F^i.

Hence, the spatial metric h is not indeterminate in the context of Newtonian kinematics. It cannot be varied arbitrarily while the other objects remain fixed; in particular, it cannot be replaced by a non-Euclidean metric h'. Accordingly, h does have unifying power in Newtonian physics. For example, one cannot unify Newtonian

[21] Note that Reichenbach also takes these cases to be paradigmatic; see [87], 374–375, [86], §36 et seq.
[22] Here I follow Glymour [47], 365–367, [45], 195–197.

298

gravitation theory (either version) with classical electrodynamics unless the same metric h appears in both theories. Moreover, if S_h ascribes the Euclidean metric h to physical space and T is the rest of Newtonian kinematics, then $\ulcorner S_h$ & $T \urcorner$ is not empirically equivalent to $\ulcorner S_{h'}$ & $T \urcorner$ for any non-Euclidean h'. If we want to replace h by h', we must revise T via (12): $\ulcorner S_h$ & $T \urcorner$ is only empirically equivalent to $\ulcorner S_{h'}$ & $T^F \urcorner$, where T^F includes the "universal" force (13).

The import of these points is as follows. First, the case of physical geometry is quite unlike the cases of absolute velocity and gravitational acceleration. The latter involve indeterminate or indistinguishable theoretical entities: *given* the laws of their respective theories, we cannot distinguish them from arbitrary variations. The former involves a perfectly determinate, well-behaved theoretical entity: *given* the laws of classical kinematics (for example, the law of motion (11)), the metric h cannot be varied arbitrarily. Second, the two different kinds of cases correspond to different types of theoretical unverifiability. Assertions about absolute velocity or gravitational acceleration are *strongly* unverifiable: they generate no empirical consequences even in conjunction with the rest of the relevant theory. Assertions about the geometry of physical space, on the other hand, are only *weakly* unverifiable: to be sure, they generate no empirical consequences in isolation, but they do generate such consequences in conjunction with the rest of the relevant theory. This is why we have to make compensatory adjustments in the latter if we want to construct empirically equivalent systems of geometry plus physics.

Reichenbach, of course, recognizes the need for such compensatory adjustments. Consider his "Theorem Θ," for example, which expresses the central principle of "the relativity of geometry":

> Given a geometry G' to which the measuring instruments conform, we can imagine a universal force F which affects the instruments in such a way that the actual geometry is an arbitrary geometry G, while the observed deviation from G is due to a universal deformation of the measuring instruments. ([86], 32)

According to Reichenbach, then, only $G + F$ has verifiable consequences. We can change G by making compensatory adjustments in F: $G + F$ is empirically equivalent to $G' + F'$, but not to $G' + F$. What Reichenbach fails to see is that this process of compensatory adjustment generates theories that are empirically equivalent but not "methodologically equivalent." The theories $\ulcorner S_h$ & $T \urcorner$ and

299

⌜$S_{h'}$ & T^F⌝, for example, are "structurally" different. The latter contains an "extra" force (13) that is itself indeterminate or indistinguishable: if h'' is any other non-Euclidean metric, then $\langle M, D', dt, h'', F', T_{\hat{a}} \rangle$ is also a model for the law of motion (12), where h'' is compatible with D' and F' is constructed from D' and $\overset{\circ}{D}$ by (13). So F has no unifying power and should be dropped via the principle of parsimony. The very same methodological considerations that lead us to reject absolute space, to prefer T_- over T_V, likewise dictate the choice of ⌜S_h & T⌝ over ⌜$S_{h'}$ & T^F⌝. Similar methodological considerations militate against the use of any other such "universal" forces.[23]

This difficulty is symptomatic of a more general problem for Reichenbach. On Reichenbach's view, one of the principal tasks of epistemology is to pinpoint the conventional elements in scientific theory, to circumscribe those sentences of science that express arbitrary definitions and, as a result, have no empirical content.[24] It is essential, on such a view, that the criteria for conventionality should be nonholistic. Some definite, particular class of sentences should be isolated; so our criteria for conventionality should apply to the sentences in this class in a way in which they do not apply to just any theoretical sentence. Hence, if we are to use considerations about unverifiability, empirical equivalence, and so on as our criteria for conventionality, the unverifiability in question must be strong unverifiability. There must be a definite, particular sentence that can vary while the rest of our theory remains fixed. For, if all we have is a case of weak unverifiability, where ⌜S & T⌝ is empirically equivalent to ⌜$(\neg S)$ & T'⌝, there is no more reason to call S conventional than to call T conventional. All theoretical sentences are equally conventional on this account.

These points are graphically illustrated in Reichenbach's discussion of geometry. At the outset it appears that there is a very particular sort of sentence, namely,

$$\text{The length of an interval} = \text{the number of times a standard rigid measuring rod can be laid off along it,} \quad (14)$$

---

[23] Of course, on Reichenbach's view, these methodological considerations involve only "descriptive simplicity" and have nothing to do with truth ([86], 35). However, as we saw in §VII.1, no significant distinction between descriptive and inductive simplicity has in fact been drawn.

[24] See, e.g., [87], 8–16.

that is being claimed to be conventional or definitional. Very soon, however, when it becomes clear that (14) is only unverifiable in the weak sense, statements about the existence of "universal" forces are also declared to be conventional ([86], 22). Unlike (14), such statements do not even look like definitions. But statements about "universal" forces have no special status either. Empirical equivalents to $G + F$ can be generated by varying F or by varying G ([86], 34). So we end up with a view according to which just about any sentence of our combined geometrical-physical theory can be a "coordinative definition." We have come far indeed from the Reichenbachian project of pinpointing and circumscribing the conventional elements of science.

We remarked above that Reichenbach's conventionalism in *The Philosophy of Space and Time* [86] is quite different from his earlier conventionalism in *The Theory of Relativity and A Priori Knowledge* [84]. The latter, as we saw, is largely motivated by Kantian themes, whereas the former rejects Kant completely.[25] Accordingly, the epistemology motivating Reichenbach's later conventionalism is fundamentally empiricist: the conventional elements of science turn out to be just the nonobservational or theoretical elements. The only facts are the observable facts; all the rest represents the "contribution of reason." By contrast, Reichenbach's earlier conventionalism is explicit antiempiricist. The conventional/factual distinction does not correspond to the theoretical/observational distinction; instead, it is drawn within the realm of the theoretical. Certain elements of theoretical structure—notably, the choice of rest system and the choice of inertial system—are arbitrary or conventional; other, equally theoretical elements—notably, the choice of metric—are not. Interestingly enough, Reichenbach's earlier Kantian conventionalism fits the actual development of relativity theory much better than his later empiricist conventionalism.

GRÜNBAUM. As we emphasized in §VII.2, there is a significant gap between empirical equivalence and full equivalence, between theoretical underdetermination and "no fact of the matter." To fill this gap we must change our theoretical ideology in such a way that the problematic elements in question, such as absolute velocity and

---

[25] See, for example, [86], xii. Reichenbach appears to have given up his alliance with neo-Kantianism between 1922 and 1924; cf. [84], xxxi. In the Introduction we saw how general relativity—especially through the idea of general covariance—contributed to this change.

gravitational acceleration, are no longer assumed to exist objectively and determinately. We must drop such problematic elements from our list of basic quantities, properties, and relations. To be sure, considerations of empirical equivalence can motivate changes in theoretical ideology; but even this is true only in very special cases where we have both empirical equivalence and "methodological equivalence." Reichenbach has not shown us how to go from empirical equivalence to full equivalence in the geometrical case. First, of course, he fails to make the crucial distinction between strong and weak unverifiability. Second, however, Reichenbach's later empiricism and verificationism prevent him from seeing that there is a gap to be filled. Since there is no notion of meaning or content besides empirical or observational content, there is no distance at all between empirical equivalence and full equivalence.

Grünbaum's work on geometry is best understood as an attempt to provide arguments for conventionalism that do not rely on a verificationist conception of meaning.[26] He recognizes that mere empirical equivalence is insufficient and that we need to consider ideological questions about what quantities and relations actually exist in the physical world. In particular, we need to ask whether, independently of our measuring procedures, there is in fact an objective physical quantity of length (relation of congruence). Grünbaum's view is that there is no such objective physical quantity: prior to the adoption of a particular measuring technique physical space simply has no metrical properties; it is "metrically amorphous."

But why does Grünbaum think that physical space lacks metrical properties? In the first instance, he argues from the assumed

---

[26] Consider the following passage, for example: "in virtue of the lack of an intrinsic metric, sameness or change of the length possessed by a body in different places and at different times consists in the ratio (relation) of that body to the conventional standard of congruence. Whether or not this ratio changes is quite independent of any human discovery of it. ... And thus the relational character of length derives, in the first instance, not from how we human beings measure length but from the failure of the continuum of physical space to possess an intrinsic metric, a failure obtaining quite independently of our measuring activities. ... Since, to begin with, there exists no property of true rigidity to be discovered by any human test, no test could possibly reveal its presence. Accordingly, the unascertainability of true rigidity by us humans is a consequence of its non-existence in physical space and evidence for that non-existence but not constitutive of it" ([51], 42).

continuity of physical space. If space were discrete or granular, the distance between two points could be defined in terms of the number of elements or "space-atoms" between them. There would then be a sense to saying that two intervals contain the same amount of space: namely, that they contain the same number of "space-atoms." Congruence would then be an "intrinsic" property of pairs of intervals. On the other hand, if physical space is in fact continuous, there is no "intrinsic" property of spatial intervals by virtue of which they can be said to be equal or unequal to each other. In a continuous space the distance between two points cannot be defined in terms of the number of points between them, for the cardinality of the set of points between any two nonidentical points is $2^{\aleph_0}$: all such sets have the same cardinality. Metrical properties are therefore not "intrinsic" to a continuous space.[27] Hence, since intervals of physical space (and time) do not stand to one another in relations of equality or inequality prior to the establishment of measuring procedures, such procedures do not ascertain relations of equality and inequality but serve instead to define them.[28] Our measuring procedures are not

[27] In Grünbaum's own words: "in the case of a discretely ordered set, the distance between two elements can be defined intrinsically in a rather natural way by the cardinality of the (least) number of intervening elements. By contrast, upon confronting the extended continuous manifolds of physical space and time ... we see that neither the cardinality of intervals nor any of their other topological properties provide a basis for an *intrinsically* defined metric. The first part of this conclusion was tellingly emphasized by Cantor's proof of the equi-cardinality of all positive intervals independently of their length. Thus, there is no *intrinsic* attribute of the space between the endpoints of a line segment AB, or any relation between these two points themselves, in virtue of which the interval AB could be said to contain the same amount of space as the space between the termini of another interval CD not coinciding with AB. Corresponding remarks apply to the time continuum. Accordingly, the continuity we postulate for physical space and time furnishes a sufficient condition for their intrinsic metric amorphousness" ([51], 9–10).

[28] In Grünbaum's own words: "Only the choice of a particular congruence standard which is *extrinsic* to the continuum itself can determine a unique congruence class, the *rigidity* or self-congruence of that standard under transport being *decreed by convention* ... the role of the spatial or temporal congruence standard cannot be construed with Newton or Russell to be the mere ascertainment of an otherwise intrinsic equality obtaining between the intervals belonging to the congruence class specified by it ... the obtaining of the congruence relation between two segments is a matter of convention, stipulation, or definition and not a factual matter concerning which empirical findings could show one to have been mistaken. And hence there can be no question at all of an empirically or factually determinate metric geometry until after a physical stipulation of congruence" ([51], 11–12).

constrained by an independently existing metric and are in this sense arbitrary or conventional.

This is indeed an ingenious argument. But what exactly does the contrast between discrete and continuous spaces really show? It shows that in the case of a discrete space metrical relations are *definable* in terms of topological properties (cardinality) and order relations; in the case of a continuous space metrical relations cannot be so defined. Does it follow that metrical relations do not objectively exist independently of measurement on a continuous space? Not at all. It only follows that in our combined theory of the topology and geometry of physical space the metric is a primitive or undefined quantity.[29] If we are to draw any conclusion about "metrical amorphousness" from the continuity of physical space, we need an additional premise: namely, that the only spatio-temporal properties and relations that objectively exist are topological properties and order relations. It follows from this claim that properties and relations not definable in terms of topological properties and order relations do not objectively exist and, consequently, that a continuous space objectively lacks metrical properties.

Thus, Grünbaum's conventionalism is actually a species of what we have called ideological relationalism. He holds that there is a privileged set of basic spatio-temporal properties and relations— roughly, topological and ordinal relations—and that all other "real" spatio-temporal properties and relations must be definable from or reducible to these privileged relations. (As we shall see later when we discuss Grünbaum's views on simultaneity, he explicitly connects his position with a "causal" theory of spatio-temporal structure.) Grünbaum is a conventionalist because he is a relationalist. For Reichenbach, on the other hand, the implication appears to go in the other direction: he holds a relationalist view of geometrical structure because he is a conventionalist. Reichenbach is convinced on epistemological grounds that there is no objective fact of the matter whether spatially separate intervals are or are not congruent. If geometrical assertions are to express objective facts at all, therefore, the predicate "congruent" must be relativized: it must contain an additional argument place filled by an arbitrarily chosen congruence standard. The predicate "length" must be introduced by some such definition as (14). This seems to be what Reichenbach

[29] This point is made by Demopoulos [18] and Fine [34]. See also Friedman [36]; Glymour [44].

304

intends by: "The objective character of the physical statement is thus shifted to a statement about relations . . . it is a statement about a relation between the universe and rigid rods" ([86], 37).

Returning to Grünbaum, it is now clear that the burden of his argument falls entirely on his version of ideological relationalism. Why should we believe that our primitive or basic spatio-temporal properties and relations include only topological and ordinal ("causal") relations? In particular, why do metrical properties and relations not belong on our list of primitives? Grünbaum cannot argue, *à la* Reichenbach, that metrical relations have no objective existence because they are conventional, for a property or relation cannot be conventional in Grünbaum's sense unless its nonexistence is already established. So how do we decide which basic properties and relations belong on our list? From the present point of view, these are just the properties and relations postulated by our most well-confirmed space-time theories: properties and relations of maximal explanatory or unifying power. Hence, from the present point of view, we can question the objectivity of a given element of theoretical ideology in only one of two ways: we can argue on internal grounds that the theory in which it is embedded is strongly unverifiable (this is what happens to our paradigmatic indeterminate entities from the development of relativity); or we can find a better theory that avoids the question altogether (this is what happens to absolute simultaneity, for example). As we have seen in our discussion of Reichenbach, the spatial metric cannot be criticized in the first way: there is nothing methodologically wrong with the Newtonian theory of h. Only the second option remains.

Grünbaum appears to feel the force of these points, for he tries to buttress the above definability argument with considerations drawn from relativity theory. Specifically, he argues that relativity theory employs alternative spatial and temporal congruences in just the sense required by conventionalism: spatial and temporal intervals can be metrized in a variety of different ways, and the choice between such alternative metrizations is arbitrary. This argument, if it were correct, would be much stronger than Grünbaum's pure definability argument—in fact, it would render the latter completely superfluous. For Grünbaum would have shown that relativity theory gives us the same kind of reason for rejecting "intrinsic" spatial and temporal congruence as we have for rejecting absolute simultaneity, say. So let us look at the examples Grünbaum offers in support of this argument: (i) the use of both proper time

305

and coordinate time in special relativity ([52], 200–205);[30] and (ii) the use of varying, time-dependent spatial metrics in general relativity ([52], 229–332).

(i) Grünbaum shows how incompatible congruences are induced by proper time and coordinate time on a rotating disk ([52], 203–204). The same point can be illustrated much more simply, however, by a single accelerating object in an inertial frame.

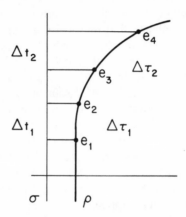

According to the coordinate time t (determined by the proper time along σ), the temporal interval e_1e_2 is congruent to the interval e_3e_4: $\Delta t_2 = \Delta t_1$. According to the proper time τ along ρ, e_1e_2 and e_3e_4 are not congruent: $\Delta \tau_2 < \Delta \tau_1$. But it is traditional relativistic practice to use either t or τ as convenience dictates. Is this not a perfect illustration of the conventionalist thesis that temporal congruence is arbitrary?

This conclusion represents a serious misunderstanding of relativity theory. In a relativistic space-time there are no such entities as temporal intervals (or spatial intervals). A given pair of events determines a unique *space-time* interval, but no unique temporal interval (or spatial interval). Instead, there are timelike and spacelike *curves*, each of which receives a unique temporal (spatial) length induced by the space-time metric. The temporal (spatial) length of

---

[30] Grünbaum uses examples of noninertial systems, which, in accordance with traditional practice, he considers in the context of *general* relativity. As he later points out ([52], 213–214), there is no reason special relativity cannot treat such examples.

any timelike (spacelike) curve is completely determinate and objective: there is no room for arbitrary choice here. Alternative temporal (spatial) metrics arise as follows: the very same *pair* of timelike (spacelike) separated events is associated with a variety of different timelike (spacelike) curves. In our example, the pairs (e_1, e_2) and (e_3, e_4) are associated with the accelerated curve ρ, the inertial curve σ (via a particular inertial reference frame), and many other curves besides. Such pairs will naturally receive different "lengths" relative to different associated curves. It should be clear, however, that this relativity of length is not conventionality. It is not a matter of one and the same temporal (spatial) manifold receiving different measures of length; rather, it is a matter of the same pair of events being associated with different temporal (spatial) manifolds or curves, each of which receives a unique and determinate measure of length.

(ii) Grünbaum's second example is the use of varying spatial metrics in general relativity. Here the idea is that with appropriate changes in the mass-energy distribution the very same spatial manifold (a succession of spacelike hypersurfaces determined by some particular reference frame) will receive alternative and incompatible geometries ([52], 231). Again, however, this does not show that the spatial metric is conventional but only that it is dynamical. At any given time, a particular spatial manifold has a unique spatial geometry, and, as time progresses, the spatial metric evolves due to changes in the distribution of mass-energy. There is never a question of arbitrary or conventional *choice* of geometry; there is a determinate geometry evolving according to determinate physical laws. Hence, although alternative geometries and "congruence definitions" are indeed possible for any given spatial manifold, the actual one depends on the matter distribution and the field equations of general relativity; it does not depend on an arbitrary choice of metrical standard.

That Grünbaum is even tempted to argue from general relativity in this way is due, I think, to a pervasive ambiguity in his thought between a conventionalist view and a relationalist view. He often seems to have the following picture in mind: an *empty* space has no "intrinsic" metric; it acquires a metric only on being filled with a particular "extrinsic" distribution of matter. This picture is relationalist but not conventionalist. The point is not that metrical relations are arbitrary but that they are constituted by (definable from?) material events and processes. (Of course, general relativity does not really vindicate this picture either, but it is at least *close*.)

Moreover, I think that it is this relationist picture, and not conventionalism, that is most clearly suggested by the passage from Riemann's inaugural address to which Grünbaum often appeals:

> while in a discrete manifold the principle of metric relations is implicit in the notion of this manifold, it must come from somewhere else in the case of a continuous manifold. Either then the actual things forming the groundwork of a space must constitute a discrete manifold, or else the basis of metric relations must be sought for outside that actuality, in colligating forces that operate upon it. ([93], 424–425)

Such "colligating forces" may, as in general relativity, have nothing whatever to do with arbitrarily chosen metrical standards.

Thus, although Grünbaum's relativistic examples do not support his conventionalism, they do provide nice illustrations of how relationist views and conventionalist views become entangled with relativity theory and with one another. Special relativity can easily suggest a conventionalist view of metrical relations, since it relativizes such relations to an arbitrary choice of reference frame. This suggestion appears especially tempting if we compare the latter relativization to our paradigmatic relativizations of velocity and gravitational acceleration, for these do implement a conventionalist strategy of "equivalent descriptions." But there is an essential difference. In relativizing velocity and gravitational acceleration, we drop the questionable entities completely: there are no elements in our new theories (§III.2 and §III.4) corresponding to absolute velocity or absolute gravitational acceleration. However, we do not simply drop metrical relations in special relativity; rather, we replace the three-dimensional spatial metric $d_S(a,b)$ and the one-dimensional temporal metric $t(a,b)$ with a single four-dimensional metric $I(a,b)$. $I(a,b)$ induces a division of the four-dimensional space-time manifold into a variety of three-dimensional spacelike hypersurfaces and one-dimensional timelike curves. Each spacelike hypersurface receives a unique and determinate spatial geometry; each timelike curve receives a unique and determinate temporal length. So there is no room for arbitrary choice of geometry on any given four-, three-, or one-dimensional manifold. What is arbitrary is the choice of any particular spacelike manifold and timelike manifold to represent three-dimensional space and one-dimensional time.

General relativity, on the other hand, can easily suggest a relationalist view of metrical relations, since such relations explicitly depend on the distribution of matter. Moreover, since conventionalism also asserts a dependence of metrical relations on "matter"—on arbitrarily chosen metrical standards or "rigid bodies"—it is tempting to conflate the two views. But this temptation should be resisted. On one view, which we might call *positive* relationalism, metrical relations do have objective existence: they are definable from "respectable" physical properties and relations. On a second view, which we might call *negative* relationalism, metrical relations do not have objective existence: they are not definable from "respectable" physical properties and relations. Only the second view leads to a conventionalist strategy of "equivalent descriptions" and arbitrary choice. Only the first view is suggested by general relativity. However, since general relativity does not implement an actual definition of metrical relations in terms of the distribution of matter, neither view is vindicated by the development of relativity theory. In relativity theory, just as in classical physics, metrical quantities are primitive and undefined—but perfectly determinate and respectable—theoretical quantities. The difference is that we have replaced three- and one-dimensional absolute quantities with a four-dimensional dynamical quantity.[31]

4. Simultaneity

We have already introduced the Reichenbach-Grünbaum thesis of the conventionality of simultaneity in §IV.7. As we emphasized there, this thesis should be carefully distinguished from the uncontroversial *relativity* of simultaneity. The point of the latter is that no unique simultaneity relation is defined on space-time; rather, each inertial frame and inertial trajectory has its own simultaneity relation. In standard formulations of special relativity, however, the relation of simultaneity-for-σ, where σ is an inertial trajectory, is unique: there is no further choice of simultaneity relation within a particular inertial frame. By contrast, the thesis of the conventionality of simultaneity asserts that an arbitrary choice of

---

[31] Many of these criticisms of Grünbaum's treatment of relativity theory are made by Hilary Putnam in "An Examination of Grünbaum's Philosophy of Geometry" ([80], chap. 6 of vol. 1).

simultaneity-relation-for-σ is possible even given a particular inertial trajectory σ. This contrast reflects the difference between the relativity of spatial and temporal congruence and the conventionality of spatial and temporal congruence that we insisted upon above. The former says that the spatial and temporal congruence relations are different in different inertial frames; the latter asserts that spatial and temporal congruence relations are subject to arbitrary choice even within a particular inertial frame (a particular "splitting" of four-dimensional space-time into three-plus-one-dimensional space-plus-time).

Once again, Reichenbach and Grünbaum argue for the very same thesis from completely different points of view. Reichenbach, arguing from an epistemological point of view, maintains that the standard simultaneity-relation-for-σ has no objective status because the relation of simultaneity-for-σ is not determinable or verifiable in a special relativistic world. There is no way (noncircularly) to compare the temporal relations of distant (spacelike separated) events. Grünbaum, on the other hand, arguing from an ideological point of view, tries to show—again on the basis of a definability argument—that no determinate simultaneity-relation-for-σ exists in a special relativistic world; consequently, events can be meaningfully said to be simultaneous or nonsimultaneous only relative to an arbitrary choice of ε. Antecedent to an arbitrary choice of ε there just are no objective temporal relations between distant (spacelike separated) events.

We should observe at the outset that there is a fundamental fact about the standard ($\varepsilon = \frac{1}{2}$) simultaneity-relation-for-σ that both Reichenbach and Grünbaum overlook: namely, in Minkowski space-time the standard relation is *explicitly definable* from the space-time metric g (in fact, from the conformal structure of g), whereas the nonstandard ($\varepsilon \neq \frac{1}{2}$) relations are not so definable.[32] To see this, recall that the standard simultaneity-relation-for-σ is generated by hypersurfaces orthogonal (in the sense of the metric g) to σ and that no nonstandard simultaneity surface can be picked out by this kind of condition. Consider, for example, the "tilted" simultaneity surfaces of pp. 171–172. One cannot pick out such a surface by requiring that it make a (nonorthogonal) fixed angle α with σ, for many distinct surfaces satisfy this condition. Thus, in two dimensions, both "plane" A and "plane" B intersect σ at the angle α. (By

---

[32] This simple, but crucial, observation is first made by David Malament in [71].

310

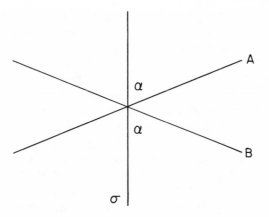

contrast, only one "plane" is orthogonal to σ.) In higher dimensions the situation is of course even worse. Hence, to pick out a unique nonstandard "tilted" surface we need to impose additional structure on Minkowski space-time. In effect, we need a distinguished temporal vector t and spatial vector s, which together determine a unique "quadrant" where the angle α is to be evaluated. To distin-

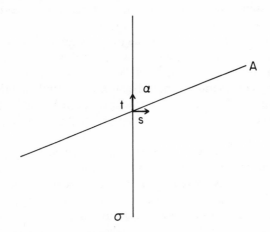

guish A from B we need a definite spatio-temporal *orientation* (such an orientation is determined by any arbitrarily chosen coordinate system). Yet Minkowski space-time, as usually understood, contains no such distinguished orientation: it is perfectly "isotropic."

The importance of this point is as follows. First, it shows that the standard $(\varepsilon = \frac{1}{2})$ simultaneity-relation-for-σ does not have the

311

defects of our paradigmatic indeterminate entities from Newtonian theory. The problem with absolute velocity and gravitational acceleration is that they are not sufficiently connected with the other quantities of Newtonian theory: they can vary arbitrarily while the other quantities remain fixed. However, not only is the standard simultaneity relation connected with the other quantities of special relativity (namely, the Minkowski metric g), but it is explicitly definable from them: two events are ($\varepsilon = \frac{1}{2}$) simultaneous-for-$\sigma$ just in case they lie on a single spacelike hypersurface orthogonal to σ. So we cannot dispense with standard simultaneity without dispensing with the entire conformal structure of Minkowski space-time. Second, it is clear that if we wish to employ a nonstandard ($\varepsilon \neq \frac{1}{2}$) simultaneity-relation-for-σ we must add further structure to Minkowski space-time. For example, to employ our "tilted" simultaneity relation we have to pick out a distinguished spatio-temporal orientation that is not itself definable from the ("isotropic") metric g. This additional structure has no explanatory power, however, and no useful purpose is served by introducing it into Minkowski space-time. Hence, the methodological principle of parsimony favors the choice of Minkowski space-time, with its "built-in" standard simultaneity, over Minkowski space-time plus any additional nonstandard simultaneity.

These considerations seem to me to undercut decisively the claim that the relation of simultaneity-for-σ is arbitrary or conventional in the context of special relativity. What is peculiar is the extent to which these considerations have been so consistently overlooked.[33] In what follows I shall explain how these points bear on the arguments for conventionality offered by Reichenbach and Grünbaum.

REICHENBACH. How does Reichenbach argue for the conventionality thesis? He considers various methods for determining distant simultaneity relative to a given trajectory, various methods for verifying statements of the form "Events e_1 and e_2 are simultaneous with respect to the given trajectory σ"; and he tries to show that none of these methods furnishes an unambiguous answer in a special relativistic world. The basic problem is this. In a Newtonian world, where there is no upper limit to the velocity of "causal" signals, we could in principle determine which event at A is simultaneous with a given event t_B at B by sending faster and faster signals

---

[33] The present author has been guilty of this oversight as well; see [39].

from A to B and back (see §IV.7). In this way we would "close in" on the unique event at A that is really simultaneous with t_B. In a special relativistic world, on the other hand, there is supposed to be an upper limit on the velocity of signals: no signal can travel faster than light. So in a special relativistic world the procedure of sending faster and faster signals from A to B and back does not determine a unique event at A. We cannot determine the event at A simultaneous with t_B any more narrowly than the range $(t'_A - t_A)$.

In the absence of arbitrarily fast signals, then, we have to make assumptions about one-way (as opposed to round-trip) velocities in order to determine distant simultaneity via "causal" signals. For example, if we assume that light rays have the same velocity from A to B as they have from B to A, the event at A simultaneous with t_B is of course given by $t_A + \frac{1}{2}(t'_A - t_A)$. But without this assumption all we know is that

$$t_B = t_A + d\left(\frac{1}{v_{\text{out}}}\right) \tag{15}$$

where d is the distance between A and B and

$$v_{\text{out}} = \frac{d}{\varepsilon(t'_A - t_A)}$$

$$(0 < \varepsilon < 1) \tag{16}$$

$$v_{\text{in}} = \frac{d}{(1 - \varepsilon)(t'_A - t_A)}.$$

Hence, we cannot use the signal method to determine simultaneity unless we already know v_{out} and v_{in}:

> Thus we are faced with a circular argument. To determine the simultaneity of distant events we need to know a velocity, and to measure a velocity we require knowledge of the simultaneity of distant events. The occurrence of this circularity proves that simultaneity is not a matter of knowledge, but of a coordinative definition, since the logical circle shows that a knowledge of simultaneity is impossible in principle. ([86], 126–127)

We cannot determine v_{out} and v_{in} unless we have already determined a value for ε in (16), unless we have already synchronized clocks at A and B.

Reichenbach has indeed shown, I think, that the signal method cannot be used to determine distant simultaneity in a special relativistic world. The circularity involved here is vicious, for, as

313

Reichenbach points out: "If we wish to determine by velocity measurements which events are simultaneous, we shall always obtain that simultaneity which has already been introduced by definition" ([86], 127). Any choice of ε in (16) will automatically "confirm" the value t_B in (15); any choice of ε in (15) will automatically "confirm" the values v_{out} and v_{in} in (16). However, as we have seen, standard simultaneity is intimately connected with the rest of the structure of Minkowski space-time. We should therefore expect that there are other methods for determining simultaneity that are free from this objectionable circularity. Consider, for example, the method of *slow-transport synchrony*.[34] In this method we synchronize a clock with A-time at A and then transport it "infinitely slowly" to B. If the reading of this clock and the original clock at B are there in agreement, A-time and B-time are in standard ($\varepsilon = \frac{1}{2}$) synchrony.

More precisely, consider an arbitrary ε_{x_1}-system adapted to a given timelike geodesic $\sigma(\tau)$. The timelike geodesic $\rho(\tau^*)$ represents

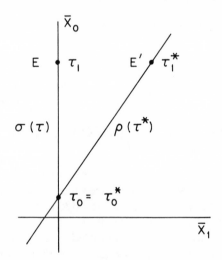

Slow-transport synchrony in an arbitrary ε_{x_1}-system (two spatial dimensions suppressed).

a clock transported with constant "velocity" $\bar{v} = d\bar{x}_1/d\bar{t}$ leaving $\sigma(\tau)$ at $\tau_0 = \tau_0^*$. Events E and E', along $\sigma(\tau)$ and $\rho(\tau^*)$ respectively,

[34] See Ellis and Bowman [32].

314

are said to be *slow-transport simultaneous* just in case

$$\lim_{\bar{v}\to 0} (\tau_1 - \tau_1^*) = 0 \tag{17}$$

where τ_1 and τ_1^* are the respective proper times of E and E'.

Equation (17) fixes the value of ε at $\frac{1}{2}$: E and E' are ε_{x_1}-simultaneous just in case $\varepsilon_{x_1} = \frac{1}{2}$. To see this, suppose that E and E' are ε_{x_1}-simultaneous: the $\bar{t} = \bar{x}_0$ coordinate of E' is just τ_1. It follows from (IV.61) that

$$(\tau_1^* - \tau_0^*) = (\tau_1 - \tau_0)\sqrt{1 + 2\delta_{x_1}\bar{v} + (\delta_{x_1}^2 - 1)\bar{v}^2}.$$

Expanding the "dilation term" in a binomial series (valid for $\bar{v}$ in a neighborhood of zero) we obtain

$$(\tau_1^* - \tau_0^*) = (\tau_1 - \tau_0)[1 + \delta_{x_1}\bar{v} + \tfrac{1}{2}(\delta_{x_1}^2 - 1)\bar{v}^2 + \cdots$$

where the rest of the series consists of second and higher powers of $\bar{v}$. Since $\tau_0^* = \tau_0$, we have

$$(\tau_1 - \tau_1^*) = (\tau_1 - \tau_0)[-\delta_{x_1}\bar{v} - \tfrac{1}{2}(\delta_{x_1}^2 - 1)\bar{v}^2 + \cdots.$$

But $(\tau_1 - \tau_0) = \bar{x}_1/\bar{v}$, where $\bar{x}_1$ is the spatial coordinate of E'. So we have

$$(\tau_1 - \tau_1^*) = \bar{x}_1[-\delta_{x_1} - \tfrac{1}{2}(\delta_{x_1}^2 - 1)\bar{v} + \cdots.$$

Letting $\bar{v} \to 0$ and substituting $2(\frac{1}{2} - \varepsilon_{x_1})$ for δ_{x_1}, we finally get

$$\lim_{\bar{v}\to 0} (\tau_1 - \tau_1^*) = \bar{x}_1(2\varepsilon_{x_1} - 1). \tag{18}$$

So ε_{x_1}-simultaneity agrees with slow-transport simultaneity just in case $\varepsilon_{x_1} = \frac{1}{2}$, just in case $\bar{x}_0$ is the coordinate time of an inertial coordinate system adapted to σ. In other words, slow-transport synchrony is just standard ($\varepsilon = \frac{1}{2}$) synchrony.

The slow-transport method allows us to determine standard ($\varepsilon = \frac{1}{2}$) simultaneity without vicious circularity. We initially synchronize the clocks at A and B according to an *arbitrary* choice of ε: we pick an arbitrary ε_{x_1}-system. We then transport clocks (synchronized with A-time at A) from A to B with successively slower and slower "velocities" as determined by our arbitrary initial choice of ε. If the differences between the readings of these transported clocks and the clock at B converge toward zero, the clocks at A and B are truly ($\varepsilon = \frac{1}{2}$) synchronized, otherwise not. So no matter what

315

value of ε is arbitrarily chosen at the beginning, we obtain $\varepsilon = \frac{1}{2}$ in the end. This procedure contrasts with the signal method, where, as we saw, any initial choice of ε is inevitably "confirmed." The point is that the slow-transport method, unlike the signal method, exploits the connections between standard simultaneity and other relativistic quantities—in this case, the connections between standard simultaneity and proper time.

Reichenbach considers the possibility of determining distant simultaneity by means of the transport of clocks, although he does not explicitly consider the slow-transport method. He points out that it will not do to say that the clocks at A and B are synchronized just in case a clock synchronized with A-time at A is synchronized with B-time when transported to B. In a special relativistic world this principle does not lead to consistent results, for the rates of clocks depend on the path and velocity of transport ([86], 133). But precisely this problem is avoided by the slow-transport method. By considering only "infinitely slow" transport, we avoid the relativistic time-dilation effects as well as question-begging assumptions about one-way velocity. So slow-transport simultaneity is a unique, well-defined simultaneity relation. What would Reichenbach say to this? Consider the following revealing passage:

> However, if relativistic physics were wrong, and the transport of clocks could be shown independent of path and velocity, this type of time comparison could not change our epistemological results, since the transport of clocks can again offer nothing but a *definition* of simultaneity. Even if the two clocks correspond when they are again brought together, how can we know whether or not both have changed in the meantime? This question is as undecidable as the question of the comparison of length of rigid rods. Again, a solution can be given only if the comparison of time is recognized as a definition. If there exists a unique transport-synchronization, it is still merely a *definition* of simultaneity. ([86], 133)

This passage is extremely interesting, for it shows that Reichenbach is willing to put aside the details of special relativity and rest his case on *a priori* arguments about the incomparability of spatially distant (and hence not jointly observable) clocks. Presumably, then, Reichenbach would reject the slow-transport method on similar grounds.

Precisely what assumptions about the rates of clocks are involved in the slow-transport method? Only this: that (ideal) clocks measure

316

the proper time along their trajectories,

$$d\tau = \sqrt{g_{ij} \frac{dx_i}{du} \frac{dx_j}{du}} \, du \qquad (19)$$

where g_{ij} is the Minkowski metric. In ε_{x_1}-systems (19) becomes

$$d\tau = \sqrt{1 + 2\delta_{x_1}\bar{v} + (\delta_{x_1}^2 - 1)\bar{v}^2} \, d\bar{t} \qquad (20)$$

which, as we have seen, entails the uniqueness of slow-transport synchrony: the fact that slow-transport simultaneity coincides with $\varepsilon = \frac{1}{2}$ simultaneity. The only way to get around this result, therefore, is to change the temporal metric on timelike curves. Thus, if we replace the metric (20) with

$$d\tau' = \sqrt{1 + 2\delta_{x_1}\bar{v} + (\delta_{x_1}^2 - 1)\bar{v}^2} \, d\bar{t} + 2\delta_{x_1}\bar{v} \, d\bar{t} \qquad (21)$$

we can eliminate the uniqueness of slow-transport synchrony. According to the metric (21)

$$\lim_{\bar{v} \to 0} (\tau_1' - \tau_1'^*) = 0$$

for *all* values of ε_{x_1}. But the important point is this. The formula (19) for the proper time metric (and its corollary (20)) is a central explanatory principle of special relativity—perhaps *the* central explanatory principle of special relativity. Equation (19) is not, of course, directly verifiable, but it is about as well-confirmed as a theoretical principle can be.

These considerations vividly illustrate how firmly the standard simultaneity relation is embedded in relativity theory. One cannot question the objectivity of this relation without also questioning significant parts of the rest of the theory. In particular, one cannot maintain that distant simultaneity is conventional without also maintaining that such basic quantities as the proper time metric are conventional as well.[35] One would need an independent argument for this latter conventionality thesis, and, as we have seen in §VII.3, the prospects for such an argument are dim indeed.

GRÜNBAUM. As in his treatment of congruence, Grünbaum tries to establish the nonexistence of an objective simultaneity relation by means of an indefinability argument. The basis of his argument is the

---

[35] This is precisely what Grünbaum does maintain in his discussion of the slow-transport method. He argues that slow-transport simultaneity is conventional *because* the temporal metric is conventional; see [53].

same as Reichenbach's, namely, the putative nonexistence of "causal" signals with velocity greater than light. In this connection, Grünbaum introduces the concept of "topological simultaneity." An event at A is topologically simultaneous with an event at B just in case the two are connectible by no "causal" signal. Grünbaum then observes that in a Newtonian world, where there is no limit to the velocity of "causal" propagation, there is a unique event at A topologically simultaneous with a given event t_B at B. The definition

e and E are simultaneous iff no signal can connect e with E (22)

picks out a unique e for every given E. In a special relativistic world, on the other hand, there are an infinity of events at A topologically simultaneous with t_B, namely, all events in the interval $[t_A, t'_A]$. Therefore, in a relativistic world "metrical simultaneity" (the ordinary relation of simultaneity) cannot be defined via (22). Metrical simultaneity does not coincide with topological simultaneity (see [51], 28–29).

But why should the relation of metrical simultaneity be definable from the relation of topological simultaneity? Why can metrical simultaneity not stand on its own feet, as it were? The answer, of course, is that Grünbaum holds a relational theory of time. He believes that all objective temporal relations are constituted by "causal" relations between events. The only temporal relations that objectively exist are those definable in terms of "causal" relations.

By maintaining that the very *existence* of *temporal* relations between non-coinciding events depends on the obtaining of some *physical* relations between them, Einstein espoused a conception of time (and space) which is *relational* by regarding them as systems of relations between physical events and things. Since time relations are first constituted by the system of physical relations obtaining among events, the character of the temporal order will be determined by the physical attributes in virtue of which events will be held to sustain relations of "simultaneous with," "earlier than," or "later than." In particular, it is a question of physical fact whether these attributes are of the kind to define temporal relations *uniquely.* ([51], 345–346)

Thus, in a world in which metrical simultaneity is not definable in terms of "causal" relations, there simply is no such physical relation. Note the similarity with Grünbaum's argument for the conventionality of congruence. The basic strategy is to select a privileged

318

set of properties and relations—topological properties and order relations in the case of congruence, "causal" relations in the case of simultaneity—and to argue that certain further properties and relations (congruence and simultaneity, respectively) are not definable from this privileged set. The strategy depends, therefore, on a prior assumption that only those properties and relations that are definable from the privileged set objectively exist.

As in our treatment of congruence in §VII.3, we can object to Grünbaum's list of primitive properties and relations. Which primitive properties and relations objectively exist cannot be settled *a priori* but only on the basis of our best confirmed space-time theories. There is a much more fundamental problem, however: Grünbaum's indefinability claim turns out to be false. Grünbaum has established only that metrical simultaneity cannot be captured by the particular "causal" definition (22). But David Malament has shown that the standard simultaneity relation is explicitly definable from "causal" relations in Minkowski space-time after all. Moreover, none of the nonstandard ($\varepsilon \neq \frac{1}{2}$) simultaneity relations is similarly definable. The standard simultaneity relation is the one and only would-be simultaneity relation that is "causally" definable in Minkowski space-time.[36]

One can get an intuitive feel for Malament's results via the following considerations. The "causal" connectibility relation on Minkowski space-time is the relation κ such that $p\kappa q$ just in case there is a timelike or null curve connecting p and q: p and q lie in or on each other's light cones. The relation κ determines the conformal structure of Minkowski space-time: all automorphisms of κ (one-one mappings f such that $p\kappa q$ iff $f(p)\kappa f(q)$) are conformal transformations. Now the conformal structure of Minkowski space-time includes the orthogonality relation, and, as we have observed, the standard ($\varepsilon = \frac{1}{2}$) simultaneity-relation-for-σ is simply definable in terms of orthogonality. On the other hand, the nonstandard ($\varepsilon \neq \frac{1}{2}$) simultaneity-relations-for-σ are not preserved by conformal transformations, and, *a fortiori*, they are not preserved by automorphisms of κ. For example, on p. 311 above, the "plane" A is mapped onto the "plane" B by a temporal inversion about their intersection. The

---

[36] See Malament [71] and chap. 2 of [69]. As Malament points out, his results are implicit in Robb [94]. Malament shows that temporal congruence (relative to a given inertial trajectory) is also uniquely definable in terms of causal relations—undermining Grünbaum's argument for the conventionality of congruence as well. See also Winnie [119].

problem again is that nonstandard simultaneity relations introduce an asymmetric orientation into otherwise "isotropic" Minkowski space-time. So nonstandard simultaneity relations cannot be defined in terms of our "isotropic" relation κ.

The central problem for both Reichenbach and Grünbaum, then, is that the standard simultaneity relation is much more intimately connected with the rest of the geometrical structure of Minkowski space-time than they have realized. In particular, they have not seen that the standard simultaneity relation is an integral part of the conformal structure of Minkowski space-time. So standard simultaneity and our paradigmatic indeterminate entities from Newtonian physics actually lie at opposite ends of the spectrum. Absolute velocity and gravitational acceleration are unusual in being quite unconnected with other quantities: they can be varied at will and are therefore not measurable or determinable. More typical quantities, like the spatial metric in Newtonian physics, are connected with other quantities by theoretical principles: they can be varied, all right, but only by making compensatory adjustments in our theory. Standard simultaneity, on the other hand, is explicitly definable from the other quantities of relativity theory: it cannot be varied without completely abandoning the basic structure of the theory.

5. *Theoretical Inference and the Evolution of Relativity Theory*

Both conventionalism and relationalism are forms of skepticism: both wish to limit the postulation of theoretical structure. Conventionalism attempts to limit our theoretical ideology, the properties and relations we hold to exist objectively and determinately. Relationalism attempts to limit our theoretical ontology, the domain of individuals (space-time points or events) over which the quantifiers of our theories are thought to range. Both views are motivated by some version of positivist or empiricist epistemology, an epistemology characterized by the rejection of empirically undecidable questions that cannot be decisively settled, even "in principle," by the accumulation of empirical data. Relationalism, for example, rejects the empirically undecidable question of "where" the material universe or set of actual events is embedded in space-time (Leibnizean indiscernibility argument); conventionalism rejects the empirically undecidable question about the "true" geometry of space or space-time (Poincaré-Reichenbach argument).

As we saw in §VII.1, this kind of empiricist epistemology conceives of scientific method as the extrapolation of observational generalizations from our finite observational data and, accordingly, tries to minimize the role of methodological criteria for choosing between empirically equivalent theories. From the present point of view, however, these criteria are essential to scientific method: all scientific inference is necessarily guided by methodological principles going beyond mere conformity to evidence. Such principles determine when it is justifiable to postulate theoretical structure and when it is not. Our task, then, is to describe the methodological criteria scientists actually employ and to show how they lead to the acceptance of some pieces of theoretical structure but not others. From this point of view, the development of relativity theory provides a central example of how our methodological principles really work. It reveals no blanket or wholesale rejection of theoretical structure: some theoretical entities (for example, absolute velocity, inertial frames) are rejected on methodological grounds; other, equally theoretical entities (for example, absolute acceleration, the space-time manifold) are accepted. An adequate philosophical account should explain how methodological considerations operate in both cases.

The elements of an account of theoretical inference and theoretical postulation have emerged above. We start with the idea of a representation in R^4: a one-one map $\phi: A \to R^4$, where A is the domain of individuals (space-time points) of our theory. The correlation ϕ supplies us with a wide range of mathematical entities—properties, relations, and functions on R^4—that are *candidates* for representing physically real items of theoretical ideology on A. A particular theory T attributes physical reality to certain of these mathematical entities, but not others, by placing conditions on the mapping ϕ that restrict it to a class of mappings Φ. Aspects of mathematical structure are physically real according to T if they are *invariant* in the class Φ; otherwise they have no physical reality and are merely representative. Moreover, we can look at this procedure in either of two ways. We can begin with a list of physically real relations $R_1, \ldots, R_n$ on A and explicitly constrain the mapping ϕ via particular representations $\phi(R_1), \ldots, \phi(R_n)$: this results in a class of mappings Φ. On the other hand, we can begin with a class of mappings Φ and ask whether

Conventionalism

particular mathematical relations $R_1, \ldots, R_n$ on R^4 are invariant under Φ: this induces a list of physically real relations $\phi^{-1}(R_1), \ldots, \phi^{-1}(R_n)$ on A.

Consider, for example, the traditional Newtonian gravitation theory of §III.3. In this theory A is an open neighborhood of the space-time manifold M, and our aim is to show how the tangent vector $T_{\hat{a}}$ of gravitationally affected particles is determined by the distribution of mass ρ on A. We postulate a gravitational potential Ψ on A and lay down the conditions

$$\frac{db^i}{d\phi_0} = -\frac{\partial(\Psi \circ \phi^{-1})}{\partial \phi_i} \qquad (i = 1,2,3)$$

$$\frac{\partial^2(\Psi \circ \phi^{-1})}{\partial \phi_1^2} + \frac{\partial^2(\Psi \circ \phi^{-1})}{\partial \phi_2^2} + \frac{\partial^2(\Psi \circ \phi^{-1})}{\partial \phi_3^2} = 4\pi k(\rho \circ \phi^{-1}) \qquad (23)$$

where $b^i = T_{\hat{a}}(\phi_i)$, on the mapping ϕ. Equation (23) constrains ϕ to the class of Newtonian inertial systems, since the covariance group of (23) is just the Galilean group. So (23) induces a relation of collinearity $R(a,b,c)$, a temporal metric $t(a,b)$, and a Euclidean spatial metric $d_S(a,b)$ on A. The postulation of Ψ and (23) commits us to $R(a,b,c)$, $t(a,b)$, and $d_S(a,b)$ even if we do not include them explicitly from the very beginning. On the other hand, (23) does not commit us to a relation $R(a,b)$ of absolute rest, for this relation is not preserved by the Galilean group.

Another interesting example is provided by the Maxwell-Lorentz equations of electrodynamics. Once again, A is an open neighborhood of M, and our aim is to show how the tangent vector $T_{\hat{a}}$ of charged particles is determined by the distribution of charge-current density J on A. We postulate an electromagnetic field F on A and lay down the conditions

$$\frac{d(T_{\hat{a}}(\phi_i))}{d\phi_0} = \frac{q}{m_0} F^i_{j\langle\phi\rangle} \frac{d\phi_j}{d\phi_0} \qquad (i = 1,2,3)$$

$$\frac{\partial F^{i0}_{\langle\phi\rangle}}{\partial \phi_0} + \frac{\partial F^{i1}_{\langle\phi\rangle}}{\partial \phi_1} + \frac{\partial F^{i2}_{\langle\phi\rangle}}{\partial \phi_2} + \frac{\partial F^{i3}_{\langle\phi\rangle}}{\partial \phi_3} = 4\pi J^i_{\langle\phi\rangle} \qquad (24)$$

$$\frac{\partial F_{ij\langle\phi\rangle}}{\partial \phi_k} + \frac{\partial F_{ik\langle\phi\rangle}}{\partial \phi_i} + \frac{\partial F_{ki\langle\phi\rangle}}{\partial \phi_j} = 0$$

on the mapping ϕ, where $F^{ij} = \eta^{ia}\eta^{jb}F_{ab}$, $F^i_j = \eta^{ia}F_{aj}$, and $(\eta^{ij}) = \mathrm{diag}(1,-1,-1,-1)$. Equation (24) constrains ϕ to the class of special

322

relativistic inertial systems, since the covariance group of (24) is just the Lorentz group. So (24) induces a relation of collinearity $R(a,b,c)$ and a Minkowskian interval $I(a,b)$ on A. The difference between classical and relativistic electrodynamics depends on the conditions we impose on the *parametrization* of our trajectories σ. If we use proper time τ, no further constraints are placed on (24), and our covariance group remains the Lorentz group. If, on the other hand, we use Newtonian time t, our covariance group is reduced to $O^3 \times T$, and (24) induces the Newtonian functions $t(a,b)$, $d_S(a,b)$, and the relation $R(a,b)$ of absolute rest as well. (Compare §IV.3,5.)

Particular attention should be paid to the relationship between covariance and invariance in these examples. The Galilean group is the covariance group of (23) and the symmetry group of $R(a,b,c)$, $t(a,b)$, and $d_S(a,b)$. Galilean transformations are not in general symmetries of the potential Ψ, of course, but they do leave the *class* of solutions to (23) invariant. Similarly, the Lorentz group is the covariance group of (24) and the symmetry group of $R(a,b,c)$ and $I(a,b)$. Lorentz transformations are not in general symmetries of the electromagnetic field F, but they do leave the *class* of solutions to (24) invariant. We should distinguish, then, between objects like Ψ and F, which are correlated with a class of functions on R^4 by means of a system of equations, and objects like $t(a,b)$, $d_S(a,b)$, and $I(a,b)$, which are correlated with particular functions on R^4 by means of conditions such as (1) and (8). An object has physical reality on A according to T just in case its correlate in R^4—a particular function or class of functions—is invariant under the class Φ of representations postulated by T.

We should also distinguish the use of R^4 as a representation of A, as in (23) and (24), from the use of R^4 as a mere coordinatization of A, as in generally covariant equations like (III.41a), (III.42a), and so forth. In the former, we postulate an isomorphism between elements of mathematical structure on R^4 and elements of geometrical structure on A. We focus attention on a privileged subclass of coordinate systems in which our representing elements are invariant. In the latter, we do not postulate an isomorphism between elements of R^4 and any particular pieces of geometrical structure on A. In effect, we are using R^4 only to represent the topological-differentiable structure of A, and, for this use, all admissible coordinate systems are equally good. Failure to distinguish these two very different uses of R^4 is central to the traditional confusions about

323

covariance and invariance; at its worst, it leads to the unfortunate idea that the space-time of a generally covariant theory lacks all geometrical structure and is "metrically amorphous."

So far we have seen how the theoretical structure postulated by a given theory T is connected with the class of representations in R^4 assumed by T: theoretical structure is real *according to* T if it is invariant in this class. Of course, our central epistemological question remains unanswered: How do we justify a particlar choice of T—along with its postulated theoretical structure—over others? When is the postulation of a particular theoretical structure $\mathscr{A} = \langle A, R_1, \ldots, R_n \rangle$ legitimate, and when is it not? Our answer to this question is based on the process of theoretical unification. We consider an initial structure $\mathscr{A}$ described by a representation of the form

$$\exists \phi \in \Phi : A \to R^4, \tag{25}$$

and ask how $\mathscr{A}$ interacts with other structures $\mathscr{A}' = \langle A, R'_1, \ldots, R'_m \rangle$ described by representations of the form

$$\exists \psi \in \Psi : A \to R^4. \tag{26}$$

In particular, we ask if we can infer the joint representation

$$\exists \chi \in \Phi \cap \Psi : A \to R^4. \tag{27}$$

A given element of structure has unifying power relative to $\mathscr{A}'$ if it facilitates the inference from (25) and (26) to (27), otherwise not. Thus, to use a by now familiar example, the relation $R(a,b)$ of absolute rest plays no role in the gravitational representation (23), nor even in the electrodynamical representation (24), but it is required by the joint representation

$$\exists \phi : A \to R^4 \text{ satisfying (23) } and \text{ (24)}. \tag{28}$$

So $R(a,b)$ does have unifying power in the context of this theoretical inference. The point, as we argued in §VI.3, is that elements of theoretical structure with unifying power are necessary if our theories are to receive *repeated* boosts of confirmation from many diverse areas of application—if, for example, our initial representation (25) is to receive confirmation from the observational success of the joint representation (27). (Of course, in the case of $R(a,b)$, the joint representation (28) happens to be an observational failure, so our theory of $R(a,b)$ in fact receives no confirmation.)

324

Now, for any particular piece of theoretical structure, we always have the choice of assigning it to the world of physical reality $\mathscr{A}$ or to the world of mathematical representation R^4. For example, we can assign the gravitational potential Ψ to the world of physical reality via (6) or to the world of mathematical representation via (7). In the former case the structure is invariant in the class of representations, in the latter not. Thus, the class of solutions to (23) is invariant in the Newtonian inertial systems but not in the Poissonian systems satisfying (7). In order to decide which option to take, we can use the following procedure. We start by assigning all elements of theoretical structure to R^4. We postulate only such observational entities as particle trajectories and source variables on A, and we let all other objects be defined on R^4 in the manner of (7). Moreover, to be thoroughly parsimonious, we can consider only the occupied events in A: we can start with $\mathscr{P} \cap A$ rather than the full neighborhood A. We then consider each element of theoretical structure in turn, asking if there are any systematic advantages to be gained—especially advantages in unifying power—by expanding our theoretical commitments beyond $\langle \mathscr{P} \cap A, T_{\hat{a}}$, source variables$\rangle$. This procedure will generate maximal insight into the methodological role of theoretical structure.

Let us begin by considering the simplest space-time theory of all: the kinematics of free particles. The key element of theoretical structure in this theory is a flat affine connection $\overset{\circ}{D}$ on A. We assert that free particles follow geodesics of $\overset{\circ}{D}$—$\overset{\circ}{D}_{T_{\hat{a}}} T_{\hat{a}} = 0$—from which it follows that in *affine* coordinate systems $\phi: A \to R^4$, such that $\overset{\circ}{\Gamma}^i_{jk} = 0$ in ϕ, we have the representation

$$\frac{d(T_{\hat{a}(u)}(\phi_i))}{du} = 0. \tag{29}$$

Suppose we try to do without the connection $\overset{\circ}{D}$ and postulate only the tangent vector field $T_{\hat{a}}$ on A ($\mathscr{P} \cap A$). In such a minimal version of kinematics we cannot derive (29) from the geodesic law and the condition $\overset{\circ}{\Gamma}^i_{jk} = 0$; instead, we lay down the representation

$$\exists \phi: A(\mathscr{P} \cap A) \to R^4 \text{ satisfying (29)} \tag{30}$$

as a primitive statement. What is wrong with this theory?

The first thing to notice is that (30) is much weaker than the geodesic law $\overset{\circ}{D}_{T_{\hat{a}}} T_{\hat{a}} = 0$: (30) picks out a much wider class of trajectories. Whereas the geodesic law picks out all and only the

325

trajectories whose absolute acceleration (according to $\overset{\circ}{D}$) vanishes, (30) picks out all the trajectories that are geodesics of *some* flat connection on A. To see the problem here, let ψ be a second co-ordinate system nonlinearly related to ϕ, and let $\overset{\circ}{D}'$ be a second flat connection such that $\overset{\circ}{\Gamma}{}'^i_{jk} = 0$ in ψ. The representation (30) does not distinguish between ϕ and ψ, between geodesics of $\overset{\circ}{D}$ and geodesics of $\overset{\circ}{D}'$. More generally, let us say that a representation of the form

$$\exists \phi \in \Phi \text{ such that } \mathscr{D}_\phi(T_{\hat{\sigma}}, \text{ source variables})$$

is *determinate* just in case the same fields $T_{\hat{\sigma}}$ are picked out by all $\phi \in \Phi$. The unrestricted representation (30) is not determinate in this sense.

Hence, if we want to formulate a determinate law of motion, we have no choice but to postulate additional structure on A ($\mathscr{P} \cap A$) If we have all of A to work with, we postulate a particular flat connection $\overset{\circ}{D}$, the geodesic law $\overset{\circ}{D}_{T_{\hat{\sigma}}} T_{\hat{\sigma}} = 0$, and the additional condition $\overset{\circ}{\Gamma}{}^i_{jk} = 0$ on our representation ϕ. If we restrict ourselves to $\mathscr{P} \cap A$, we postulate a primitive quantity $a^i(\sigma)$ on $\mathscr{P} \cap A$, the law $a^i(\sigma) = 0$ for free particles, and the additional condition

$$a^i(\sigma) = \frac{d(T_{\hat{\sigma}(u)}(\phi_i))}{du} \tag{31}$$

on ϕ. Unlike (30), (31) does pick out a determinate class of trajectories on $\mathscr{P} \cap A$, namely, trajectories whose absolute acceleration really vanishes. Note that this is essentially the same point, made from a somewhat more abstract perspective, that we argued in §IV.2: we need an affine connection *or* a primitive property of absolute acceleration to formulate an accurate law of motion.

Kinematics has no determinate content unless we postulate an affine connection $\overset{\circ}{D}$ on A (a quantity $a^i(\sigma)$ on $\mathscr{P} \cap A$). By contrast, we can formulate determinate dynamical laws, in the style of (23) and (24), without explicitly introducing $\overset{\circ}{D}$. Instead, we start with *force variables*, like Ψ and F, and lay down field equations in addition to a law of motion. These equations pick out a determinate class of linearly related coordinate systems and thereby induce a flat affine connection $\overset{\circ}{D}$. But we can also reverse this procedure. We can start by postulating $\overset{\circ}{D}$ rather than force variables and pick out a determinate class of linearly related coordinate systems by the conditions $\overset{\circ}{\Gamma}{}^i_{jk} = 0$. Restricting our field equations to this class of

326

coordinate systems then induces a determinate class of force variables (recall our dual viewpoint on covariance from §II.2). These different ways of generating affine structure reflect the familiar fact that force and acceleration are two sides of the same coin. We can postulate one or the other, but we cannot do without both.

On the other hand, suppose we want to unify two different dynamical theories. Suppose, for example, we want a joint representation satisfying (23) and (24). For this purpose we need a guarantee that the affine coordinate systems induced by (23) and the affine coordinate systems induced by (24) are in fact the same. Nothing we can say about Ψ or F individually will supply such a guarantee, however; our only option is to introduce $\overset{\circ}{D}$ into both theories explicitly and to place the condition $\overset{\circ}{\Gamma}{}^i_{jk} = 0$ on ϕ (alternatively, we can explicitly introduce $a^i(\sigma)$ into both theories and place the condition (31) on ϕ).[37] Hence, the affine connection (or primitive quantity of absolute acceleration) plays two key roles in the evolution of our space-time theories. In the context of any single space-time theory it generates (either explicitly or implicitly) a determinate class of linearly related coordinate systems and thereby gives determinate content to the law of motion of that theory. Second, and perhaps even more important, it must be explicitly assumed for the purpose of unifying any two space-time theories, for we need a guarantee that the privileged, linearly related coordinate systems of any two such theories coincide.

Next, consider temporal structure. This structure also plays two roles. First, it supplies additional determinate content to laws of motion by picking out a distinguished class of *timelike* trajectories. In a Newtonian context this role is performed by the co vector field dt (function $t(a,b)$), in a relativistic context by the metric g (function $I(a,b)$). In either case, we restrict the range of admissible parameters in our laws of motion to a narrower class than the class of affine parameters. In a Newtonian theory we restrict ourselves to the time-functions t, in a relativistic context to functions linearly related to the proper time τ (§IV.3). We also single out a distinguished ϕ_0-coordinate in our class of admissible representations and thereby put additional constraints on this class. Thus, for example, although

---

[37] In general, of course, the affine connection is not *sufficient* for this kind of unification: for example, the joint representation satisfying (23) and (24) requires the field V as well. The point here is that an affine connection is always *necessary*, for we always need a privileged class of affine coordinate systems. We may need narrower classes—inertial, rest, and so on—as well.

the law of motion (29) is preserved by all linear transformations

$$\psi_i = a_{ij}\phi_j + b_i \tag{32}$$

the gravitational law (23) puts the additional condition

$$\psi_0 = \phi_0 + b_0 \tag{33}$$

on our representation-preserving transformations (of course, (23) induces further constraints as well, to which we shall return in a moment).

Once again, we can turn this procedure around in dynamical theories like (23) and (24). Instead of explicitly postulating temporal structure, we can start by postulating force variables and field equations that then induce temporal structure. The class of representations (23) induces a temporal metric $t(a,b)$; the class of representations (24) induces a Minkowskian interval $I(a,b)$. Again, however, if we want to unify two such theories, we have no choice but to postulate temporal structure explicitly. In unifying (23) and (24), for example, we need a guarantee that the ϕ_0-coordinates of both representations are Newtonian time-functions. (In this case, interestingly enough, the introduction of the temporal structure dt is already sufficient for our desired unification. Constraining the Lorentz transformations by (33) immediately restricts us to the group $O^3 \times T$, which then induces $d_S(a,b)$ and $R(a,b)$. So we do not need to introduce $R(a,b)$ explicitly after all.)

What about spatial structure? This structure has three distinguishable elements: the planes of simultaneity (either relativized or absolute), the spatial metric $d_S(a,b)$ defined on these planes, and (in a Newtonian context) the absolute space vector field V or relation $R(a,b)$. The first element is really an aspect of temporal structure, so we have in effect considered it already. Moreover, in a relativistic context, the spatial metric $d_S(a,b)$ (properly relativized, of course) and the temporal metric $t(a,b)$ (also properly relativized) are both defined from the spatio-temporal interval $I(a,b)$, which we have also considered already. It remains, then, to consider the Newtonian objects $d_S(a,b)$ and $R(a,b)$. $R(a,b)$ is easy. It obviously adds no determinate content to the kinematical law of motion (29), nor is it induced by the dynamical representations (23) and (24). But the joint representation satisfying (23) and (24) does induce $R(a,b)$ (covariance group $O^3 \times T$), so $R(a,b)$ is required in classical electrodynamics.

The Newtonian spatial metric $d_S(a,b)$ (tensor field h) is more interesting. The metric $d_S(a,b)$ adds no determinate content to the kinematical law (29) either, but it is induced by the dynamical representation (23). This is because (23) imposes the additional constraint of spatial orthogonality

$$a_{ik} a_{jk} = \delta_{ij} \qquad (i,j,k = 1,2,3) \qquad (34)$$

on a linear transformation satisfying (32) and (33). So $d_S(a,b)$ does have physical reality in the gravitation theory (23). We know, however, that something is wrong with this version of gravitation theory. Equation (23) attributes physical reality to the potential Ψ, and this is a mistake. Yet if we deny physical reality to the potential Ψ in (23), our group of representation-preserving transformations is widened, and the number of representation-induced objects is correspondingly decreased.

Thus, suppose we assign Ψ to the world of mathematical representation: we postulate only the objects $T_{\bar{a}}$ and ρ on A $(\mathscr{P} \cap A)$, and we replace (23) with

There exists a function $\Psi: R^4 \to R$ such that

$$\frac{d(T_{\bar{a}}(\phi_i))}{d\phi_0} = -\frac{\partial \Psi}{\partial \phi_i} \qquad (i = 1,2,3)$$

$$\frac{\partial^2 \Psi}{\partial \phi_1^2} + \frac{\partial^2 \Psi}{\partial \phi_2^2} + \frac{\partial^2 \Psi}{\partial \phi_3^2} = 4\pi k(\rho \circ \phi^{-1}). \tag{35}$$

Equation (35) is not yet a determinate law of motion, for there can be distinct nonflat connections D and D' such that $\Gamma^i_{00} = \Psi_{,i}$ in ϕ, while $\Gamma'^i_{00} = \Phi_{,i}$ in ϕ, for some second function Φ on R^4. To achieve determinacy we have to postulate a particular nonflat connection D and add the condition $\Gamma^i_{00} = \Psi_{,i}$ to (35). Once this is done, (35) picks out a definite class of trajectories, namely, geodesics of D.

Consider the class of representations satisfying (35) (with $\Psi_{,i} = \Gamma^i_{00}$). This is just the class of *Poissonian* coordinates, any two members of which are related by the nonlinear transformation

$$\psi_0 = \phi_0 + b_0$$

$$\psi_i = a_{ij}\phi_j + b_i(t) \qquad (i,j = 1,2,3) \tag{36}$$

where b_0 and the a_{ij} are constants, and $b_i(t)$ is time-dependent. The transformations (36) have several notable features. First, they do not preserve the nonflat connection D. If we want D to be invariant, we must add the condition

$$\ddot{b}_i(t) = \Gamma^i_{00} \tag{37}$$

where the Γ^i_{jk} are the components of D in ϕ—in other words, we have to restrict the "acceleration" $\ddot{b}_i(t)$ to gravitational "acceleration." The need for (37) reflects the breakdown of the standard connections between covariance and invariance in a nonflat space-time. In a flat space-time the covariance group of the class of representations is also the invariance group of the objects used to specify this class. For example, the Galilean group is the covariance group of (24) and the invariance group of $\mathring{D}$, dt, and h; the Lorentz group is the covariance group of (24) and the invariance group of $\mathring{D}$ and g. In the present nonflat gravitation theory, however, the covariance group of the class of representations (35) (with $\Psi_{,i} = \Gamma^i_{00}$, of course) is (36), and the invariance group of D requires the additional condition (37).

Second, the transformations (36), even with the condition (37), do not preserve the spatial metric h. The a_{ij} must be constants, but they do not necessarily satisfy the orthogonality condition (34). Thus, the Poissonian coordinate systems generated by (36) exceed the Galilean systems; the *local affine* coordinate systems generated by (36) and (37) exceed the local *inertial* systems. Therefore, although the class of representations (35) does induce a temporal metric $t(a,b)$, it does not induce a spatial metric $d_S(a,b)$. In this sense, the spatial metric lacks physical reality in the gravitation theory (35).[38] (Another way to see this point is to compare equations (III.62a) and (III.63a) with the corresponding equations (III.41a) and (III.42a). The latter contain the metric h^{ij} explicitly; the former contain only the time t_i and the nonflat connection Γ^i_{jk}.)

We do not need a spatial metric h to give determinate content to a gravitational law of motion; the only elements of theoretical structure we need are D and dt. But this does not mean that the

---

[38] This fact captures whatever truth there is in the thesis of the conventionality of geometry. Interestingly enough, it arises in the context of a theory in which the only real "universal force"—gravitational force—is eliminated. I have no idea what the significance of this is, however. I should also note that the indeterminacy in h due to the nonorthogonal character of the a_{ij} in (36) is quite limited. It is true that different metrics (not even related by a scale factor) are induced by different representations satisfying (35). Nevertheless, since the a_{ij} in (36) are constants, these different metrics are all Euclidean: no choice between Euclidean and non-Euclidean geometry is left open by (35).

spatial metric lacks physical reality *simpliciter*, for we also have to consider the unifying power of h. Do we need to postulate h in order to unify the gravitation theory (35) with other dynamical theories? Here, of course, the answer is yes: we need h to unify (35) with electrodynamics. In particular, the electrodynamic tensor g^* must take the form $V \otimes V - h$ if the class of gravitational representations (35) is to have a nonempty intersection with the class of electrodynamic representations. So both h and V have physical reality in this unification. (This case requires some additional complications. For, in accordance with the principle of equivalence, we should reformulate electrodynamics using the nonflat connection D rather than the flat connection $\overset{\circ}{D}$. When this is done, additional terms involving $\Gamma^i_{00} = \Psi_{,i}$ appear in the covariant derivatives in (III.80a)–(III.82a) and, therefore, in the electrodynamic representation (24). Nevertheless, this revised representation still requires h and V for the purpose of unification with (35).)

This completes our survey of the role of theoretical ideology in the development of relativity theory. We start by postulating only source variables and particle trajectories on A ($\mathscr{P} \cap A$), and we assign all other elements of theoretical ideology to the world of representation R^4: we obtain a *minimal representation*, in the style of (29) and (30), containing only particle trajectories and source variables explicitly. We then ask two questions. What pieces of theoretical structure are required in order to give determinate content to our laws of motion? What pieces of theoretical structure are required in order to unify our different space-time theories, to guarantee that the distinct classes of representations generated by different space-time theories overlap?

Our results are these. Affine structure is necessary for giving determinate content to all laws of motion; it is also necessary for unifying any two space-time theories. However, although a flat affine connection $\overset{\circ}{D}$ is needed for the kinematical law of motion (29), only the nonflat connection D is needed for giving determinate content to the gravitational law of motion (35). Further, in accordance with the principle of equivalence, it is D rather than $\overset{\circ}{D}$ that is required for unifying any two space-time theories. Temporal structure adds content to the laws of motion by picking out a class of timelike curves (here is where the characteristic differences between classical and relativistic laws appear), and it too is necessary for theoretical unification. In the case of Newtonian spatial structure, on the other hand, these two roles come apart (in a relativistic context, of course, spatial

331

structure is built in to the temporal structure $I(a,b)$). Neither $d_S(a,b)$ nor $R(a,b)$ adds content to any of our laws of motion, but both are necessary for unifying the gravitational law of motion with the electrodynamic law of motion.

These considerations also provide additional insight into the methodological role of the relativity principle (R2). They show us what is wrong with theories containing indeterminate quantities, with theories whose indistinguishability groups exceed their symmetry groups. Picture the situation as follows. In the kinematics of §III.1 we have three overlapping classes of privileged coordinate

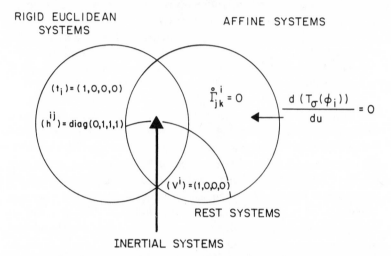

RIGID EUCLIDEAN SYSTEMS

AFFINE SYSTEMS

$(t_i) = (1,0,0,0)$

$(h^{ij}) = \mathrm{diag}(0,1,1,1)$

$\overset{\circ}{\Gamma}{}^i_{jk} = 0$

$$\frac{d\,(T_\sigma(\phi_i))}{du} = 0$$

$(v^i) = (1,0,0,0)$

REST SYSTEMS

INERTIAL SYSTEMS

systems. That the class of rest systems is properly included in the class of affine systems reflects the fact that v^i is not needed for a determinate law of motion. That the class of inertial rest systems is properly included in the class of inertial systems reflects the fact that v^i is not needed for unification in our combined theory of space-time geometry and kinematics. Similarly, in the gravitation theory of §III.3 we also have three overlapping classes of privileged coordinate systems. That the class of affine systems is properly included in the class of Poissonian systems reflects the fact that $\overset{\circ}{\Gamma}{}^i_{jk}$ is not needed for a determinate law of motion. That the class of inertial systems is properly included in the class of Galilean systems reflects the fact that $\overset{\circ}{\Gamma}{}^i_{jk}$ is not needed for unification in our combined theory of space-time geometry and gravitation. Hence, dropping v^i from

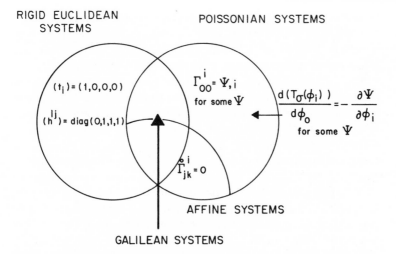

$$\exists \phi \in \Phi : A(\mathscr{P} \cap A) \rightarrow R^4 \text{ such that } \mathscr{D}_\phi(T_{\hat{\sigma}}, \text{ source variables}). \quad (39)$$

§III.1 and $\overset{\circ}{\Gamma}{}^i_{jk}$ from §III.3 leaves us with just enough objects for determinate laws of motion and theoretical unification. This is the full methodological force of the relativity principle (R2).

The methodological moves we have been discussing so far are exclusively ideological. We have been concerned with the acceptance of "good" elements of theoretical ideology, like D, dt, and h, and the rejection of "bad" elements of theoretical ideology, like V and $\overset{\circ}{D}$. The question of theoretical ontology, the choice between the full neighborhood A and the occupied subpart $\mathscr{P} \cap A$, was the subject of §VI.4. There we argued that the postulation of the full set A is in fact necessary for theoretical unification: key theoretical inferences, like the inference from (VI.39) and (VI.40) to (VI.36), do not go through unless the domain of definition of our representation ϕ includes all of A.

It is instructive to compare the inferential advantage of A over $\mathscr{P} \cap A$ with the inferential advantages of theoretical ideology. The point of theoretical ideology is to place constraints on the class Φ of mappings in a representation

$$\exists \phi \in \Phi : A(\mathscr{P} \cap A) \rightarrow R^4 \text{ such that } \mathscr{D}_\phi(T_{\hat{\sigma}}, \text{ source variables}). \quad (39)$$

These constraints should be just strong enough to ensure that a determinate class of trajectories is picked out by (39) and that the class Φ has a nonempty intersection with other classes of mappings assumed by other space-time theories. By contrast, the point of the full set A concerns the strength of the *existential* statement

$$\exists \phi : A(\mathscr{P} \cap A) \to R^4 \text{ such that } \mathscr{D}_\phi(T_{\hat{\sigma}}, \text{ source variables}). \quad (40)$$

This statement is typically much stronger if the domain of definition of ϕ is A rather than $\mathscr{P} \cap A$—in the extreme case where $\mathscr{P} \cap A$ is just a single trajectory σ, for example, (40) will say nothing at all. As we saw in §VI.4, such extra strength plays a crucial role in important "conjunctive" inferences.

This difference between ontological strength and ideological strength also explains why the Leibnizean indiscernibility argument does not have the methodological import of the relativity principle (R2). At first sight this can appear puzzling. Both the Leibnizean argument and (R2) apply to states of affairs that are indistinguishable *even assuming all of our theoretical apparatus.* Why should (R2) lead to the rejection of such states of affairs and the Leibnizean argument not? The explanation is that theories violating (R2) postulate too much ideology: they contain more vocabulary than is required for the class Φ in (39) to have the desired properties. Dropping these excess elements of vocabulary leaves us with just the constraints we need on Φ. By contrast, the extra strength of A over $\mathscr{P} \cap A$ has noting to do with vocabulary and nothing to do with constraints on Φ; rather, it involves the power of the simple existential claim (40). Dropping the unoccupied events from A does not leave us with the same inferential power. Moreover, since the problem addressed by (R2) is a problem about vocabulary, theories violating (R2) contain sentences that are unverifiable even given the rest of the relevant theory, namely, the sentences S_v ascribing absolute velocity in the context of §III.1 and sentences $S_{\hat{b}_i}$ ascribing gravitational acceleration in the context of §III.3. Theories subject to the Leibnizean argument, on the other hand, contain no such unverifiable sentences: the indistinguishable states of affairs in question are not distinguished by theoretical vocabulary.

6. *Concluding Remarks*

The above, unhappily brief sketch represents our first approximation to an adequate picture of scientific method. The development of relativity theory embodies key methodological moves, which, on the present account, are best understood in terms of the process of theoretical unification. From this point of view, the guiding aim of scientific method is the development of maximally well-confirmed laws about the observable world $\mathscr{B}$—in the present case, about the minimal structure $\langle \mathscr{P} \cap A, T_{\hat{a}}, \text{source variables}\rangle$. We find, however, that in order to do this it is necessary to embed the observable world $\mathscr{B}$ in a theoretical structure $\mathscr{A}$—in the present case, we embed the structure $\langle \mathscr{P} \cap A, T_{\hat{a}}, \text{source variables}\rangle$ in the structure $\langle A, \text{theoretical quantities}, T_{\hat{a}}, \text{source variables}\rangle$. The idea is that unless we literally embed $\mathscr{B}$ in $\mathscr{A}$, the lawlike assertions we wish to make about $\mathscr{B}$ will not receive *repeated* boosts in confirmation; without the unifying power generated by $\mathscr{A}$ our theories about $\mathscr{B}$ will be much less well-confirmed. By the same token, however, there are limits on the amount of theoretical structure that is to be postulated in $\mathscr{A}$: namely, no more than is required for maximally unified, and therefore maximally well-confirmed, theories about $\mathscr{B}$. Any candidates for theoretical structure going beyond this amount are to be assigned a purely representative role; they are not to be taken literally.

This picture of scientific method, crude as it is, appears to make good sense of the evolution of relativity theory. In particular, it provides a plausible rationale for the operation of traditional relativity principles. The parsimonious principle (R2) explains both the elimination of absolute space and velocity via the special or restricted principle of relativity and the elimination of gravitational acceleration via the principle of equivalence. (R2), in turn, is best understood as a mechanism for ensuring that we postulate no more theoretical structure than the process of theoretical unification requires: that is, we have just enough constraints on the class Φ of mappings in a representation

$$\exists \phi \in \Phi : A \to R^4$$

to guarantee a nonempty intersection with the class Ψ of mappings in a second representation

$$\exists \psi \in \Psi : A \to R^4.$$

We need $\Phi \cap \Psi$ to be nonempty if theoretical unification is to work, but we need no further constraints on Φ. In our two paradigm cases from the evolution of relativity the principle (R2) functions to ensure precisely this.

Our account also illuminates such traditional methodological criteria as the principle of parsimony and the identity of indiscernibles. These criteria are also best understood as mechanisms for limiting the process of theoretical postulation to just those entities needed for theoretical unification. Thus, for example, faced with a choice between the parsimonious kinematics T_- and the absolute space theories T_V, our methodological criteria recommend T_-. Faced with a choice between the parsimonious gravitation theory T_D and the gravitational potential theories T_Ψ, our methodological criteria recommend T_D. In both cases we choose the "simpler" theory, the theory postulating fewer theoretical entities. Moreover, as we have emphasized many times, these methodological choices are basic to a proper understanding of relativity theory. Combine the first choice with the "empirical" discovery of the invariance of the velocity of light and one obtains special relativity; combine this "empirical" discovery with the second choice and one obtains general relativity. Since the principle of parsimony and the identity of indiscernibles do play leading roles in the evolution of relativity, it is not surprising that positivist philosophers have tried to extend their application to all aspects of geometrical structure via the skeptical arguments of Leibniz and Poincaré.

What saves us from general skepticism about theoretical structure is that the entities rejected in our two paradigm applications of the principle of parsimony have a very special status. In both cases the entities in question fail to be determined by the rest of the relevant theory; they can vary arbitrarily while the other entities postulated by the theory remain fixed. This, as we have seen, is why we can easily eliminate such entities without disturbing the explanatory or unifying power of the theories in which they are embedded. The theory T_- is equal in unifying power to the theories T_V; the theory T_D is equal in unifying power to the theories T_Ψ. By contrast, we cannot drop just any theoretical entity—for example, the spatial metric h or absolute time dt—without producing a serious loss of unifying power. The principle of parsimony is itself constrained by the process of theoretical unification: entities can be eliminated if and *only if* no loss of unifying power results. If we abstract the notion of parsimony from this wider methodological context, it becomes genuinely puzzling why parsimony has led to the elimination of absolute space and inertial frames but not to the elimination of all other theoretical entities as well.

The special nature of the entities involved in our two paradigm examples goes hand in hand with a special kind of empirical equivalence. Since the entities in question can be varied arbitrarily while the rest of our theoretical structure remains fixed, we can obtain classes of empirically equivalent theories by systematically varying one particular sentence while leaving the rest of the relevant theory fixed. Thus, the empirically equivalent absolute velocity theories take the form $T_V = \ulcorner S_v \ \& \ T_- \urcorner$, where S_v is an ascription of the absolute velocity v to some particular object; the empirically equivalent gravitational potential theories take the form $T_\Psi = \ulcorner S_{\ddot{b}_i} \ \& \ T_D \urcorner$, where $S_{\ddot{b}_i}$ is an ascription of the gravitational acceleration $\ddot{b}_i = -\Psi_{,i}$ to some particular object. Because these classes of empirically equivalent theories are generated by such limited variations, the theories in question are not only empirically equivalent; they are "structurally identical" as well. So we cannot use methodological criteria like parsimony to decide between them. On the other hand, when we eliminate the dubious entities, we obtain theories empirically equivalent to all the theories in the above classes but not "structurally identical" to them. The theory T_- is empirically equivalent to but "structurally simpler" than all the theories $\ulcorner S_v \ \& \ T_- \urcorner$; the theory T_D is empirically equivalent to but "structurally simpler" than all the theories $\ulcorner S_{\ddot{b}_i} \ \& \ T_D \urcorner$.

We can also generate empirically equivalent theories involving more typical theoretical entities, of course, but there is an all-important difference. Since such entities are typically not indeterminate relative to the other entities of their respective theories, we have to make *compensatory adjustments* in the rest of the theory in question: we cannot vary one sentence arbitrarily while leaving the remainder fixed. In general, then, the empirically equivalent theories with which we are faced will be "structurally different," and methodological criteria will dictate a choice between them. For example, one cannot vary the Euclidean spatial metric of Newtonian theory without introducing compensatory "universal" forces into the law of motion, but such "universal" forces should be (and in actual practice are) eliminated by the principle of parsimony. Similarly, one cannot introduce a nonstandard simultaneity relation into special relativity without adding a privileged spatio-temporal orientation to the structure of Minkowski space-time; the principle of parsimony once again militates against such alterations of theory. In the context of the debate over conventionalism the principle of parsimony is therefore a double-edged sword. When applied to certain special examples of empirically equivalent theories (the absolute space theories and the gravitational potential theories), it yields the desired skeptical conclusions about the controversial entities in question, and it leads to new theories based on a strategy of equivalent descriptions. However, when the same principle is applied to more typical examples of empirically equivalent theories (including those involved in the central cases of congruence and simultaneity), it does not lead to such skeptical methodological moves; instead, it justifies our preference for the "simpler" (more parsimonious) standard theory.

In other words, the principle of parsimony functions as an *upper limit* on the process of theoretical postulation, a process that also has a *lower limit* generated by the need for theoretical unification. This is why considerations about empirical equivalence, theoretical underdetermination, and so on can only motivate the elimination of entities that truly lack unifying power. Such considerations have no general antitheoretical force. Hence, the generalized conventionalist argument of Poincaré and Reichenbach fails. For the same reason, the Leibnizean argument for a relationalist interpretation of space-time fails as well. The full space-time manifold—including, if need be, unoccupied space-time points—plays an essential role in the process of theoretical unification. It must be admitted, however, that the Leibnizean indiscernibility argument is stronger than the Poincaré-Reichenbach argument: the former, unlike the latter, does apply to states of affairs that can vary independently of the rest of the relevant theory. The kind of unverifiability to which Leibniz appeals is *strong* unverifiability. Nevertheless, since the debate about relationalism concerns ontological strength rather than ideological strength, even an appeal to strong unverifiability falls short. One simply cannot drop the unoccupied events from the domain of our spatio-temporal representations without a genuine loss of inferential power.[39]

The present conception of scientific method involves no general distinction between factual statements on the one side and conventions or arbitrary definitions on the other. Using Reichenbach's terminology, we can discern no interesting distinction between "principles of coordination" and "principles of connection." All elements of theoretical apparatus—ontology, ideology, and laws—are in the

---

[39] If this point is correct, one cannot use determinacy (relative to the rest of our theoretical apparatus) as the criterion for a "good" piece of theoretical structure and indeterminacy (relative to the rest of our theoretical apparatus) or indistinguishability (relative to the observable world) as the criterion for a "bad" piece of theoretical structure. Indeterminacy and indistinguishability are only symptoms of the real problem: lack of unifying power. Indeterminacy and indistinguishability coincide with lack of unifying power only in the case of theoretical ideology; in the case of theoretical ontology they come apart.

same boat. They are all subject to confirmation and disconfirmation, and in essentially the same way, that is, by a process of theoretical unification that looks for repeated boosts in confirmation. It is futile, then, to attempt to distinguish "constitutive" principles that are supposed to provide a framework for empirical theorizing (and are therefore not subject to confirmation and disconfirmation) from ordinary empirical laws. In particular, principles like

Rigid rods retain their length under transport

and

The speed of light is the same in all directions

are not to be distinguished from other empirical laws. Similarly, there is no general distinction between "inductive simplicity" and "descriptive simplicity." There is only one kind of "simplicity": that required by the proper operation of the principle of parsimony.

It is worth noting, however, that the development of relativity theory does reveal a limited or "local" counterpart of the conventional/factual distinction, namely, the distinction between *extrinsic* and *intrinsic*. Intrinsic features of a space-time reflect those aspects of geometrical structure that objectively characterize the space-time. Accordingly, such features are invariant in the class of coordinate representations and are in this sense independent of "our representation." Extrinsic features, on the other hand, vary from one coordinate representation to another and are therefore not objective or factual; they are mere artifacts of "our representation." But this relativistic counterpart of the conventional/factual distinction does not enable us to sidestep questions about empirical confirmation and disconfirmation. On the contrary, it raises such questions in a new and more acute form. What is our rationale for assigning some pieces of structure to the realm of physical reality and others to the realm of mathematical representation? Why do we use some aspects of structure as part of our intrinsic characterization of space-time and regard others as merely extrinsic? This question, as we have seen, is just the question concerning which theoretical structures are legitimately postulated and which are not. Our rough picture of scientific method is meant to provide the beginnings of an answer.

Appendix:
Differential Geometry

1. Differentiable Manifolds

Let R^n be the n-fold Cartesian product of the set of real numbers with itself, considered as a topological space, that is, endowed with the usual metric topology induced by the metric:

$$d[(a_1, \ldots, a_n),(b_1, \ldots, b_n)] = \left[\sum_{i=1}^{n} (b_i - a_i)^2 \right]^{1/2}.$$

A real-valued function $f: A \to R$, where A is an open subset of R^n, is C^r iff (if and only if) f possesses continuous partial derivatives of all orders $\leq r$ on A (if $r = \infty$, f possesses continuous partial derivatives of all orders on A). In what follows u_i will be the projection function on R^n: $u_i(a_1, \ldots, a_n) = a_i$.

A differentiable manifold M is a topological space with two additional properties: M can be coordinatized in a "nice" way by R^n (and so M is n-dimensional); there is a meaningful notion of *differentiability* for functions defined on M (unlike a simple topological space, which has a notion of continuity but not of differentiability). One's first thought is that a topological space M is "coordinatizable" iff there is a homeomorphism $\phi: M \to R^n$. However, there are many spaces we want to count as n-dimensional manifolds even though they cannot be mapped homeomorphically into R^n. The surface of a sphere, for example, cannot be mapped homeomorphically into R^2. Thus, the familiar stereographic projection mapping s on the surface of the sphere onto p on the plane, leaves out the "north pole" N. But, although the sphere cannot be coordinatized by a single homeomorphism into R^2, it can be by a pair of homeomorphisms: for example, one mapping everything but the "north pole" into R^2 by a stereographic projection, and a second mapping an open set around N into R^2 by a stereographic projection originating at the "south pole" Furthermore, these two

340

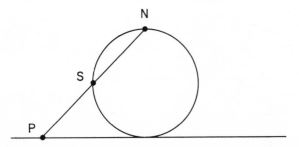

homeomorphisms fit together smoothly: if ϕ is the first and ψ the second, then $u_1 \circ \phi \circ \psi^{-1}$ and $u_2 \circ \phi \circ \psi^{-1}$ are differentiable maps from R^2 into R where ϕ and ψ overlap. Thus, for M to be coordinatizable we do not demand the existence of a single homeomorphism $\phi: M \to R^n$ but merely the existence of a family of homeomorphisms that jointly cover M and fit together smoothly where they overlap.

More precisely: if M is a topological space, an *n-chart* on M is a homeomorphism ϕ from an open subset $A \subseteq M$ onto an open subset of R^n. Two *n*-charts ϕ and ψ on M, with domains A and B respectively, are *C^r-related* iff $u_i \circ \phi \circ \psi^{-1}$ is C^r for each i on $A \cap B$.

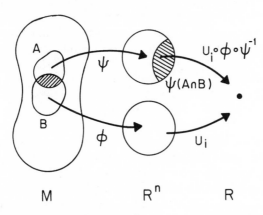

C^r-related *n*-charts on $M: u_i \circ \phi \circ \psi^{-1}$ is a C^r map from an open subset $\psi(A \cap B) \subseteq R^n \to R$.

A family of C^r-related *n*-charts is a C^r *n-atlas* for M iff the union of their domains covers M. A C^r *n*-atlas is *maximal* iff every *n*-chart that is C^r-related to all *n*-charts in the atlas is also in the atlas. Finally, a C^r *n-manifold* is a topological space M together with a

341

maximal C^r n-atlas for M. For simplicity I shall assume that $r = \infty$, although we rarely need to consider orders of differentiability greater than 2 in physics. In what follows, then, a *differentiable manifold* (or just *manifold*) will be a C^∞ n-manifold.

A differentiable manifold comes equipped with a natural notion of differentiability. Let f be a real-valued function defined on an open subset A of a manifold M. The function f is C^r iff $f \circ \phi^{-1}$ is C^r on $A \cap \mathrm{dom}(\phi)$ for every chart ϕ on M. If f maps A into a second manifold N, then f is C^r iff f is continuous, and for every real-valued C^r function g on N, $g \circ f$ is C^r on A. For any chart ϕ on a manifold M, the *coordinate functions* $x_i = u_i \circ \phi$ are clearly C^∞. From now on we shall almost always specialize to four-dimensional manifolds or *space-time manifolds*, and, following tradition, we let i range from 0 to 3. So our coordinates will be quadruples $\langle x_0, x_1, x_2, x_3 \rangle$.

2. Vectors on a Manifold

For motivation consider vectors at a given point p in Euclidean three-space. Associated with each vector $\vec{v}$ at p is a mapping X from functions differentiable around p into R such that X takes f to $\vec{v} \cdot \vec{\nabla} f$, where $\vec{\nabla}$ is the vector operator $(\partial/\partial x, \partial/\partial y, \partial/\partial z)$ and $\cdot$ is the "dot" product (thus X is the *directional derivative* of f in the direction $\vec{v}$). This mapping is easily seen to have the following properties:

$$
\begin{aligned}
X(f + g) &= Xf + Xg \\
X(bf) &= bXf \\
X(fg) &= f(p)Xg + g(p)Xf
\end{aligned}
\tag{1}
$$

where b is in R. That is, X is a *linear derivation* on the set of functions differentiable around p. The collection of all such mappings is itself a vector space over R with the obvious operations $(X + Y)f = Xf + Yf, (bX)f = b(Xf)$. So we can identify the vector $\vec{v}$ with the mapping $X : f \to \vec{v} \cdot \vec{\nabla} f$. This is precisely what we do in order to define vectors on an arbitrary manifold.

Let p be a point on a manifold M. Let $C^\infty (p)$ be the set of real-valued functions that are C^∞ on a neighborhood of p. A *tangent vector* or *contra vector* at p is a real-valued linear derivation on $C^\infty(p)$, namely, a real-valued function $X : C^\infty(p) \to R$ satisfying (1), where f and g are in $C^\infty(p)$. The vector space T_p of all such objects is called the *tangent space* at p.

If $\langle x_i \rangle = \langle u_i \circ \phi \rangle$ is a coordinate system around p, the *coordinate vectors* $\partial/\partial x_i(p)$ defined by

$$\left(\frac{\partial}{\partial x_i}\right)_p [f] = \frac{\partial(f \circ \phi^{-1})}{\partial u_i}\bigg|_{\phi(p)}$$

form a basis for the tangent space at p. Each vector in the tangent space can be written in the form[1]

$$X = \sum_{i=0}^{3} (Xx_i)\left(\frac{\partial}{\partial x_i}\right)_p. \tag{2}$$

The numbers (Xx_i) are called the *components* of X in the coordinate system $\langle x_i \rangle$. If $\langle y_j \rangle$ is a second coordinate system around p, then by (2)

$$\left(\frac{\partial}{\partial x_i}\right)_p = \left(\frac{\partial y_j}{\partial x_i}\right)_p \left(\frac{\partial}{\partial y_j}\right)_p.$$

(Note the use of the *summation convention*: repeated indices are summed over.) So if

$$X = a^i \left(\frac{\partial}{\partial x_i}\right)_p = b^j \left(\frac{\partial}{\partial y_j}\right)_p$$

where $a^i = (Xx_i)$ and $b^j = (Xy_j)$, then

$$b^j = a^i \left(\frac{\partial y_j}{\partial x_i}\right)_p. \tag{3}$$

Equation (3) is the classical "transformation law" for contra vectors.

Let T_p be the tangent space at p. Since T_p is a vector space, it possesses a *dual space* T_p^* consisting of linear functions $w: T_p \to R$ (with the obvious pointwise operations). Elements of T_p^* are called *cotangent* vectors or *co vectors*. The dual space of T_p^*, T_p^{**}, is naturally (independent of basis) isomorphic to T_p if we let $X \in T_p$ correspond to the element $w^* \in T_p^{**}$ such that $w^*(w) = w(X)$ for all $w \in T_p^*$. So we do not distinguish between T_p^{**} and T_p.

If $f \in C^\infty(p)$, there is an element df of T_p^* such that $df(X) = (Xf)$ for all $X \in T_p$. The object df is called the *differential* of f. Thus, if $\langle x_i \rangle$ is a coordinate system around p, the *coordinate differentials* dx_i are in T_p^*. The dx_i are in fact a basis for T_p^* since $dx_i(\partial/\partial x_j)(p) = \partial x_i/\partial x_j(p) = \delta_{ij}(\delta_{ij} = 1$ if $i = j$; $\delta_{ij} = 0$ if $i \neq j)$, and every element

---

[1] See Hicks [59], 7–8; Bishop and Goldberg [5], 52–53.

343

of T_p^* has the form

$$w = w\left(\frac{\partial}{\partial x_i}\right)_p dx_i. \tag{4}$$

If $X = a^i(\partial/\partial x_i)(p), w(X) = a^i w(\partial/\partial x_i)(p)$ by linearity. But $a^i = (Xx_i)$ and $(Xx_i) = dx_i(X)$. So $w(X) = w(\partial/\partial x_i)(p) \, dx_i(X)$. In particular, $df = \partial f/\partial x_i(p) \, dx_i$. The numbers $w(\partial/\partial x_i)(p)$ are the *components* of the co vector w in the coordinate system $\langle x_i \rangle$. If $\langle y_j \rangle$ is a second coordinate system around p, then by (4)

$$dx_i = dx_i\left(\frac{\partial}{\partial y_j}\right)_p dy_j = \left(\frac{\partial x_i}{\partial y_j}\right)_p dy_j.$$

So if $w = a_i \, dx_i = b_j \, dy_j$, where $a_i = w(\partial/\partial x_i)(p)$ and $b_j = w(\partial/\partial y_j)(p)$, then

$$b_j = a_i\left(\frac{\partial x_i}{\partial y_j}\right)_p. \tag{5}$$

Equation (5) is the classical "transformation law" for co vectors.

We can now define contra vector fields and co vector fields. A *contra vector field* on a subset $A \subseteq M$ is a mapping that assigns a tangent vector $X(p)$ at p to every point $p \in A$. A contra vector field is C^∞ on A iff A is open and the function (Xf) mapping p onto $X(p)[f]$ is C^∞ on $A \cap B$ for every C^∞ function f with domain B. Thus, a C^∞ contra vector field can also be regarded as a mapping that assigns a C^∞ function (Xf) to every C^∞ function f. If $A \subseteq M$ a *base field* on A is a quadruple of contra vector fields $\langle X_i \rangle$ such that the $X_i(p)$ are linearly independent for all $p \in A$. Clearly, the coordinate vector fields $\partial/\partial x_i$ form a C^∞ base field on the domain of their coordinate system $\langle x_i \rangle$. Therefore, any C^∞ contra vector field on the domain of the chart ϕ can be written as $X = a^i(\partial/\partial x_i)$, where each a^i is a C^∞ function equal to (Xx_i) and $x_i = u_i \circ \phi$. The "transformation law" (3) then holds for contra vector fields. If ϕ and ψ are two charts with $x_i = u_i \circ \phi$, $y_j = u_j \circ \psi$, and if X is a C^∞ field on $\text{dom}(\phi) \cap \text{dom}(\psi)$ such that

$$X = a^i \frac{\partial}{\partial x_i} = b^j \frac{\partial}{\partial y_j}$$

then on $\text{dom}(\phi) \cap \text{dom}(\psi)$

$$b^j = a^i \frac{\partial y_j}{\partial x_i}. \tag{6}$$

A *co vector field* on a subset $A \subseteq M$ is a mapping that assigns a co vector $w(p)$ at p to every point $p \in A$. A co vector field is C^∞ on A iff A is open and the function (wX) mapping p onto $w(p)[X(p)]$ is C^∞ on $A \cap B$ for every C^∞ contra vector field X with domain B. Thus, a C^∞ co vector field can also be regarded as a mapping that assigns a C^∞ function (wX) to every C^∞ contra vector field X. Co vector base fields are defined analogously to contra vector base fields. The coordinate differentials dx_i form a C^∞ base field on the domain of their coordinate system $\langle x_i \rangle$, and every C^∞ co vector field on the domain of the chart ϕ can be written as $w = a_i \, dx_i$, where each a_i is a C^∞ function equal to $w(\partial/\partial x_i)$ and $x_i = u_i \circ \phi$. The "transformation law" corresponding to (6) then obviously holds as well.

3. Tensors on a Manifold

Let T_p and T_p^* be the tangent space and its dual space at p, and let V be an arbitrary vector space. A *V-valued tensor of type* (r,s) at p is an $(r + s)$-linear mapping $\Theta : T_p^{*r} \times T_p^s \to V$. Most tensors used here are *real-valued*—that is, $V = R$—but other kinds of tensors occasionally appear. The collection of *real-valued* tensors of type (r,s) at p forms a vector space $T^{(r,s)}(p)$. Elements of $T^{(r,0)}(p)$ are *r-contra tensors*; elements of $T^{(0,s)}(p)$ are *s-co tensors*; all others are *mixed tensors*. (So $T^{(1,0)}(p)$ is just T_p; $T^{(0,1)}(p)$ is just T_p^*; $T^{(0,0)}(p)$ is just R.) In giving definitions for arbitrary tensors in $T^{(r,s)}(p)$, it is convenient to use representatives from $T^{(2,2)}(p)$, say. Thus, for example, if $\Theta, \Lambda \in T^{(2,2)}(p)$, their *tensor product* $\Theta \otimes \Lambda$ is the element of $T^{(4,4)}(p)$ such that

$$\Theta \otimes \Lambda(w,w',v,v';X,X',Y,Y') = \Theta(w,w';X,X')\Lambda(v,v';Y,Y').$$

This definition generalizes in the obvious way for $\Theta \in T^{(r,s)}(p)$ and $\Lambda \in T^{(m,n)}(p)$.

Let $\langle x_i \rangle$ be a coordinate system around p and let Θ be in $T^{(2,2)}(p)$. Since any co vector w is equal to $a_k \, dx_k$ by (4) and any contra vector X is equal to $b^i(\partial/\partial x_i)$ by (2), we have

$$\Theta(w,w';X,X') = a_k a_l' b^i b'^j \Theta\left(dx_k, dx_l; \frac{\partial}{\partial x_i}, \frac{\partial}{\partial x_j}\right)$$

by linearity. It follows that

$$\Theta = \Theta\left(dx_k, dx_l; \frac{\partial}{\partial x_i}, \frac{\partial}{\partial x_j}\right)\frac{\partial}{\partial x_k} \otimes \frac{\partial}{\partial x_l} \otimes dx_i \otimes dx_j.$$

Appendix

Therefore, the $(\partial/\partial x_k) \otimes (\partial/\partial x_l) \otimes dx_i \otimes dx_j$ form a basis for $T^{(2,2)}(p)$. The numbers

$$\Theta_{ij}^{kl} = \Theta\left(dx_k, dx_l; \frac{\partial}{\partial x_i}, \frac{\partial}{\partial x_j}\right)$$

are called the *components* of Θ in the coordinate system $\langle x_i \rangle$. The k,l are called *contra indices*; the i,j are called *co indices*. If $\langle y_j \rangle$ is a second coordinate system around p and $\bar{\Theta}_{rs}^{mn}$ are the components of Θ in $\langle y_j \rangle$, it follows from (3) and (5) that

$$\bar{\Theta}_{rs}^{mn} = \frac{\partial y_m}{\partial x_k}\frac{\partial y_n}{\partial x_l}\frac{\partial x_i}{\partial x_r}\frac{\partial x_j}{\partial x_s}\Theta_{ij}^{kl}. \tag{7}$$

Equation (7) is the classical "transformation law" for tensors of type (2,2).

A tensor is *symmetric* in a pair of arguments iff its value remains the same when these arguments are interchanged. For example, if $\Theta \in T^{(2,2)}(p)$ and

$$\Theta(w,w';X,X') = \Theta(w,w';X',X)$$

then Θ is symmetric in its first two co arguments. A tensor is symmetric in a pair of arguments iff its *components* are unchanged when the corresponding *indices* are interchanged. Thus, if Θ is as above, then

$$\Theta_{ij}^{kl} = \Theta_{ji}^{kl}.$$

A tensor is *antisymmetric* in a pair of arguments iff its value reverses in sign when these arguments are interchanged. So if Φ is antisymmetric in its first two contra arguments, say, we have

$$\Theta(w,w';X,X') = -\Theta(w',w;X,X')$$

and

$$\Theta_{ij}^{kl} = -\Theta_{ij}^{lk}.$$

A tensor is *symmetric (antisymmetric) simpliciter* iff it is symmetric (antisymmetric) in all its arguments.

Now, let Θ be in $T^{(1,4)}(p)$, for example, so the components of Θ have the form Θ_{jklm}^i. The *symmetrization* of Θ with respect to the second through fourth co variables, say, is the tensor

$$\Theta_{2}s_4 = \frac{1}{3!}[\Theta(v,X,Y,Z,W) + \Theta(v,X,Y,W,Z) + \Theta(v,X,W,Y,Z)$$

$$+ \Theta(v,X,W,Z,Y) + \Theta(v,X,Z,W,Y) + \Theta(v,X,Z,Y,W)].$$

346

This tensor is clearly symmetric in its second through fourth co variables, and its components are

$$\Theta^i_{j(klm)} = \frac{1}{3!}\left[\Theta^i_{jklm} + \Theta^i_{jkml} + \Theta^i_{jmkl} + \Theta^i_{jmlk} + \Theta^i_{jlmk} + \Theta^i_{jlkm}\right].$$

The *antisymmetrization* of Θ with respect to the second through fourth co variables, say, is the tensor

$$\Theta_2 a_4 = \frac{1}{3!}\left[\Theta(v,X,Y,Z,W) - \Theta(v,X,Y,W,Z) + \Theta(v,X,W,Y,Z)\right.$$

$$\left. - \Theta(v,X,W,Z,Y) + \Theta(v,X,Z,W,Y) - \Theta(v,X,Z,Y,W)\right].$$

This tensor is clearly antisymmetric in its second through fourth co variables, and its components are

$$\Theta^i_{j[klm]} = \frac{1}{3!}\left[\Theta^i_{jklm} - \Theta^i_{jkml} + \Theta^i_{jmkl} - \Theta^i_{jmlk} + \Theta^i_{jlmk} - \Theta^i_{jlkm}\right].$$

The notions of symmetrization and antisymmetrization for contra variables and contra indices are defined analogously.

A further operation on tensors called *contraction* is defined as follows. Suppose again that $\Theta \in T^{(2,2)}(p)$, then $tr^{12}\Theta$ is in $T^{(1,1)}(p)$ and

$$tr^{12}\Theta(w,X) = \sum_{k=0}^{3} \Theta\left(dx_k, w; X, \frac{\partial}{\partial x_k}\right)$$

where $\langle x_k \rangle$ is any coordinate system around p (note that $tr^{12}\Theta$ is independent of the choice of $\langle x_k \rangle$). Similarly, $tr^{22}\Theta$, for example, is also in $T^{(1,1)}(p)$ and

$$tr^{22}\Theta(w,X) = \sum_{k=0}^{3} \Theta\left(w, dx_k; X, \frac{\partial}{\partial x_k}\right).$$

In terms of components we have

$$(tr^{12}\Theta)^l_i = \Theta^{kl}_{ik}$$

and

$$(tr^{12}\Theta)^l_i = \Theta^{lk}_{ik}.$$

Addition, scalar multiplication, tensor product, symmetrization and antisymmetrization, and contraction are the basic algebraic operations on tensors.

347

A *tensor field* of type (r,s) on a subset $A \subseteq M$ is a mapping that assigns a tensor in $T^{(r,s)}(p)$ to every point $p \in A$. A tensor field Θ of type (2,2), say, is C^{∞} on A iff A is open and the function $\Theta(w,v;X,Y)$ mapping p onto $\Theta_p(w(p),v(p);X(p),Y(p))$ is C^{∞} on A for all quadruples $\langle w,v,X,Y \rangle$ of C^{∞} vector fields. Thus, a C^{∞} tensor field (of type (2,2)) can also be regarded as a mapping that assigns a C^{∞} function $\Theta(w,v;X,Y)$ to every quadruple of vector fields $\langle w,v,X,Y \rangle$. *Base fields* and *components* are defined analogously to the case of vector fields. The $(\partial/\partial x_k) \otimes (\partial/\partial k_l) \otimes dx_i \otimes dx_j$ form a base field on the domain A of the coordinate system $\langle x_i \rangle$; the components of Θ with respect to $\langle x_i \rangle$ are C^{∞} functions; and the obvious "transformation law" then holds for these components as expressed in a second coordinate system.

4. Curves on a Manifold

A *curve* on M is a C^{∞} map $\sigma: U \to M$, where U is an open interval in R. If σ is a curve with domain U, then for every $t \in U$ there is a vector $T_\sigma(t)$ at $\sigma(t) \in M$, called the *tangent vector to σ at t*, such that for every $f \in C^{\infty}(\sigma(t))$

$$T_\sigma(t)[f] = \frac{d(f \circ \sigma)}{du}\bigg|_t.$$

Clearly, $T_\sigma(t)$ is in $T_{\sigma(t)}$. If $\langle x_i \rangle$ is a coordinate system around $\sigma(t)$, the components of $T_\sigma(t)$ with respect to $\langle x_i \rangle$ are

$$T_\sigma(t)[x_i] = \frac{d(x_i \circ \sigma)}{du}\bigg|_t$$

so

$$T_\sigma(t) = \frac{dx_i}{du}\bigg|_t \left(\frac{\partial}{\partial x_i}\right)\bigg|_{\sigma(t)} \tag{8}$$

(where I have abbreviated $x_i \circ \sigma$ by x_i). Hence, associated with each curve σ with domain U is a C^{∞} contra vector *field* with domain $\sigma(U)$. Conversely, if X is a C^{∞} contra vector field, then σ is an *integral curve* of X iff whenever $\sigma(t)$ is in the domain of X we have

$$T_\sigma(t) = X(\sigma(t)).$$

In these circumstances we say that the curve σ *fits* the field X. Let X be a contra vector field and let p be in dom(X). Let $\langle x_i \rangle$ be a

348

coordinate system around p whose domain is included in $\text{dom}(X)$. Finally, let σ be a curve through p: that is, for some $t \in \text{dom}(\sigma)$, $\sigma(t) = p$. The condition that σ fits X on the domain of $\langle x_i \rangle$ is just

$$\frac{d(x_i \circ \sigma)}{du} = a^i \circ \sigma \qquad (9)$$

where the a^i are the components of X: $a^i = (Xx_i)$. Therefore, integral curves always exist locally.[2]

5. *Affine Connections*

Differentiable manifolds possess considerable topological structure but no intrinsic geometrical structure. They cannot in general be said to be "flat" or "curved," Euclidean or non-Euclidean. We can introduce such structure by specifying which C^∞ curves on M are *geodesics*. Now, geodesics can be regarded either as "shortest" curves or as "straightest" curves. We can give a precise definition of "shortest" curve by introducing a *metric* on M, but there is a conceptually distinct notion of "straightest" curve that can be introduced independently of a metric. We can think of a "straight" curve as one whose tangent vector field is "constant": the tangent vectors to the curve at different points all have the same "direction" or are all "parallel" to one another, Moreover, we can make the idea of a "constant" vector field precise by defining a generalized *derivative* for vector fields: the derivative of a vector field in a given direction.

For motivation think of curves and vector fields in ordinary Euclidean three-space. As we observed, associated with each vector $\vec{v}$ is a directional derivative defined on real-valued functions: $D_{\vec{v}} f = \vec{v} \cdot \vec{\nabla} f$. We can extend this operation to a derivative on *vector fields* componentwise: if $\vec{a} = (a_1, a_2, a_3)$, then

$$D_{\vec{v}} \vec{a} = (\vec{v} \cdot \vec{\nabla} a_1, \vec{v} \cdot \vec{\nabla} a_2, \vec{v} \cdot \vec{\nabla} a_3).$$

This operation satisfies the conditions:

$$D_{\vec{v}}(\vec{a} + \vec{s}) = D_{\vec{v}} \vec{a} + D_{\vec{v}} \vec{s}$$
$$D_{\vec{v} + \vec{a}}(\vec{s}) = D_{\vec{v}} \vec{s} + D_{\vec{a}} \vec{s}$$
$$D_{f\vec{v}} \vec{a} = f D_{\vec{v}} \vec{a}$$
$$D_{\vec{v}} f \vec{a} = \vec{v} \cdot \vec{\nabla} f \vec{a} + f D_{\vec{v}} \vec{a}$$

---

[2] See Hicks [59], 12; Bishop and Goldberg [5], 122.

(The last condition is the general "Leibniz product rule" for derivatives.) So if $\vec{r}(t)$ is a curve in Euclidean three-space with tangent vector $\vec{t}(t) = d\vec{r}/dt$, and if $\vec{a} = (a_1, a_2, a_3)$ is a vector field, then $\vec{a}$ is constant along $\vec{r}(t)$ iff $D_{\vec{t}}\vec{a} = 0$ iff $da_i/dt = 0$ ($i = 1, 2, 3$). So $\vec{r}(t)$ is a straight line or geodesic iff $D_{\vec{t}}\vec{t} = 0$ iff $d^2x_i/dt^2 = 0$ ($i = 1, 2, 3$), where $x_i = u_i(\vec{r}(t))$.

Generalizing all this to arbitrary manifolds, we say that an *affine connection* is a mapping D that assigns a C^∞ vector field $D_X Y$ to each pair of C^∞ vector fields X and Y and satisfies:

$$
\begin{aligned}
D_X(Y + Z) &= D_X Y + D_X Z \\
D_{X+Y} Z &= D_X Z + D_Y Z \\
D_{fX} Y &= f D_X Y \\
D_X f Y &= (Xf)Y + f D_X Y
\end{aligned}
\tag{10}
$$

where Z is an arbitrary C^∞ vector field and f is a C^∞ function. Now let σ be a curve with tangent vector field T_σ. If Y is a C^∞ vector field on σ, Y is *parallel* along σ (*constant* along σ) iff $D_{T_\sigma} Y = 0$. The curve σ is a *geodesic* iff T_σ is itself parallel along σ—that is, $D_{T_\sigma} T_\sigma = 0$.

Let $\langle x_i \rangle$ be a coordinate system and let $\langle \partial/\partial x_i \rangle$ be the natural base field on the domain of this coordinate system. Define the functions Γ^k_{ij} by

$$
D_{\partial/\partial x_j} \frac{\partial}{\partial x_i} = \Gamma^k_{ij} \frac{\partial}{\partial x_k}.
$$

The functions Γ^k_{ij} completely determine D. For if $X = a^j(\partial/\partial x_j)$ and $Y = b^i(\partial/\partial x_i)$, it follows from (10) that

$$
D_X Y = \left[a^j \frac{\partial b^k}{\partial x_j} + a^j b^i \Gamma^k_{ij} \right] \frac{\partial}{\partial x_k}.
\tag{11}
$$

The Γ^k_{ij} are the *components* of D in $\langle x_i \rangle$. Let $\langle y_j \rangle$ be a second coordinate system with overlapping domain. The components of D in $\langle y_j \rangle$ are $\bar{\Gamma}^k_{ij}$, where

$$
D_{\partial/\partial y_j} \frac{\partial}{\partial y_i} = \bar{\Gamma}^k_{ij} \frac{\partial}{\partial y_k}.
$$

But $\partial/\partial y_j = (\partial x_m/\partial y_j)(\partial/\partial x_m)$ and $\partial/\partial y_i = (\partial x_n/\partial y_i)(\partial/\partial x_n)$. Therefore, via (11) and the "chain rule" we have

$$
\bar{\Gamma}^k_{ij} = \left[\frac{\partial^2 x_l}{\partial y_j \partial y_i} + \frac{\partial x_m}{\partial y_j} \frac{\partial x_n}{\partial y_i} \Gamma^l_{nm} \right] \frac{\partial y_k}{\partial x_l}.
\tag{12}
$$

Equation (12) is the classical "transformation law" for the affine connection.

In terms of components, then, the condition that Y be parallel along $\sigma(u)$ becomes

$$\frac{db^k}{du} + b^i \frac{dx_j}{du} \Gamma^k_{ij} = 0 \qquad (13)$$

where $Y = b^i(\partial/\partial x_i)$ and $T_\sigma = (dx_j/du)(\partial/\partial x_j)$ by (8). So the condition for $\sigma(u)$ to be a geodesic is

$$\frac{d^2 x_k}{du^2} + \frac{dx_i}{du}\frac{dx_j}{du} \Gamma^k_{ij} = 0. \qquad (14)$$

In this case u is called an *affine parameter*. If we change to a new parameter $\lambda = \lambda(u)$, where λ is a monotonic and differentiable function of u, (14) becomes

$$\frac{d^2 x_k}{d\lambda^2} + \frac{dx_i}{d\lambda}\frac{dx_j}{d\lambda} \Gamma^k_{ij} = -\frac{d^2\lambda}{du^2}\frac{dx_k}{d\lambda}\frac{1}{\left(\dfrac{d\lambda}{du}\right)^2}. \qquad (15)$$

Therefore, λ is an affine parameter iff $d^2\lambda/du^2 = 0$ iff $\lambda = au + b$.

If X and Y are C^∞ contra vector fields, their *Lie Bracket* is a C^∞ vector field $[X,Y]$ such that $[X,Y]_p(f) = X_p(Yf) - Y_p(Xf)$. With the help of the Lie Bracket we can define several important tensors associated with an affine connection. First, the *torsion tensor* of an affine connection D is a vector-valued tensor field assigning a vector $\text{Tor}(X_p, Y_p)$ at p to each pair of vectors X_p, Y_p at p, where

$$\text{Tor}(X_p, Y_p) = (D_X Y - D_Y X - [X,Y])_p \qquad (16)$$

and X and Y are C^∞ *fields* such that $X(p) = X_p$ and $Y(p) = Y_p$.[3] A connection is *symmetric* iff $\text{Tor}(X_p, Y_p) = 0$ for all X_p, Y_p and all p. Second, the *curvature tensor* of an affine connection D is a linear-transformation-valued tensor field assigning a mapping $R(X_p, Y_p)$: $T_p \to T_p$ to each pair of vectors X_p, Y_p at p, where

$$R(X_p, Y_p)Z_p = (D_X D_Y Z - D_Y D_X Z - D_{[X,Y]}Z)_p \qquad (17)$$

and X, Y, Z are C^∞ *fields* such that $X(p) = X_p$, $Y(p) = Y_p$, $Z(p) = Z_p$. (Again, this is independent of the fields.) Finally, the *Riemann-Christoffel curvature tensor* is a C^∞ real-valued tensor field of type

---

[3] The definition is independent of the choice of fields. See Hicks [59], 59.

(1,3) such that

$$K_p(w_p; X_p, Y_p, Z_p) = w_p[R(Y_p, Z_p)X_p].$$ (18)

In terms of a coordinate system $\langle x_i \rangle$ the components of $\mathrm{Tor}(X, Y)$ are given by

$$a^i b^j [\Gamma_{ij}^k - \Gamma_{ji}^k]$$

where $X = a^i(\partial/\partial x_i)$ and $Y = b^j(\partial/\partial x_j)$. Hence, D is symmetric iff

$$\Gamma_{ij}^k = \Gamma_{ji}^k.$$

The components of the Riemann-Christoffel curvature tensor are

$$R_{jkl}^i = K\left(dx_i; \frac{\partial}{\partial x_j}, \frac{\partial}{\partial x_k}, \frac{\partial}{\partial x_l}\right) = \frac{\partial \Gamma_{jl}^i}{\partial x_k} - \frac{\partial \Gamma_{jk}^i}{\partial x_l} + \Gamma_{jl}^m \Gamma_{mk}^i - \Gamma_{jk}^m \Gamma_{ml}^i.$$ (19)

Equation (19) is the classical definition of the curvature tensor.

The significance of the curvature tensor is as follows. It can be shown that there is a coordinate system around $p \in M$ in which the components of the affine connection vanish iff the curvature tensor vanishes at p.[4] In such a coordinate system the condition for geodesics (14) becomes

$$\frac{d^2 x_k}{du^2} = 0.$$

Therefore, if the curvature tensor vanishes at p, there is a neighborhood of p on which the connection behaves like the ordinary Euclidean "directional derivative." So a region $A \subseteq M$ on which the curvature tensor vanishes is called *flat* or *semi-Euclidean*.

An affine connection D can be extended to a general directional derivative for tensors of any type. If Θ is in $T^{(r,s)}(A)$—the collection of all C^∞ tensor fields of type (r,s) on A—and X is in $T^{(1,0)}(A)$, then our directional derivative $\bar{D}_X \Theta$ is in $T^{(r,s)}(A)$. If $f \in T^{(0,0)}(A)$, then $\bar{D}_X f = (Xf)$; if $Y \in T^{(1,0)}(A)$, then $\bar{D}_X Y = D_X Y$; if $w \in T^{(0,1)}(A)$, then

$$(\bar{D}_X w)Y = X(w(Y)) - w(D_X Y).$$

---

[4] The implication from left to right follows trivially from (19). For the other direction see, e.g., Laugwitz [66], 109–110.

In general, if $\Theta \in T^{(2,2)}(A)$, say, then

$$\bar{D}_X\Theta = X(\Theta(w,v;Y,Z)) - \Theta(\bar{D}_Xw,v;Y,Z) - \Theta(w,\bar{D}_Xv;Y,Z) -$$
$$\Theta(w,v;\bar{D}_XY,Z) - \Theta(w,v;Y,\bar{D}_XZ). \tag{20}$$

By incorporating the field X, we can construct a mapping from $T^{(2,2)}(A)$ into $T^{(2,3)}(A)$:

$$(\bar{D}\Theta)(w,v;Y,Z,X) = (\bar{D}_X\Theta)(w,v;Y,Z). \tag{21}$$

Equation (21) defines the *general covariant derivative*. Using (20), (21), and the definition of Γ^i_{jk}, we find the components of $(\bar{D}\Theta)$ are given by

$$\Theta^{ij}_{kl;m} = \frac{\partial\Theta^{ij}_{kl}}{\partial x_m} + \Gamma^i_{nm}\Theta^{nj}_{kl} + \Gamma^j_{nm}\Theta^{in}_{kl} - \Gamma^n_{km}\Theta^{ij}_{nl} - \Gamma^n_{lm}\Theta^{ij}_{kn}. \tag{22}$$

Equation (22) is the classical definition of the covariant derivative. Note also that, whereas we represent covariant differentiation by a semicolon in (22), it is customary to represent ordinary differentiation by a comma:

$$\Theta^{ij}_{kl,m} = \frac{\partial\Theta^{ij}_{kl}}{\partial x_m}.$$

Finally, we can use the covariant derivative (21) to define a generalized *divergence* mapping $T^{(2,2)}(A)$ into $T^{(1,2)}(A)$:

$$\text{div}(\Theta) = tr^{23}(\bar{D}\Theta). \tag{23}$$

So the components of div(Θ) are given by

$$\Theta^{im}_{jk;m}$$

If div(Θ) = 0, Θ is said to be *conservative*.

6. *Metric Tensors and Semi-Riemannian Manifolds*

In §A.5 we saw how to introduce geometrical structure on a manifold by means of an affine connection. In this section we shall construct geometrical structure on the basis of a distinguished C^∞ co tensor field called the *metric tensor*. A *semi-Riemannian metric tensor* is a C^∞ tensor field of type (0,2) with the properties

$$g(X,Y) = g(Y,X) \qquad \text{(symmetry)}$$
$$g(X,Y) = 0 \text{ for all } Y \text{ iff } X = 0 \qquad \text{(nonsingularity)}. \tag{24}$$

353

If, in addition

$$g(X,X) > 0 \text{ for all } X \neq 0 \qquad \text{(positive-definiteness)}$$

then g is a *Riemannian* metric tensor. (So a Riemannian metric tensor is a real-valued *inner product* on the tangent space at each point $p \in M$.) A manifold with a semi-Riemannian (Riemannian) metric tensor defined on it is called a semi-Riemannian (Riemannian) manifold. If $\langle x_i \rangle$ is a coordinate system on $A \subseteq M$, we know that g can be written in the form $g_{ij}dx_i \otimes dx_j$ on A, where $g_{ij} = g(\partial/\partial x_i, \partial/\partial x_j)$. The g_{ij} are therefore the *components* of the metric tensor in $\langle x_i \rangle$.

Now let h_p be an arbitrary symmetric (but not necessarily nonsingular) tensor of type (0,2) at p: that is, h_p is a bilinear form on T_p. There exist coordinate systems $\langle x_i \rangle$ around p such that the matrix $(h_{ij}) = (h_p(\partial/\partial x_i, \partial/\partial x_j))$ is diagonalized with diagonal elements $+1$, -1, or 0 at p. Further, the number of positive, negative, and zero diagonal elements is the same for every such coordinate system.[5] Let n_+ be the number of positive diagonal elements, n_- the number of negative diagonal elements, n_0 the number of zero diagonal elements. The triple $\langle n_+, n_-, n_0 \rangle$ is called the *signature* of h_p. If h_p is nonsingular—so h_p is a *semi-Riemannian* metric—then all diagonal elements of (h_{ij}) are nonvanishing and h_p has signature $\langle n_+, n_- \rangle$. If h_p is positive-definite—so h_p is a *Riemannian* metric—then $h_{ij} = \delta_{ij}$. If h is a C^∞ symmetric *field* of type (0,2), then h has signature $\langle n_+, n_-, n_0 \rangle$ iff h_p has this signature at every point $p \in M$. Finally, note that all this remains true if h is a symmetric tensor of type (2,0), that is, if h_p is a bilinear form on T_p^*. So we can talk about the *signatures* of such objects as well.

Since a *semi-Riemannian* metric tensor g is nonsingular, g induces a nonsingular map G from T_p onto T_p^* at each p such that if $X, Y \in T_p$, $G(X)Y = g_p(X,Y)$. Since G is nonsingular, there is an inverse map G_* from T_p^* onto T_p such that if $w \in T_p^*$, $X \in T_p$, $g_p(G_*(w),X) = w(X)$. G therefore induces a tensor g^* of type (2,0) on $T_p^* \times T_p^*$ such that if $w,v \in T_p^*$, $g_p^*(w,v) = g_p(G_*(w),G_*(v))$. Hence, if the components of g_p are $g_p(\partial/\partial x_i, \partial/\partial x_j) = g_{ij}$, the components of g_p^* are $g_p^*(dx_i,dx_j) = g_p(G_*(dx_i),G_*(dx_j)) = g^{ij}$, where $g^{ik}g_{kj} = \delta_{ij}$. Consequently, the matrix (g^{ij}) is the inverse of the matrix (g_{ij}).

---

[5] See Bishop and Goldberg [5], 104–106.

The map G induces mappings from $T^{(r,s)}(p)$ into $T^{(r-1,s+1)}(p)$ such that, for example, if $\Theta \in T^{(2,2)}(p)$, then

$$G^2\Theta(w;X,Y,Z) = \Theta(w,G(Z);X,Y).$$

Similarly, the map G_* induces the maps $G_*^m : T^{(r,s)}(p) \to T^{(r+1,s-1)}(p)$ defined analogously. In general, then, G^n is the operation of "lowering" the nth contra index, and G_*^m is the operation of "raising" the mth co index. Thus, for example, the components of $G^2\Theta$ above are given by

$$g_{jm}\Theta_{kl}^{im}$$

whereas the components of $G_*^1\Theta$, say, are given by

$$g^{km}\Theta_{ml}^{ij}.$$

G_* allows us to define the *gradient* and *Laplacian* of a C^∞ function thus: $\mathrm{grad}(f) = G_*(df)$ and $\mathrm{del}(f) = \mathrm{div}(\mathrm{grad}(f))$.

An affine connection D on a semi-Riemannian manifold is said to be semi-Riemannian or *compatible* with the metric tensor g iff D is symmetric and $\bar{D}g = 0$: that is, the covariant derivative of g with respect to D vanishes. By the definition of the covariant derivative (20), (21), this means that for all C^∞ vector fields X,Y,Z:

$$Z[g(X,Y)] = g(D_Z X,Y) + g(X,D_Z Y). \tag{24}$$

Equation (24) implies that there is a unique connection compatible with a given semi-Riemannian metric tensor g and that its components in any coordinate system $\langle x_i \rangle$ are given by[6]

$$\Gamma_{jk}^i = \tfrac{1}{2}g^{im}\left[\frac{\partial g_{mk}}{\partial x_j} + \frac{\partial g_{mj}}{\partial x_k} - \frac{\partial g_{jk}}{\partial x_m}\right]. \tag{25}$$

Equation (25) is the classical definition of a semi-Riemannian connection. The numbers on the right-hand side of (25) are usually denoted by the *Christoffel symbols* "$\{_{jk}^{\;i}\}$."

Associated with a semi-Riemannian connection are several additional important tensors. First, there is the *Ricci tensor*, $\mathrm{Ric} = tr^{13}K$, with components $R_{jk} = R^m{}_{jkm}$. Second, there is the *curvature scalar*, $R = tr^{11}G_*^1\mathrm{Ric}$, or $R = g^{mn}R_{mn}$. Finally, there is the *Einstein tensor*,

$$\mathscr{G} = G_*^1 G_*^2 \mathrm{Ric} - \tfrac{1}{2}Rg^*. \tag{26}$$

---

[6] See Hicks [59], 72.

with components $\mathcal{G}^{ij} = g^{im}g^{jn}R_{mn} - \frac{1}{2}g^{ij}R$. It follows from the *second Bianchi identity*—$R^i_{jkl;m} + R^i_{jmk;l} + R^i_{jlm;k} = 0^7$—by contraction that $\mathcal{G}$ is *conservative*: div($\mathcal{G}$) = 0.

It makes sense to talk about the *length* of a vector and the *angle* between two vectors on a semi-Riemannian manifold. Define the *square* of the length of X by $|X|^2 = g(X,X)$. (This quantity can be *negative* if the manifold is not Riemannian.) A vector is *timelike* if $g(X,X) > 0$, *spacelike* if $g(X,X) < 0$, and *null* if $g(X,X) = 0$. Define the length of a timelike vector by $|X| = \sqrt{g(X,X)}$, the length of a spacelike vector by $|X| = \sqrt{-g(X,X)}$, and the angle between X and Y by $\cos(\alpha) = g(X,Y)/|X||Y|$. A *curve* is timelike if its tangent vector field is everywhere timelike, spacelike if its tangent vector field is everywhere spacelike, and null if its tangent vector field is everywhere null. Thus, we can define the length of a timelike or spacelike curve by integrating its tangent. That is, if $[a,b] \subseteq \text{dom}(\sigma)$ and if $\sigma(a) = p$ and $\sigma(b) = q$, define the length of σ from p to q by

$$|\sigma|_p^q = \int_a^b |T_\sigma(u)|\, du. \tag{27}$$

The length $|\sigma|_p^q$ is independent of the particular parameter u.[8] On a Riemannian manifold we have a well-defined distance function or metric:

$$d(p,q) = \inf\{|\sigma|_p^q : \sigma \text{ a curve through } p \text{ and } q\}. \tag{28}$$

(Equation (28) assumes that our manifold is sufficiently *connected*, so that there exist "enough" curves through p and q.)

We can use the length of a curve between p and q as a new parameter s between p and q, where

$$s(u) = \int_a^u |T_\sigma(u)|\, du. \tag{29}$$

It follows from (29) that $ds/du = |T_\sigma(u)|$, so $|T_\sigma(s)| = 1$: tangent vectors to curves parametrized by their length s have unit length. On a semi-Riemannian manifold we have $\bar{D}g = 0$. So if $X_\sigma(u)$ is any vector field along σ, then $(\bar{D}_{T_\sigma}g)(X_\sigma(u),X_\sigma(u)) = 0$, and if $X_\sigma(u)$ is parallel along σ, then $d(g(X_\sigma(u),X_\sigma(u)))/du = 0$: parallel fields have constant length. In particular, in the case of a geodesic we have $d(g(T_\sigma(u),T_\sigma(u)))/du = 0$. Thus, if s is the length of σ, $d^2s/du^2 = 0$ and s is an affine parameter. On a semi-Riemannian manifold the

---

[7] Ibid., 95.

[8] Ibid., 70. See also Bishop and Goldberg [5], 209.

condition for affine geodesics therefore becomes

$$\frac{d^2 x_i}{ds^2} + \Gamma^i_{jk} \frac{dx_j}{ds} \frac{dx_k}{ds} = 0. \tag{30}$$

On a Riemannian manifold we also have a notion of *metric geodesic*: a curve of "shortest" (or extreme) length. Such a curve is one for which the integral

$$\int_a^b |T_\sigma(u)| \, du$$

takes on an extreme value. Hence,

$$h(x_i(s); dx_i/ds) = \left(g_{ij} \frac{dx_j}{ds} \frac{dx_i}{ds} \right)^{\frac{1}{2}}$$

must satisfy the Euler-Lagrange equations

$$\frac{\partial h}{\partial x_i} - \frac{d}{ds} \left(\frac{\partial h}{\partial x_i'} \right) = 0 \tag{31}$$

where $x_i' = dx_i/ds$. Substituting $(g_{ij}(dx_j/ds)(dx_i/ds))^{1/2}$ for h in (31), we obtain

$$\frac{d^2 x_i}{ds^2} + \{^i_{jk}\} \frac{dx_j}{ds} \frac{dx_k}{ds} = 0 \tag{32}$$

as the condition for a metric geodesic. Comparing (32) and (30), and noting that for a Riemannian manifold $\Gamma^i_{jk} = \{^i_{jk}\}$, we see that metric geodesics and affine geodesics are identical on a Riemannian manifold.

On a semi-Remannian manifold things are more complicated. In general, affine geodesics are not extremal curves. Consider, for example, spacelike geodesics in Minkowski spacetime—a flat, four-dimensional, semi-Riemannian manifold of signature $\langle 1,3 \rangle$. There are nearby longer spacelike curves and nearby shorter spacelike curves. On a semi-Riemannian manifold we have the following facts. (i) Although affine geodesics are not extremal curves, they are *critical* curves: although their lengths are not necessarily maxima or minima, they are "points of inflection."[9] (ii) If the signature is $\langle 1, n - 1 \rangle (\langle n - 1, 1 \rangle)$—where M is n-dimensional—then timelike (spacelike) geodesics are extremal. (So in Minkowski space-time *timelike* geodesics are always extremal.) (iii) If we confine ourselves

---

[9] See Bishop and Goldberg [5], 249.

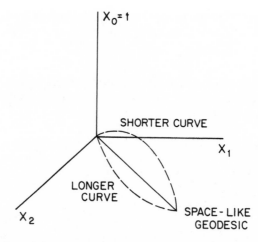

Spacelike geodesics in Minkowski space-time (one spatial dimension suppressed).

to a particular timelike (spacelike) submanifold N—that is, all curves in N are timelike (spacelike)—then timelike (spacelike) geodesics are extremal in N.

Suppose that for some $p \in M$ there is a coordinate system around p such that the components g_{ij} of the metric tensor are constants. Then $g_{ij,k} = 0$, and it follows from (25) that the Γ^i_{jk} vanish on a neighborhood of p. Conversely, if the Γ^i_{jk} vanish, it follows from (22) that $g_{ij;k} = g_{ij,k}$ and—since $g_{ij;k} = 0$—that the components of the metric are constants. Therefore, a semi-Riemannian manifold is *flat* or *semi-Euclidean* at $p \in M$ iff there is a coordinate system around p for which the components of the metric are constants. If g has signature $\langle n_+, n_- \rangle$, then by a linear transformation we can find a coordinate system around p in which the matrix (g_{ij}) is constant and diagonalized, with n_+ diagonal elements equal to $+1$ and n_- diagonal elements equal to -1.

7. Differentiable Transformations and Symmetry Groups

Let M and N be two manifolds. A *diffeomorphism* is a one-one C^∞ mapping h from an open subset $A \subseteq M$ onto an open subset $hA \subseteq N$ such that h^{-1} is also C^∞. A diffeomorphism h on A induces a mapping from $T^{(r,s)}(A)$ onto $T^{(r,s)}(hA)$. If $f \in T^{(0,0)}(A)$, then $hf(hp) = f(p)$; if $X \in T^{(1,0)}(A)$, then $hX_{hp}(hf) = X_p(f)$; if

$w \in T^{(0,1)}(A)$, then $hw_{hp}hX_{hp} = w_p X_p$; and, in general, if $\Theta \in T^{(2,2)}(A)$, say, then

$$h\Theta_{hp}(hw,hv;hX,hY) = \Theta_p(w,v;X,Y) \qquad (33)$$

for all $p \in A$. Further, if D is an affine connection on A, h induces a connection hD on hA such that $(hD_{hx}hY)_{hp}hf = (D_X Y)_p f$: that is, $h(D_X Y) = hD_{hx}hY$. Since h is one-one, these objects are all well-defined. So if Θ is a *geometrical object* on A (that is, a tensor field or affine connection on A), it follows from (33) that the components of Θ with respect to $\langle x_i \rangle$ at $p \in A$ are equal to the components of $h\Theta$ with respect to $\langle hx_i \rangle$ at $hp \in hA$.

Of particular importance is the case in which both A and hA are neighborhoods of some point $p \in M$. Such an h is called a *differentiable transformation* around p. Via (33) a differentiable transformation h induces a transformation from the set of geometrical objects on A onto the set of geometrical objects on $B = hA$. If the induced transformation maps Θ onto itself on B (wherever Θ is defined), then h is said to be a *symmetry* of Θ or to leave Θ *invariant*. From the above, it follows that h is a symmetry of Θ iff the components of Θ with respect to $\langle x_i \rangle$ at p are equal to the components of Θ with respect to $\langle hx_i \rangle$ at hp.

Let h be a differentiable transformation around p and let ϕ be a chart around p. Suppose that $A \subseteq \text{dom}(\phi)$ and $B = hA \subseteq \text{dom}(\phi)$, so that hq remains in $\text{dom}(\phi)$ for all $q \in \text{dom}(h)$. Then we can represent the action of h by a system of equations

$$x_i^* = f_i(x_0,x_1,x_2,x_3) \qquad (34)$$

where $f_i = x_i \circ h \circ \phi^{-1}$ and $x_i = u_i \circ \phi$. Equation (34) gives the coordinates of hq in $\langle x_i \rangle$ as a function of the coordinates of q in $\langle x_i \rangle$. We can also use (34) to define a *coordinate transformation* from the old coordinate system $\langle x_i \rangle$ to a new coordinate system $\langle y_j \rangle$, where

$$y_j(q) = x_j(hq). \qquad (35)$$

So $x_i = hy_i$. Now suppose that Θ is defined on $A \cup B$ and that for all $q \in A \cup B$ the components of Θ with respect to $\langle x_i \rangle$ at q equal the components of Θ with respect to $\langle x_i \rangle$ at hq (this will happen, for example, if the components of Θ in $\langle x_i \rangle$ are *constants* on $A \cup B$). The transformation h will be a symmetry of Θ iff the components of Θ with respect to $\langle x_i \rangle$ at q equal the components of Θ with

359

respect to $\langle y_j \rangle$ at q. In such a situation we can associate a *coordinate* transformation (35) that leaves the components of Θ invariant with a *manifold* transformation (34) that leaves Θ itself invariant.

Our ultimate aim is to construct certain *groups* of transformations leaving certain geometrical objects invariant: we want to find the *symmetry groups* of various geometrical objects. However, from the purely *local* point of view we have been using so far, this turns out to be rather messy. Thus, for example, the set of all differentiable transformations around p leaving Θ invariant will not in general form a group because of awkward problems with domains. In this connection, then, it is simplest to adopt a *global* point of view: we consider only differentiable transformations h such that $A =$ dom(h) $= M =$ .hA. In other words, we consider only diffeomorphisms from the entire manifold M onto itself. Furthermore, we must assume that M has a "nice" global topology, that M is globally homeomorphic to R^4 for example. Otherwise, there will not exist "enough" diffeomorphisms h$: M \xrightarrow{onto} M$; for example, if M has irregular "holes" or deletions, then no nontrivial global diffeomorphisms will exist. When we talk about the symmetry group of a particular geometrical object of a particular space-time theory, therefore, we mean a group of global diffeomorphisms on a topologically "nice" model of the theory.[10]

First, we need some more definitions. A *Lie group* is a set $\mathscr{L}$ such that

(i) $\mathscr{L}$ is a group
(ii) $\mathscr{L}$ is a C^∞ n-manifold
(iii) The mapping $\circ: \mathscr{L} \times \mathscr{L} \to \mathscr{L}$ determined by the group operation $\circ$ is C^∞ on the product manifold $|\mathscr{L} \times \mathscr{L}|$

(The *product* manifold $|\mathscr{L} \times \mathscr{L}|$ is the set of pairs $\mathscr{L} \times \mathscr{L}$ endowed with the product topology such that if $\langle x_1, \ldots, x_n \rangle$ is a chart around $p \in \mathscr{L}$ and $\langle y_1, \ldots, y_n \rangle$ is a chart around $q \in \mathscr{L}$, then $\langle x_1, \ldots, x_n, y_1, \ldots, y_n \rangle$ is a chart around $\langle p, q \rangle \in \mathscr{L} \times \mathscr{L}$.) Some important examples of Lie groups are the following. First, consider the group of all mappings h$: R^3 \to R^3$ taking $\langle a_1, a_2, a_3 \rangle \in R^3$ onto

[10] It is possible to construct symmetry groups on arbitrary manifolds using the concept of a *local group*; see Cohn [16], 41–44, 48–51; Bishop and Goldberg [5], 126. However, the details are too complicated to go into here. I am indebted to David Malament and Roger Jones for emphasizing the problems with a purely local approach.

$\langle b_1 b_2, b_3 \rangle \in R^3$ such that

$$b_i = \alpha_{ij} a_j + \beta_i \qquad (36)$$

where the α_{ij} and β_i are constants and $\alpha_{ij}\alpha_{kj} = \delta_{ik}$. This is the group of all *orthogonal* transformations or *Euclidean rigid motions*. It is a six-dimensional Lie group—three dimensions of rotations and three of translations. Second, consider the group of all mappings $h: R^4 \to R^4$ taking $\langle a_0, a_1, a_2, a_3 \rangle \in R^4$ onto $\langle b_0, b_1, b_2, b_3 \rangle \in R^4$ such that

$$\begin{aligned} b_0 &= a_0 + \beta_0 \\ b_i &= \alpha_{ij} a_j + \beta_i \quad (i = 1,2,3) \end{aligned} \qquad (37)$$

where the α_{ij} and β_i are constants and $\alpha_{ij}\alpha_{kj} = \delta_{ik} \, (i,j,k = 1,2,3)$. This is the (inhomogeneous) *Galilean* group—a ten-dimensional Lie group consisting of the orthogonal group (36) plus one dimension of temporal translations and three dimensions of Galilean transformations. Finally, consider the group of mappings $h: R^4 \to R^4$ such that

$$b_i = \alpha_{ij} a_j + \beta_i \qquad (38)$$

where the α_{ij} and β_i are constants and, if $(\eta_{ik}) = \text{diag}(1, -1, -1, -1)$, then $\alpha_{ij}\alpha_{kl}\eta_{ik} = \eta_{jl}$. This is the (inhomogeneous) *Lorentz* group—a ten-dimensional Lie group consisting of the orthogonal group (36) plus one dimension of temporal translations and three dimensions of Lorentz transformations

A Lie group $\mathscr{L}$ is said to *act* on a manifold M, or to form a *Lie group of differentiable transformations on M*, iff to each pair of elements $p \in M, 1 \in \mathscr{L}$ there corresponds an element $1(p) \in M$ such that

(i) The mapping $\langle 1, p \rangle \to 1(p)$ of $\mathscr{L} \times M$ into M is C^∞
(ii) For fixed $1 \in \mathscr{L}$, $1(p)$ is a (global) differentiable transformation of M onto itself $\qquad (39)$
(iii) $1(p) = p$ for all $p \in M$ iff $1 = e$, the identity element of $\mathscr{L}$
(iv) $k(1(p)) = (k \circ 1)p$ for all $p \in M, 1, k \in \mathscr{L}$.

For example, if M is a topologically well-behaved 4-manifold, then both the Galilean group (37) and the Lorentz group (38) *act* on M: if ϕ is a global coordinate system and 1 is in (37) or (38), then $\phi^{-1} \circ 1 \circ \phi$ is a global diffeomorphism on M. If $\mathscr{L}$ acts on M and the dimension of $\mathscr{L}$ is d, $\mathscr{L}$ is said to be a *d-parameter* group of differentiable transformations on M. Any d-parameter group of differentiable transformations is a subgroup of the group $\mathscr{M}$ of all (global) differentiable transformations on M with composition as the group operation. For if $l \in \mathscr{L}$, there is an $h \in \mathscr{M}$ such that $l(p) = h(p)$ for all p, and by (39.iv) the group operation $\circ$ on $\mathscr{L}$ is composition of functions on M.

Let $\mathscr{L}$ be a d-parameter group of differentiable transformations on an n-manifold M. Let $\langle y_1, \ldots, y_d \rangle$ be a coordinate system around $e \in \mathscr{L}$ and let $\langle x_1, \ldots, x_n \rangle$ be a coordinate system around $p \in M$. By (39) we can represent the action of $\mathscr{L}$ on M by the relations

$$x_i(l(q)) = f_i(y_1(l), \ldots, y_d(l); x_1(q), \ldots, x_n(q))$$

where the f_i are C^∞ for l in a neighborhood of e and q in a neighborhood A of p. Form the $d + n$ functions

$$g^i_j = \frac{\partial f_i}{\partial y_j}\bigg|_{l=e}$$

and the d vector fields defined on A

$$X_j = g^i_j \frac{\partial}{\partial x_i}. \tag{40}$$

The fundamental theorem of the theory of Lie groups of differentiable transformations is that the d vector fields (40) are independent and span a d-dimensional subspace of $T^{(1,0)}(A)$. This d-dimensional subspace is closed under the Lie Bracket operation $[X, Y]$ and is, accordingly, called the *Lie Algebra* of $\mathscr{L}$ on A. Conversely, every such Lie Algebra determines a local d-parameter Lie group.[11] Hence the vector fields (40) are called the *generators* of the Lie group $\mathscr{L}$.

Using the Lie Algebra we can decompose a d-parameter group $\mathscr{L}$ into d 1-parameter subgroups generated by the d vector fields (40). Such 1-parameter groups are especially easy to work with. First, it can be shown that there is always a coordinate system y on a neighborhood B around e on a 1-parameter Lie group $\{h_\alpha\}$ such that for all $l, k \in B$, $y(l \circ k) = y(l) + y(k)$.[12] This chart is called the *canonical* chart. Thus, 1-parameter Lie groups are usually thought of as indexed by a parameter $t \in R$ such that $h_{t+s} = h_t \circ h_s$; they are thought of as parametrized by the canonical chart. Second, asso-

---

[11] See Cohn [16], chap. 5 and pp. 66–68.

[12] Ibid., 54–55.

ciated with a 1-parameter group $\{h_t\}$ of differentiable transformations on M is a family of curves on M: if $p \in M$, define a curve through p by

$$\sigma_p(t) = h_t(p). \tag{41}$$

The curve defined by (41) is called the *orbit of p* generated by $\{h_t\}$. The set of orbits induces a vector field X_h on M such that

$$X_h(p)[f] = \frac{d(f \circ \sigma_p)}{dt}\Bigg|_{t=0} = \frac{\partial(f \circ h_t)}{\partial t}\Bigg|_{\substack{t=0 \\ p}}. \tag{42}$$

So X_h is tangent to every orbit of $\{h_t\}$. In any coordinate system $\langle x_i \rangle$ around p the components of X_h are $X_h x_i = \partial(x_i \circ h_t)/\partial t(t = 0)$; so X_h is the *generator* of $\{h_t\}$ as defined by (40). Conversely, then, every vector field X determines a local 1-parameter group.

We have now reached the goal of our discussion, for looking at the 1-parameter subgroups of a Lie group $\mathcal{L}$ enables us to determine the *invariants* of $\mathcal{L}$: the geometrical objects left invariant by every transformation in $\mathcal{L}$. An object Θ will be left invariant by $\mathcal{L}$ iff Θ is left invariant by every 1-parameter subgroup of $\mathcal{L}$; and, associated with every 1-parameter subgroup $\{h_t\}$, we have an operation $L_{X_h}\Theta$ indicating whether Θ is left invariant by $\{h_t\}$. If Θ is a tensor field on A and X is a vector field on A, then $L_X\Theta$—the *Lie derivative* of Θ with respect to X—is a tensor field of the same type as Θ, defined as follows. If $f \in T^{(0,0)}(A)$, then $L_X f = Xf$; if $Y \in T^{(1,0)}(A)$, then $L_X Y = [X,Y]$; if $w \in T^{(0,1)}(A)$, then $(L_X w)Y = X(wY) - w(L_X Y)$; and, in general, if $\Theta \in T^{(2,2)}(A)$, say then

$$\begin{aligned}(L_X\Theta)(w,v;Y,Z) = {} & L_X(\Theta(w,v;Y,Z)) - \Theta(L_X w,v;Y,Z) \\ & - \Theta(w,L_X v;Y,Z) - \Theta(w,v;L_X Y,Z) \\ & - \Theta(w,v;Y,L_X Z). \end{aligned} \tag{43}$$

(Note the similarity of (43) to (20), the *covariant* derivative.) Hence, if $\Theta \in T^{(2,2)}(A)$, the components of $L_X\Theta$ are given by

$$\begin{aligned}(L_X\Theta)^{ij}_{kl} = {} & a^m \frac{\partial\Theta^{ij}_{kl}}{\partial x_m} + \frac{\partial a^m}{\partial x_k}\,\Theta^{ij}_{ml} + \frac{\partial a^m}{\partial x_l}\,\Theta^{ij}_{km} \\ & - \frac{\partial a^i}{\partial x_m}\,\Theta^{mj}_{kl} - \frac{\partial a^j}{\partial x_m}\,\Theta^{im}_{kl} \end{aligned} \tag{44}$$

where the a^k are the components of X.

The significance of the Lie derivative is as follows. Let $\{h_t\}$ be a 1-parameter group of transformations on M with generator X_h and let Θ be in $T^{(2,2)}(M)$. If $\langle x_i \rangle$ is a coordinate system around $p \in M$, then the components of Θ with respect to $\{h_t x_i\}$ at $h_t p$ are C^∞ functions of t. Let these functions be

$$\Theta^*(h_t p)^{ij}_{kl}.$$

It can be shown[13] that

$$((L_{X_h}\Theta)(h_t p))^{ij}_{kl} = \frac{d(\Theta^*(h_t p)^{ij}_{kl})}{dt}.$$

But we know that h_t leaves Θ invariant iff the components of Θ at p in $\langle x_i \rangle$ equal the components of Θ at $h_t p$ in $\langle h_t x_i \rangle$. Hence, *every* $h_t \in \{h_t\}$ leaves Θ invariant iff the functions $\Theta^*(h_t p)^{ij}_{kl}$ are constants, that is, iff $L_{X_h}\Theta = 0$.

We can therefore use the Lie derivative to find an explicit form for the 1-parameter groups that leave Θ invariant. If $\langle x_i \rangle$ is a coordinate system around $p \in M$ we set $L_{Xh}\Theta = 0$ and use (44) to determine the components a^i of a generator X_h of a 1-parameter invariance group $\{h_t\}$ of Θ. We then solve the equations

$$\frac{\partial(x_i \circ h_t)}{\partial t} = a^i \circ h_t \tag{45}$$

subject to the initial conditions $x_i \circ h_t = x_i$ for $t = 0$, to find the functions $x_i \circ h_t$. This gives us an explicit representation around p of $\{h_t\}$. If the full invariance group of Θ is a d-parameter group, we can use this procedure to find d-independent generators for d-independent 1-parameter subgroups.

A particularly important type of invariance group is the invariance group of a semi-Riemannian metric tensor. Such a group is called a group of *metric automorphisms* or *isometries*. If g is a semi-Riemannian metric, the generators of the 1-parameter invariance groups of g must satisfy

$$L_X g = 0$$

which, in terms of components in a coordinate system $\langle x_i \rangle$, becomes

$$a^r \frac{\partial g_{ij}}{\partial x_r} + g_{rj}\frac{\partial a^r}{\partial x_i} + g_{ir}\frac{\partial a^r}{\partial x_j} = 0. \tag{46}$$

---

[13] See Bishop and Goldberg [5], 128–131.

Let $a_i = g_{ir}a^r$; that is, the a_i are the components of the "lowering" of X by g. Using (22), we find

$$a_{i;j} = g_{ir}\frac{\partial a^r}{\partial x_j} + a^r\frac{\partial g_{ir}}{\partial x_j} - g_{kr}\Gamma_{ij}^k a^r.$$

Using (25) and the fact that $g_{rk}g^{km} = \delta_{rm}$:

$$a_{i;j} = g_{ir}\frac{\partial a^r}{\partial x_j} + a^r\frac{\partial g_{ij}}{\partial x_j} - \frac{1}{2}\left[\frac{\partial g_{rj}}{\partial x_i} + \frac{\partial g_{ri}}{\partial x_j} - \frac{\partial g_{ij}}{\partial x_r}\right]a^r. \qquad (47)$$

Interchanging i and j in (47) and adding:

$$a_{i;j} + a_{j;i} = a^r\frac{\partial g_{ij}}{\partial x_r} + g_{rj}\frac{\partial a^r}{\partial x_i} + g_{ir}\frac{\partial a^r}{\partial x_j}. \qquad (48)$$

Comparing (46) and (48), we see that X generates a 1-parameter invariance group of g iff

$$a_{i;j} + a_{j;i} = 0. \qquad (49)$$

Equation (49) is called *Killing's Equation*, and any vector field satisfying (49) is called a *Killing Field*.

As an example, let us find the invariance group of ordinary Euclidean three-space E^3. E^3 is a flat C^∞ 3-manifold with a Riemannian metric tensor g of signature $\langle 3,0\rangle$. Around every point $p \in E^3$ there is a coordinate system $\langle x_1,x_2,x_3\rangle$ in which $(g_{ij}) = \mathrm{diag}(1,1,1)$: $g_{ij} = \delta_{ij}$. We know that a vector field X with components a^i in $\langle x_1,x_2,x_3\rangle$ generates an isometry of E^3 iff a^i satisfies (49) with $a_i = \delta_{ik}a^k = a^i$. Since covariant differentiation reduces to ordinary differentiation in our (Cartesian) coordinate system, we also have

$$a_{i;\,j;k} - a_{i;k;j} = 0. \qquad (50)$$

Permuting indices in (49) and (50), we obtain

$$a_{i;\,j;k} = 0. \qquad (51)$$

In our coordinate system (49) and (51) become

$$\frac{\partial a^i}{\partial x_j} + \frac{\partial a^j}{\partial x_i} = 0$$

and

$$\frac{\partial^2 a^i}{\partial x_k \partial x_j} = 0$$

respectively.

Hence, the a^i must take the form

$$a^i = b_{ij}x_j + c_i \qquad (52)$$

where the b_{ij} and c_i are constants and $b_{ij} = -b_{ji}$. Equation (52) contains six undetermined constants, so we can find six independent generators by successively setting one of the undetermined constants equal to 1 and the rest equal to 0. The b_{ij} so obtained from an anti-symmetric matrix

$$(b_{ij}) = \begin{pmatrix} 0 & 1 & 1 \\ -1 & 0 & 1 \\ -1 & -1 & 0 \end{pmatrix}$$

and we find that the independent generators are of two types.

Translations	Rotations
(1,0,0)	$(x_2, -x_1, 0)$
(0,1,0)	$(0, x_3, -x_2)$
(0,0,1)	$(x_3, 0, -x_1)$

We can now use (45) to find the six independent 1-parameter subgroups generated by these vector fields. For example, the generator (1,0,0) yields the equations

$$\frac{\partial x_1^*}{\partial t} = 1 \qquad \frac{\partial x_2^*}{\partial t} = 0 \qquad \frac{\partial x_3^*}{\partial t} = 0$$

where $x_i^* = x_i \circ h_t$. So the 1-parameter group generated by (1,0,0) takes the explicit form

$$x_1^* = x_1 + t$$
$$x_2^* = x_2$$
$$x_3^* = x_3.$$

That is, it represents a translation along the x_1-axis. Similarly, the generator $(x_2, -x_1, 0)$ yields the equations

$$\frac{\partial x_1^*}{\partial t} = x_2^* \qquad \frac{\partial x_2^*}{\partial t} = -x_1^* \qquad \frac{\partial x_3^*}{\partial t} = 0.$$

So the 1-parameter group generated by $(x_2, -x_1, 0)$ takes the form

$$x_1^* = x_1 \cos(t) + x_2 \sin(t)$$
$$x_2^* = x_2 \cos(t) - x_1 \sin(t) \qquad (53)$$
$$x_3^* = x_3.$$

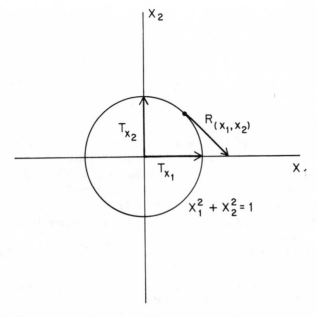

FIGURE 8. The Symmetry Group of Euclidean Space (one dimension suppressed)

$$T_{x_1} = (1,0,0) \text{ generates a translation along the } x_1\text{-axis}$$
$$T_{x_2} = (0,1,0) \text{ generates a translation along the } x_2\text{-axis}$$
$$R_{(x_1,x_2)} = (x_2, -x_1, 0) \text{ generates a rotation of the } (x_1,x_2) \text{ plane}$$

That is, it represents a rotation around the x_3-axis. Hence, the invariance group of Euclidean three-space is just the 6-parameter Lie group O^3 (36) as expected (Fig. 8).[14]

---

[14] I have taken this technique for finding an invariance group largely from Robertson and Noonan [95], 321–323.

Bibliography

1. Adler, R., Bazin, M., and Schiffer, M. *Introduction to General Relativity*. New York: McGraw-Hill, 1965.
2. Alexander, H. G., ed. *The Leibniz-Clarke Correspondence*. Manchester: Manchester University Press, 1956. (Original edition published by Clarke in 1717.)
3. Anderson, J. L. *Principles of Relativity Physics*. New York: Academic Press, 1967.
4. Bergmann, P. G. *Introduction to the Theory of Relativity*. Englewood Cliffs: Prentice-Hall, 1942.
5. Bishop, R. I., and Goldberg, S. I. *Tensor Analysis on Manifolds*. New York: Macmillan, 1968.
6. Boyd, R. *Realism and Scientific Epistemology*. Cambridge: Cambridge University Press, forthcoming.
7. Bunge, M. *Foundations of Physics*. New York: Springer, 1967.
8. Carnap, R. *The Logical Structure of the World*. Translated by R. A. George. Berkeley: University of California Press, 1967. (Original German edition published in 1928.)
9. Carnap, R. *The Logical Syntax of Language*. Translated by A. Smeaton. Patterson: Littlefield, Adams & Co., 1959. (Original German edition published in 1934.)
10. Carnap, R. "Testability and Meaning." *Philosophy of Science* 3 (1936): 419–471; and 4 (1937): 1–40.
11. Carnap, R. *Logical Foundations of Probability*. Chicago: University of Chicago Press, 1950.
12. Carnap, R. *An Introduction to the Philosophy of Science*. Edited by M. Gardner. New York: Basic Books, 1966.
13. Cartan, E. "Sur les variétés à connexion affine et la théorie de la relativité generalisée." *Ann. Ecole Norm. Sup.* 40 (1923): 325–412; and 41 (1924): 1–25.
14. Cassirer, E. *Substance and Function* and *Einstein's Theory of Relativity*. Translated by W. C. Swabey and M. C. Swabey.

New York: Dover, 1953. (Original German edition of *Einstein's Theory of Relativity* published in 1921.)

15. Castellan, G. W. *Physical Chemistry.* Reading: Addison-Wesley, 1964.

16. Cohn, P. M. *Lie Groups.* Cambridge: Cambridge University Press, 1968.

17. d'Abro, A. *The Evolution of Scientific Thought from Newton to Einstein.* 2d ed. New York: Dover, 1950.

18. Demopoulos, W. "On the Relation of Topological to Metrical Structure." In M. Radner and S. Winokur, eds., *Minnesota Studies in the Philosophy of Science,* vol. 4. Minneapolis: University of Minnesota Press, 1970.

19. Earman, J. "Space-Time, or How to Solve Philosophical Problems and Dissolve Philosophical Muddles Without Really Trying." *Journal of Philosophy* 67 (1970): 259–277.

20. Earman, J. "Who's Afraid of Absolute Space?" *Australasian Journal of Philosophy* 48 (1970): 287–319.

21. Earman, J. "Implications of Causal Propagation outside the Null Cone." *Australasian Journal of Philosophy* 50 (1972): 222–237.

22. Earman, J. "Covariance, Invariance and the Equivalence of Frames." *Foundations of Physics* 4 (1974): 267–289.

23. Earman, J., and Friedman, M. "The Meaning and Status of Newton's Law of Inertia and the Nature of Gravitational Forces." *Philosophy of Science* 40 (1973): 329–359.

24. Earman, J., and Glymour, C. "Lost in the Tensors: Einstein's Struggles with Covariance Principles, 1912–1916." *Studies in History and Philosophy of Science* 9 (1978): 251–278.

25. Earman, J., Glymour, C., and Stachel, J., eds. *Minnesota Studies in the Philosophy of Science,* vol. 8. Minneapolis: University of Minnesota Press, 1977.

26. Einstein, A. "On the Electrodynamics of Moving Bodies." In *The Principle of Relativity* [78]. (Original German version published in *Annalen der Physik* 17 [1905].)

27. Einstein, A. "On the Influence of Gravitation on the Propagation of Light." In *The Principle of Relativity* [78]. (Original German version published in *Annalen der Physik* 35 [1911].)

28. Einstein, A. "The Foundation of the General Theory of Relativity." In *The Principle of Relativity* [78]. (Original German version published in *Annalen der Physik* 49 [1916].)

29. Einstein, A. *The Meaning of Relativity.* 5th ed. Princeton:

369

Princeton University Press, 1953. (The Stafford Little Lectures of Princeton University, May 1921.)

30. Einstein, A. "Autobiographical Notes." In *Albert Einstein: Philosopher-Scientist* [96] (1949).

31. Ellis, B. "The Origin and Nature of Newton's Laws of Motion." In R. G. Colodny, ed., *Beyond the Edge of Certainty.* Pittsburgh: Pittsburgh University Press, 1965.

32. Ellis, B., and Bowman, P. "Conventionality in Distant Simultaneity." *Philosophy of Science* 34 (1967): 116–136.

33. Field, H. *Science Without Numbers: A Defense of Nominalism.* Princeton: Princeton University Press, 1980.

34. Fine, A. "Reflections on a Relational Theory of Space." In *Space Time and Geometry* [105] (1973).

35. Fock, V. *The Theory of Space, Time and Gravitation.* Translated by N. Kemner. Oxford: Pergamon, 1964. (Original Russian edition published in 1955.)

36. Friedman, M. "Grünbaum on the Conventionality of Geometry." In *Space, Time and Geometry* [105] (1973).

37. Friedman, M. "Relativity Principles, Absolute Objects, and Symmetry Groups." In *Space, Time and Geometry* [105] (1973).

38. Friedman, M. "Explanation and Scientific Understanding." *Journal of Philosophy* 71 (1974): 5–19.

39. Friedman, M. "Simultaneity in Newtonian Mechanics and Special Relativity." In *Minnesota Studies*, vol. 8 [25] (1977).

40. Friedman, M. "Truth and Confirmation." *Journal of Philosophy* 76 (1979): 361–382.

41. Geroch, R. *General Relativity from A to B.* Chicago: University of Chicago Press, 1978.

42. Geroch, R., and Jang, P. S. "Motion of a Body in General Relativity." *Journal of Mathematical Physics* 16 (1975): 65–67.

43. Glymour, C. "Theoretical Realism and Theoretical Equivalence." In R. C. Buck and R. S. Cohen, eds., *Boston Studies in the Philosophy of Science*, vol. 8. Dordrecht: Reidel, 1971.

44. Glymour, C. "Physics by Convention." *Philosophy of Science* 39 (1972): 322–340.

45. Glymour, C. "Topology, Cosmology and Convention." In *Space, Time and Geometry* [105] (1973).

46. Glymour, C. "Indistinguishable Space-Times and the Fundamental Group." In *Minnesota Studies*, vol. 8 [25] (1977).

370

47. Glymour, C. *Theory and Evidence*. Princeton: Princeton University Press, 1980.
48. Gödel, K. "An Example of a New Type of Cosmological Solutions of Einstein's Field Equations of Gravitation." *Reviews of Modern Physics* 21 (1949): 447–450.
49. Goldstein, H. *Classical Mechanics*. Reading: Addison-Wesley, 1950.
50. Goodman, N. *Fact, Fiction, and Forecast*. New York: Bobbs-Merrill, 1955.
51. Grünbaum, A. *Philosophical Problems of Space and Time*. New York: Knopf, 1963.
52. Grünbaum, A. *Geometry and Chronometry in Philosophical Perspective*. Minneapolis: University of Minnesota Press, 1968.
53. Grünbaum, A. "Simultaneity by Slow Clock Transport in the Special Theory of Relativity." *Philosophy of Science* 36 (1969): 5–43.
54. Hanson, N. R. "Newton's First Law; A Philosopher's Door Into Natural Philosophy." In R. G. Colodny, ed., *Beyond the Edge of Certainty*. Pittsburgh: Pittsburgh University Press, 1965.
55. Havas, P. "Four-Dimensional Formulations of Newtonian Mechanics and Their Relations to the Special and General Theory of Relativity." *Reviews of Modern Physics* 36 (1964): 938–965.
56. Havas, P. "Foundation Problems in General Relativity." In M. Bunge, ed., *Delaware Seminar in the Foundations of Physics*. New York: Springer, 1967.
57. Hawking, S. W., and Ellis, G.F.R. *The Large Scale Structure of Space-Time*. Cambridge: Cambridge University Press, 1973.
58. Hempel, C. G. *Aspects of Scientific Explanation*. New York: Free Press, 1965.
59. Hicks, N. J. *Notes on Differential Geometry*. Princeton: Van Nostrand, 1965.
60. Horwich, P. "On Some Alleged Paradoxes of Time Travel." *Journal of Philosophy* 72 (1975): 432–444.
61. Horwich, P. "On the Existence of Time, Space and Space-Time." *Nous* 12 (1978): 397–419.
62. Jones, R. "The Special and General Principles of Relativity." In P. Barker and C. G. Shugart, eds., *After Einstein:*

371

Proceedings of the Einstein Centenary Conference. Memphis: Memphis State University Press, 1981.

63. Kneale, W. *Probability and Induction*. Oxford: Oxford University Press, 1949.

64. Kretschmann, E. "Über den physikalischen Sinn der Relativitätspostulate. A. Einsteins neue und seine ursprungliche Relativitätstheorie." *Annalen der Physik* 53 (1917): 575–614.

65. Kreyszig, E. *Introduction to Differential Geometry and Riemannian Geometry*. Toronto: University of Toronto Press, 1967.

66. Laugwitz, D. *Differential and Riemannian Geometry*. New York: Academic Press, 1965.

67. Lorentz, H. A. "Electromagnetic Phenomena in a System Moving with any Velocity less than that of Light." In *The Principle of Relativity* [78]. (Original English version published in *Proceedings of the Academy of Sciences of Amsterdam* 6 [1904].)

68. Mach, E. *The Science of Mechanics: A Critical and Historical Account of Its Development*. Translated by T. J. McCormak. La Salle: Open Court, 1960. (Original German edition published in 1883.)

69. Malament, D. "Does the Causal Structure of Space-Time Determine its Geometry?" Ph.D. diss., Rockefeller University, 1975.

70. Malament, D. "Observationally Indistinguishable Space-Times." In *Minnesota Studies*, vol. 8 [25] (1977).

71. Malament, D. "Causal Theories of Time and the Conventionality of Simultaneity." *Nous* 11 (1977): 293–300.

72. Malament, D. "Newtonian Gravity, Limits, and the Geometry of Space." In R. G. Colodny, ed., *Pittsburgh Studies in the Philosophy of Science*. Pittsburgh: Pittsburgh University Press, forthcoming.

73. Minkowski, H. "Space and Time." In *The Principle of Relativity* [78]. (Address delivered at the 80th Assembly of German Natural Scientists and Physicians, 1908.)

74. Misner, C. W., Thorne, K. S., and Wheeler, J. A. *Gravitation*. San Francisco: W. H. Freeman, 1973.

75. Newton, I. *Principia*. Translated by A. Motte (1729); revised by A. Cajori. Berkeley: University of California Press, 1934. (Original Latin edition published in 1687.)

76. Noll, W. "Space-Time Structures in Classical Mechanics." In

M. Bunge, ed., *Delaware Seminar in the Foundations of Physics*. New York: Springer, 1967.

77. Ozsváth, I., and Schücking, E. L. "The Finite Rotating Universe." *Annals of Physics* 55 (1969): 166–204.

78. Perrett, W., and Jeffery, G. B., trans. *The Principle of Relativity: A Collection of Original Memoirs on the Special and General Theory of Relativity*. New York: Dover, 1923.

79. Poincaré, H. *Science and Hypothesis*. Translated by W. J. Greensreet. New York: Dover, 1952. (Original French edition published in 1902.)

80. Putnam, H. *Philosophical Papers*. 2 vols. Cambridge: Cambridge University Press, 1975.

81. Quine, W. V. *The Ways of Paradox and Other Essays*. New York: Random House, 1966.

82. Quine, W. V. *Ontological Relativity and Other Essays*. New York: Columbia University Press, 1969.

83. Quine, W. V. "On the Reasons for Indeterminacy of Translation." *Journal of Philosophy* 67 (1970): 178–183.

84. Reichenbach, H. *The Theory of Relativity and A Priori Knowledge*. Translated by M. Reichenbach. Berkeley: University of California Press, 1960. (Original German edition published in 1920.)

85. Reichenbach, H. *Axiomatization of the Theory of Relativity*. Translated by M. Reichenbach. Berkeley: University of California Press, 1969. (Original German edition published in 1924.)

86. Reichenbach, H. *The Philosophy of Space and Time*. Translated by M. Reichenbach and J. Freund. New York: Dover, 1957. (Original German edition published in 1928.)

87. Reichenbach, H. *Experience and Prediction*. Chicago: University of Chicago Press, 1938.

88. Reichenbach, H. *Philosophic Foundations of Quantum Mechanics*. Berkeley: University of California Press, 1944.

89. Reichenbach, H. "The Philosophical Significance of the Theory of Relativity." In *Albert Einstein: Philosopher-Scientist* [96] (1949).

90. Resnick, R. *Introduction to Special Relativity*. New York: Wiley, 1968.

91. Ricci, G., and Levi-Civita, T. "Méthodes de calcul différentiel absolu et leurs applications." *Math. Ann.* 54 (1901): 125–201.

92. Ricketts, T. "Rationality, Translation, and Epistemology Naturalized." *Journal of Philosophy* 79 (1982): 117–136.

93. Riemann, B. "On the Hypotheses Which Lie at the Foundations of Geometry." Translated by H. S. White. In D. E. Smith, ed., *A Source Book in Mathematics.* New York: Dover, 1959. (Original German version delivered in 1854.)

94. Robb, A. A. *A Theory of Time and Space.* Cambridge: Cambridge University Press, 1914.

95. Robertson, H. and Noonan, T. *Relativity and Cosmology.* Philadelphia: Saunders, 1968.

96. Schilpp, P. A., ed. *Albert Einstein: Philosopher-Scientist.* New York: Harper, 1949.

97. Schlick, M. *Space and Time in Contemporary Physics.* Translated by H. L. Brose. Oxford: Oxford University Press, 1920. (Original German edition published in 1917.)

98. Schlick, M. *General Theory of Knowledge.* Translated by A. E. Blumberg. New York: Springer, 1974. (Original German edition published in 1918.)

99. Sellars, W. *Science, Perception and Reality.* London: Routledge, 1963.

100. Sklar, L. *Space, Time, and Spacetime.* Berkeley: University of California Press, 1974.

101. Sklar, L. "Inertia, Gravitation and Metaphysics." *Philosophy of Science* 43 (1976): 1–23.

102. Stein, H. "Newtonian Space-Time." *Texas Quarterly* 10 (1967): 174–200.

103. Stein, H. "On Einstein-Minkowski Space-Time." *Journal of Philosophy* 65 (1968): 5–23.

104. Stein, H. "Some Philosophical Prehistory of General Relativity." In *Minnesota Studies*, vol. 8 [25] (1977).

105. Suppes, P., ed. *Space, Time and Geometry.* Dordrecht: Reidel, 1973.

106. Taylor, A. E. *Advanced Calculus.* New York: Blaisdell, 1955.

107. Taylor, E. F., and Wheeler, J. A. *Spacetime Physics.* San Francisco: W. H. Freeman, 1966.

108. Trautman, A. "Sur la Théorie Newtonienne de la Gravitation." *Comptes Rendus* A257 (1963): 617–620.

109. Trautman, A. "Foundations and Current Problems of General Relativity." In S. Deser and K. W. Ford, eds., *Lectures on General Relativity.* Englewood Cliffs: Prentice-Hall, 1965.

110. Trautman, A. "Comparison of Newtonian and Relativistic

Theories of Space-Time." In B. Hoffmann, ed., *Perspectives in Geometry and Relativity*. Bloomington: Indiana University Press, 1966.

111. Tuller, A. *A Modern Introduction to Geometries*. Princeton: Van Nostrand, 1967.

112. van Fraassen, B. C. *An Introduction to the Philosophy of Time and Space*. New York: Random House, 1970.

113. van Fraassen, B. C. "Earman on the Causal Theory of Time" in *Space, Time and Geometry* [105] (1973).

114. van Fraassen, B. C. *The Scientific Image*. Oxford: Oxford University Press. 1980.

115. Weinberg, S. *Gravitation and Cosmology*. New York: Wiley. 1972.

116. Weyl, H. *Space-Time-Matter*. Translated by H. L. Brose. New York: Dover. 1952. (Original German edition published in 1918.)

117. Whewell, W. *The Philosophy of the Inductive Sciences, Founded upon their History*. 2 vols. London: J. W. Parker, 1840.

118. Winnie, J. A. "Special Relativity Without One-Way Velocity Assumptions." *Philosophy of Science* 37 (1970): 81–99; 223–238.

119. Winnie, J. A. "The Causal Theory of Space-Time" in *Minnesota Studies*, vol. 8 [25] (1977).

120. Zeeman, E. C. "Causality implies the Lorentz group." *Journal of Mathematical Physics* 5 (1964): 490–493.

Index

absolute acceleration, 5, 12, 14, 66, 97ff., 225ff.; four-dimensional nature of, 16–17, 111–112, 228ff., 232; and gravitation, 28, 97ff., 192ff., 279; in general relativity, 209ff.; as primitive quantity, 232ff., 326; and unification, 255ff., 260ff.

absolute motion: vs. relative motion, 5, 12, 17, 65ff., 226ff.; observable significance of, 14; and four-dimensionality, 34, 232; and evolution of relativity, 223ff.; and absolute position, 227–228; definability of, 230

absoluteness, senses of, 62ff., 108ff., 211, 232

absolute objects, 56ff., 64, 109, 154; lack of in general relativity, 56, 61, 64, 214; definition of, 60

absolute rotation, 14, 17, 27, 66ff., 192–193, 225ff., 229ff.; four-dimensional nature of, 16–17, 111–112, 232; in general relativity, 209ff.; and unification, 260ff.

absolute space, 12–14, 176, 237–238, 258–259; in electrodynamics, 14–15, 105ff., 115, 156ff., 249–250, 284, 290, 292–293, 323, 328, 331; in Newtonian space-time, 74ff., 108ff.; and laws of motion, 119; lack of unifying power of, 248–249, 290, 332; and observational/theoretical distinction, 275; and conventionalist strategy, 278ff.

absolute time, 72; in Newtonian space-time, 108ff.; elimination of in special relativity, 158

absolute velocity, 5, 14; three-dimensional nature of, 16–17, 228; indistinguishability of, 113ff., 152, 275, 278–279, 291ff.; vs. acceleration, 227

acceleration. *See* absolute acceleration

acceleration four-vector: in Newtonian space-time, 81, 88; in special relativity, 134; in general relativity, 182

action-at-a-distance theories, 123–124

Adler, R., 180n, 181n, 207, 210n

aether, 110, 151, 158n, 292, 293. *See also* absolute space

affine connection: informal explanation of, 38ff.; as correction terms in covariant derivative, 54; and problem of motion, 67ff.; in theoretical inference, 325ff.; precise definition of, 350. *See also* affine structure

affine structure, 11; informal explanation of, 38ff.; and laws of motion, 115ff.; and Lorentz transformations, 139ff., 163ff.; and unification, 252ff., 331ff.

analyticity, 288n

Anderson, J. L., 105n, 123n, 159n, 160n, 162n, 180n; on covariance, 52n; on absolute/dynamical distinction, 56ff., 64, 109; on invariance or symmetry, 56ff., 154; on general principle of relativity, 214ff.

a prioricity, 7, 18, 285ff. *See also* Kant

automorphism, 11, 47, 52, 56, 164, 240, 319, 364

axioms of connection, 286ff.

axioms of coordination, 286ff.

Bazin, M., 180n, 181n, 207, 210n

Bergmann, P. G., 47n, 117n, 149n, 150n, 180n, 184n

Bishop, R. I., 343n, 349n, 354n, 356n, 357n, 360n, 364n

black holes, 187, 261

Bowman, P., 314n

Boyd, R., 242n, 246n

Boyle-Charles law, 239

Bunge, M., 117n, 159n

379

Library of Congress Cataloging in Publication Data

Friedman, Michael, 1947–
    Foundations of space-time theories.

    Bibliography: p.
    Includes index.
    1. Relativity (Physics)   2. Space and time.
3. Science—Philosophy.   I. Title.
QC173.55.F74   1983   530.1'1   82-61362
ISBN 0-691-07239-6